Reactive Extrusion

Reactive Extrusion

Principles and Applications

Edited by Günter Beyer and Christian Hopmann

Editors

Dr. Günter Beyer
Head of Department
Chem. -phys. Laboratories
Kabelwerk Eupen AG
Malmedyer Straße 9
4700 Eupen
Belgium

Prof. Dr. Christian Hopmann
RWTH Aachen University
Institute of Plastics Processing (IKV)
52056 Aachen
Germany

■ All books published by **Wiley-VCH** are carefully produced. Nevertheless, authors, editors, and publisher do not warrant the information contained in these books, including this book, to be free of errors. Readers are advised to keep in mind that statements, data, illustrations, procedural details or other items may inadvertently be inaccurate.

Library of Congress Card No.: applied for

British Library Cataloguing-in-Publication Data
A catalogue record for this book is available from the British Library.

Bibliographic information published by the Deutsche Nationalbibliothek
The Deutsche Nationalbibliothek lists this publication in the Deutsche Nationalbibliografie; detailed bibliographic data are available on the Internet at <http://dnb.d-nb.de>.

© 2018 Wiley-VCH Verlag GmbH & Co. KGaA, Boschstr. 12, 69469 Weinheim, Germany

All rights reserved (including those of translation into other languages). No part of this book may be reproduced in any form – by photoprinting, microfilm, or any other means – nor transmitted or translated into a machine language without written permission from the publishers. Registered names, trademarks, etc. used in this book, even when not specifically marked as such, are not to be considered unprotected by law.

Print ISBN: 978-3-527-34098-9
ePDF ISBN: 978-3-527-80153-4
ePub ISBN: 978-3-527-80155-8
Mobi ISBN: 978-3-527-80156-5
oBook ISBN: 978-3-527-80154-1

Cover Design Schulz Grafik-Design, Fußgönheim, Germany
Typesetting SPi Global, Chennai, India
Printing and Binding C.O.S. Printers Pte Ltd Singapore

Printed on acid-free paper

Contents

Preface *xiii*
List of Contributors *xv*

Part I   Introduction *1*

1   Introduction to Reactive Extrusion *3*
Christian Hopmann, Maximilian Adamy, and Andreas Cohnen
References *9*

Part II   Introduction to Twin-Screw Extruder for Reactive Extrusion *11*

2   The Co-rotating Twin-Screw Extruder for Reactive Extrusion *13*
Frank Lechner
2.1   Introduction *13*
2.2   Development and Key Figures of the Co-rotating Twin-Screw Extruder *14*
2.3   Screw Elements *16*
2.4   Co-rotating Twin-Screw Extruder – Unit Operations *22*
2.4.1   Feeding *23*
2.4.2   Upstream Feeding *23*
2.4.3   Downstream Feeding *24*
2.4.4   Melting Mechanisms *24*
2.4.5   Thermal Energy Transfer *24*
2.4.6   Mechanical Energy Transfer *25*
2.4.7   Mixing Mechanisms *25*
2.4.8   Devolatilization/Degassing *25*
2.4.9   Discharge *26*
2.5   Suitability of Twin-Screw Extruders for Chemical Reactions *26*
2.6   Processing of TPE-V *27*
2.7   Polymerization of Thermoplastic Polyurethane (TPU) *29*

2.8	Grafting of Maleic Anhydride on Polyolefines 31
2.9	Partial Glycolysis of PET 32
2.10	Peroxide Break-Down of Polypropylene 33
2.11	Summary 35
	References 35

Part III Simulation and Modeling 37

3 Modeling of Twin Screw Reactive Extrusion: Challenges and Applications 39
Françoise Berzin and Bruno Vergnes

3.1	Introduction 39
3.1.1	Presentation of the Reactive Extrusion Process 39
3.1.2	Examples of Industrial Applications 40
3.1.3	Interest of Reactive Extrusion Process Modeling 41
3.2	Principles and Challenges of the Modeling 41
3.2.1	Twin Screw Flow Module 42
3.2.2	Kinetic Equations 44
3.2.3	Rheokinetic Model 44
3.2.4	Coupling 45
3.2.5	Open Problems and Remaining Challenges 45
3.3	Examples of Modeling 46
3.3.1	Esterification of EVA Copolymer 46
3.3.2	Controlled Degradation of Polypropylene 50
3.3.3	Polymerization of ε-Caprolactone 55
3.3.4	Starch Cationization 59
3.3.5	Optimization and Scale-up 61
3.4	Conclusion 65
	References 66

4 Measurement and Modeling of Local Residence Time Distributions in a Twin-Screw Extruder 71
Xian-Ming Zhang, Lian-Fang Feng, and Guo-Hua Hu

4.1	Introduction 71
4.2	Measurement of the Global and Local RTD 72
4.2.1	Theory of RTD 72
4.2.2	In-line RTD Measuring System 73
4.2.3	Extruder and Screw Configurations 75
4.2.4	Performance of the In-line RTD Measuring System 76
4.2.5	Effects of Screw Speed and Feed Rate on RTD 77
4.2.6	Assessment of the Local RTD in the Kneading Disk Zone 79
4.3	Residence Time, Residence Revolution, and Residence Volume Distributions 81
4.3.1	Partial RTD, RRD, and RVD 82
4.3.2	Local RTD, RRD, and RVD 86

4.4	Modeling of Local Residence Time Distributions	*88*
4.4.1	Kinematic Modeling of Distributive Mixing	*88*
4.4.2	Numerical Simulation	*89*
4.4.3	Experimental Validation	*92*
4.4.4	Distributive Mixing Performance and Efficiency	*93*
4.5	Summary	*97*
	References	*98*

5	**In-process Measurements for Reactive Extrusion Monitoring and Control**	*101*
	José A. Covas	
5.1	Introduction	*101*
5.2	Requirements of In-process Monitoring of Reactive Extrusion	*103*
5.3	In-process Optical Spectroscopy	*111*
5.4	In-process Rheometry	*116*
5.5	Conclusions	*125*
	Acknowledgment	*126*
	References	*126*

Part IV   Synthesis Concepts   *133*

6	**Exchange Reaction Mechanisms in the Reactive Extrusion of Condensation Polymers**	*135*
	Concetto Puglisi and Filippo Samperi	
6.1	Introduction	*135*
6.2	Interchange Reaction in Polyester/Polyester Blends	*138*
6.3	Interchange Reaction in Polycarbonate/Polyester Blends	*143*
6.4	Interchange Reaction in Polyester/Polyamide Blends	*148*
6.5	Interchange Reaction in Polycarbonate/Polyamide Blends	*155*
6.6	Interchange Reaction in Polyamide/Polyamide Blends	*159*
6.7	Conclusions	*166*
	References	*167*

7	***In situ* Synthesis of Inorganic and/or Organic Phases in Thermoplastic Polymers by Reactive Extrusion**	*179*
	Véronique Bounor-Legaré, Françoise Fenouillot, and Philippe Cassagnau	
7.1	Introduction	*179*
7.2	Nanocomposites	*179*
7.2.1	Synthesis of *in situ* Nanocomposites	*181*
7.2.2	Some Specific Applications	*183*
7.2.2.1	Antibacterial Properties of PP/TiO_2 Nanocomposites	*183*
7.2.2.2	Flame-Retardant Properties	*184*
7.2.2.3	Protonic Conductivity	*186*
7.3	Polymerization of a Thermoplastic Minor Phase: Toward Blend Nanostructuration	*188*

7.4	Polymerization of a Thermoset Minor Phase Under Shear *196*
7.4.1	Thermoplastic Polymer/Epoxy-Amine Miscible Blends *197*
7.4.2	Examples of Stabilization of Thermoplastic Polymer/Epoxy-Amine Blends *202*
7.4.3	Blends of Thermoplastic Polymer with Monomers Crosslinking via Radical Polymerization *202*
7.5	Conclusion *203*
	References *204*

8 Concept of (Reactive) Compatibilizer-Tracer for Emulsification Curve Build-up, Compatibilizer Selection, and Process Optimization of Immiscible Polymer Blends *209*

Cai-Liang Zhang, Wei-Yun Ji, Lian-Fang Feng, and Guo-Hua Hu

8.1	Introduction *209*
8.2	Emulsification Curves of Immiscible Polymer Blends in a Batch Mixer *210*
8.3	Emulsification Curves of Immiscible Polymer Blends in a Twin-Screw Extruder Using the Concept of (Reactive) Compatibilizer *213*
8.3.1	Synthesis of (Reactive) Compatibilizer-Tracers *213*
8.3.2	Development of an In-line Fluorescence Measuring Device *214*
8.3.3	Experimental Procedure for Emulsification Curve Build-up *216*
8.3.4	Compatibilizer Selection Using the Concept of Compatibilizer-Tracer *219*
8.3.5	Process Optimization Using the Concept of Compatibilizer-Tracer *220*
8.3.5.1	Effect of Screw Speed *220*
8.3.5.2	Effects of the Type of Mixer *221*
8.3.6	Section Summary *221*
8.4	Emulsification Curves of Reactive Immiscible Polymer Blends in a Twin-Screw Exturder *222*
8.4.1	Reaction Kinetics between Reactive Functional Groups *222*
8.4.2	(Non-reactive) Compatibilizers Versus Reactive Compatibilizers *223*
8.4.3	An Example of Reactive Compatibilizer-Tracer *224*
8.4.4	Assessment of the Morphology Development of Reactive Immiscible Polymer Blends Using the Concept of Reactive Compatibilizer *225*
8.4.5	Emulsification Curve Build-up in a Twin-Screw Extruder Using the Concept of Reactive Compatibilizer-Tracer *229*
8.4.6	Assessment of the Effects of Processing Parameters Using the Concept of Reactive Compatibilizer-Tracer *233*
8.4.6.1	Effect of the Reactive Compatibilizer-Tracer Injection Location *233*
8.4.6.2	Effect of the Blend Composition *235*
8.4.6.3	Effect of the Geometry of Screw Elements *238*
8.5	Conclusion *241*
	References *241*

Part V Selected Examples of Synthesis *245*

9 Nano-structuring of Polymer Blends by *in situ* Polymerization and *in situ* Compatibilization Processes *247*
Cai-Liang Zhang, Lian-Fang Feng, and Guo-Hua Hu
9.1 Introduction *247*
9.2 Morphology Development of Classical Immiscible Polymer Blending Processes *248*
9.2.1 Solid–Liquid Transition Stage *249*
9.2.2 Melt Flow Stage *251*
9.2.3 Effect of Compatibilizer *253*
9.3 *In situ* Polymerization and *in situ* Compatibilization of Polymer Blends *255*
9.3.1 Principles *255*
9.3.2 Classical Polymer Blending Versus *in situ* Polymerization and *in situ* Compatibilization *255*
9.3.3 Examples of Nano-structured Polymer Blends by *in situ* Polymerization and *in situ* Compatibilization *257*
9.3.3.1 PP/PA6 Nano-blends *257*
9.3.3.2 PPO/PA6 Nano-blends *264*
9.3.3.3 PA6/Core–Shell Blends *264*
9.4 Summary *267*
References *268*

10 Reactive Comb Compatibilizers for Immiscible Polymer Blends *271*
Yongjin Li, Wenyong Dong, and Hengti Wang
10.1 Introduction *271*
10.2 Synthesis of Reactive Comb Polymers *272*
10.3 Reactive Compatibilization of Immiscible Polymer Blends by Reactive Comb Polymers *274*
10.3.1 PLLA/PVDF Blends Compatibilized by Reactive Comb Polymers *274*
10.3.1.1 Comparison of the Compatibilization Efficiency of Reactive Linear and Reactive Comb Polymers *274*
10.3.1.2 Effects of the Molecular Structures on the Compatibilization Efficiency of Reactive Comb Polymers *278*
10.3.2 PLLA/ABS Blends Compatibilized by Reactive Comb Polymers *282*
10.4 Immiscible Polymer Blends Compatiblized by Janus Nanomicelles *289*
10.5 Conclusions and Further Remarks *293*
References *293*

11		**Reactive Compounding of Highly Filled Flame Retardant Wire and Cable Compounds** *299*
		Mario Neuenhaus and Andreas Niklaus
11.1		Introduction *299*
11.2		Formulations and Ingredients *300*
11.2.1		Typical Formulation and Variations for the Evaluation *300*
11.2.2		Principle of Silane Crosslinking by Reactive Extrusion *301*
11.2.3		Production of Aluminum Trihydroxide (ATH) *301*
11.2.4		Mode of Action of Aluminum Trihydroxide *302*
11.2.5		Selection of Suitable ATH Grades *303*
11.3		Processing *306*
11.3.1		Compounding Line *306*
11.3.2		Compounding Process for Cross Linkable HFFR Products *308*
11.3.2.1		Two-Step Compounding Process *308*
11.3.2.2		One-Step Compounding Process *309*
11.3.2.3		Advantages and Disadvantages of the Two Process Concepts (Two-Step vs One-Step) *313*
11.4		Evaluation and Results on the Compound *314*
11.4.1		Crosslinking Density *314*
11.4.2		Mechanical Properties *315*
11.4.3		Aging Performance *315*
11.4.4		Fire Performance on Laboratory Scale *317*
11.4.5		Results of the Non-Polar Compounds *318*
11.5		Cable Trials *322*
11.5.1		Fire Performance of Electrical Cables According to EN 50399 *322*
11.5.2		Burning Test on Experimental Cables According to EN 50399 *323*
11.6		Conclusions *328*
		References *329*
12		**Thermoplastic Vulcanizates (TPVs) by the Dynamic Vulcanization of Miscible or Highly Compatible Plastic/Rubber Blends** *331*
		Yongjin Li and Yanchun Tang
12.1		Introduction *331*
12.2		Morphological Development of TPVs from Immiscible Polymer Blends *333*
12.3		TPVs from Miscible PVDF/ACM Blends *334*
12.4		TPVs from Highly Compatible EVA/EVM Blends *338*
12.5		Conclusions and Future Remarks *342*
		References *342*

Part VI Selected Examples of Processing 345

13   Reactive Extrusion of Polyamide 6 with Integrated Multiple Melt Degassing 347
Christian Hopmann, Eike Klünker, Andreas Cohnen, and Maximilian Adamy
13.1   Introduction 347
13.2   Synthesis of Polyamide 6 347
13.2.1   Hydrolytic Polymerization of Polyamide 6 347
13.2.2   Anionic Polymerization of Polyamide 6 348
13.3   Review of Reactive Extrusion of Polyamide 6 in Twin-Screw Extruders 352
13.4   Recent Developments in Reactive Extrusion of Polyamide 6 in Twin-Screw Extruders 354
13.4.1   Reaction System and Experimental Setup 354
13.4.2   Influence of Number of Degassing Steps and Activator Content on Residual Monomer Content and Molecular Weight 356
13.4.3   Influence of Amount and Type of Entrainer on Residual Monomer Content and Molecular Weight 365
13.4.4   Influence of Polymer Throughput on Residual Monomer Content 367
13.5   Conclusion 368
References 369

14   Industrial Production and Use of Grafted Polyolefins 375
Inno Rapthel, Jochen Wilms, and Frederik Piestert
14.1   Grafted Polymers 375
14.2   Industrial Synthesis of Grafted Polymers 376
14.2.1   Melt Grafting Technology 377
14.2.2   Solid State Grafting Technology 378
14.3   Main Applications 380
14.3.1   Use as Coupling Agents 380
14.3.2   Grafted Polyolefins for Polymer Blending 392
14.3.2.1   Reactive Blending of Polyamides 392
14.3.3   Grafted TPE's for Overmolding Applications 400
14.4   Conclusion and Outlook 403
References 404

Index 407

Preface

One elegant possibility to improve properties of polymers is by chemical modifications of polymers by reactive extrusion steps. Reactive extrusion is often used in industry to improve polymers for many reasons such as:

improved adhesion to other polymers and also to metals
improved compatibilization of fillers to a polymer matrix
improved compatibilization of polymers in blends.

To understand the underlying chemical, physical and processing steps during reactive extrusion in detail, many analytical tools are available. This book describes in detail the chemical reactions that occur during the reactive extrusion step, the possibilities to investigate and measure the kinetic reaction parameters, and also the modification of screws to improve residence times and mixing steps. Authors, both from industry and academics, from Germany, Italy, France, Portugal, Switzerland, and China, present overviews for these topics. Our hope is that this new book on reactive extrusion is helpful for scientists and engineers to understand better the principles of reactive extrusion and provides inspiration for new ideas as well as answers to existing questions.

Both editors want to thank all authors for their contributions. We want to also thank Wiley for their support that enabled this book.

September 2017,
Aachen

Günter Beyer
Christian Hopmann

List of Contributors

Maximilian Adamy
RWTH Aachen University
Institute of Plastics Processing (IKV)
Seffenter Weg 201
D-52074 Aachen
Germany

Françoise Berzin
Université de Reims
Champagne-Ardenne
UMR 0614 Fractionnement des
AgroRessources et Environnement
2 Esplanade Roland Garros
BP 224, 51686 Reims Cedex
France

Véronique Bounor-Legaré
Université de Lyon
69003 Lyon
France

and

Université de Lyon 1
Ingénierie des Matériaux Polymères
CNRS UMR 5223
69622 Villeurbanne
France

Philippe Cassagnau
Université de Lyon
69003 Lyon
France

and

Université de Lyon 1
Ingénierie des Matériaux Polymères
CNRS UMR 5223
69622 Villeurbanne
France

Andreas Cohnen
RWTH Aachen University
Institute of Plastics Processing (IKV)
Seffenter Weg 201
D-52074 Aachen
Germany

José A. Covas
University of Minho
Institute for Polymers
Composites/I3N, 4800-058 Guimarães
Portugal

Wenyong Dong
Hangzhou Normal University
College of Material, Chemistry and
Chemical Engineering
No. 16 Xuelin Road
Hangzhou 310036
P.R. China

Lian-Fang Feng
Zhejiang University
State Key Laboratory of Chemical Engineering
College of Chemical and Biochemical Engineering
Hangzhou 310027
China

Françoise Fenouillot
Université de Lyon
69003 Lyon
France

and

INSA de Lyon
Ingénierie des Matériaux Polymères
CNRS UMR 5223
69621 Villeurbanne
France

Christian Hopmann
RWTH Aachen University
Institute of Plastics Processing (IKV)
Seffenter Weg 201
D-52074 Aachen
Germany

Guo-Hua Hu
Université de Lorraine-CNRS
Laboratoire Réactions et Génie des Procédés
1 rue Grandville, BP 20451
54001 Nancy
France

Wei-Yun Ji
Zhejiang University
State Key Laboratory of Chemical Engineering
College of Chemical and Biochemical Engineering
Hangzhou 310027
China

E. Klünker
3M Deutschland GmbH
Carl-Schurz-Str. 1
D-41453 Neuss
Germany

Frank Lechner
Coperion GmbH
Process Technology
Theodorstrasse 10
D-70469 Stuttgart
Germany

Yongjin Li
Hangzhou Normal University
College of Material, Chemistry and Chemical Engineering
No. 16 Xuelin Road
Hangzhou 310036
P.R. China

Mario Neuenhaus
Martinswerk GmbH
Kölner Strasse 110
50127 Bergheim
Germany

Andreas Niklaus
Buss AG
Hohenrainstrasse 10
4133 Pratteln
Switzerland

Frederik Piestert
BYK-Chemie GmbH
Abelstrasse 45
46483 Wesel
Germany

Concetto Puglisi
Composites and Biomaterials (IPCB)
U.O.S. di Catania
CNR, Institute for Polymers
Via Gaifami 18
95126 Catania
Italy

Inno Rapthel
BYK-Chemie GmbH
Value Park Y 42
06258 Schkopau
Germany

Filippo Samperi
Composites and Biomaterials (IPCB)
U.O.S. di Catania
CNR, Institute for Polymers
Via Gaifami 18
95126 Catania
Italy

Yanchun Tang
Hangzhou Normal University
College of Material, Chemistry and
Chemical Engineering
No. 16 Xuelin Road
Hangzhou 310036
P.R. China

Bruno Vergnes
MINES ParisTech.
Centre de Mise en Forme des
Matériaux (CEMEF)
UMR CNRS 7635
CS 10207, 06904 Sophia Antipolis
Cedex
France

Hengti Wang
Hangzhou Normal University
College of Material, Chemistry and
Chemical Engineering
No. 16 Xuelin Road
Hangzhou 310036
P.R. China

Jochen Wilms
BYK-Chemie GmbH
Abelstrasse 45
46483 Wesel
Germany

Cai-Liang Zhang
Zhejiang University
State Key Laboratory of Chemical
Engineering
College of Chemical and Biochemical
Engineering
Hangzhou 310027
China

Xian-Ming Zhang
Zhejiang Sci-Tech University
Key Laboratory of Advanced Textile
Materials and Manufacturing
Technology
Ministry of Education
Hangzhou 310018
China

Part I

Introduction

1

Introduction to Reactive Extrusion

Christian Hopmann, Maximilian Adamy, and Andreas Cohnen

RWTH Aachen University, Institute of Plastic Processing (IKV), Seffenter Weg 201, D-52074 Aachen, Germany

Engineering plastics have become indispensable to our daily lives. For several decades, the development of plastics technology to expand the scope of applications for plastics has been marked by "tailor-made" materials as well as by new processing techniques. Today, plastics are used in many applications ranging from plastics packaging and the automotive sector to medical products and they are increasingly substituting metal and ceramic. The success of this class of material is due to its characteristics such as good processability, low density, reasonable price, and especially its adjustable material properties.

The processing chain usually starts with the synthesis of the polymers, which is carried out by mostly petroleum-based monomers. In the field of polymer synthesis, a distinction is made between different polymerization reactions. These include, among others, mass polymerization, solution polymerization, and polymerization in gas phase [10]. Mass polymerization, which takes place in the melt, achieves high throughputs and high purities. However, the reaction rate, which is typically quite slow, results in the sedimentation of polymer chains with high molecular weight. Solution polymerization enables a higher homogeneity of the molecular weight distribution in the mixture because of the use of solvents. However, the solvent has to be removed by an energy-intensive process following polymerization. In addition, the presence of the solvent may produce undesirable side reactions. During gas phase polymerization, the monomers are transferred into a fluid state by an inert gas flow prior to polymerization [10].

Polymerization can be carried out in three different ways: batchwise, semi-continuous, and continuous. The choice of the reaction process influences both the material properties, such as viscosity or achievable molecular weight and the reaction kinetics and the resulting heat dissipation. Although polymerization in a batchwise process allows the production of small amounts of special polymers the batchwise process is less suitable for the mass production of a polymer due to the restricted achievable viscosity. Semi-continuous processes are often used for polycondensate reactions, when either low molecular weight fractions have to be removed to shift the chemical equilibrium or when additives have to be added. The continuous process is characterized by high throughputs

Reactive Extrusion: Principles and Applications, First Edition. Edited by Günter Beyer and Christian Hopmann.
© 2018 Wiley-VCH Verlag GmbH & Co. KGaA. Published 2018 by Wiley-VCH Verlag GmbH & Co. KGaA.

with simultaneously good heat dissipation. The process used influences the residence time as well as the residence time distribution. Typical reactors are the stirring tank reactor, stirred tank reactors in series, and the tube reactor [8].

After the synthesis, the properties of polymers frequently do not meet the requirements for the production of component parts. Therefore, the material properties are specifically adjusted by adding fillers and additives. This takes place in a separate processing step that follows polymer synthesis. Mostly, a co-rotating and closely intermeshing twin-screw extruder is used for this compounding task, in which the polymers are melted and mixed with additives and functional fillers [12]. The additives and fillers can be mixed dispersively and distributively into the melt to achieve homogeneous material properties. The processing during compounding as well as the material composition influence the material properties. The produced material is usually formed as granules, and goes through further processes such as injection molding or extrusion [12]. Alternatively, direct compounding of sheets or profiles is also possible. This saves energy, since the material does not have to be melted again. In addition, the material experiences a lower thermal stress, which enhances the material properties.

To summarize, the supply chain of plastics processing usually consists of three steps: polymerization, compounding, and further processing. In industrial applications, these three steps are separated in time and space, as each process step is usually performed by a different company with clearly defined areas of responsibilities. At the beginning of the supply chain, the production of polymers is already characterized by inflexibility, as the chemical reactors are generally designed for the production of large amounts of polymers. A typical stirring tank reactor can handle a volume of up to $50\,\mathrm{m}^3$ [10]. Furthermore, many reactor types are not designed for the handling and mixing of high-viscosity liquids. In such cases, the synthesis of polymers takes place in the presence of a solution, suspension, or emulsion. The removal of the processing aids is complex and reduces the economic efficiency. Therefore, the continuous processing of specialty polymers for specific applications in small amounts ensures efficiency. The raw material manufacturers cannot react rapidly and flexibly to the changing market demands. Thus, it is not possible to satisfy individual customer requirements for small volumes. Therefore, raw material manufacturers produce only few variations in large quantities. This restriction at the beginning of the process chain hinders use in plastics applications that demand more specific characteristics profiles than are currently available. A higher flexibility during polymer synthesis could therefore create a wider range of scope of plastics applications.

In this context, reactive extrusion offers a more flexible alternative to polymerization in the presence of a solution, suspension, or emulsion and subsequent compounding. In reactive extrusion, the co-rotating, intermeshing twin-screw extruder that is traditionally used to melt, homogenize, and pump polymers through dies for compounding, is used for the synthesis of polymers. The extruder, which conveys the reactant oligomer or monomer in solid, liquid, or molten state represents a horizontal chemical reactor. As such, reactive extrusion includes the backward integration of the polymer synthesis into the compounding process. In principle, all reaction types of polymer synthesis can be realized in the twin-screw extruder, as long as the reaction rate is adapted to the residence time and distribution of the material in the processing machine [4, 7]. The residence time depends on the throughput, the length of the extruder, and the screw design and speeds. An increase of the residence time is possible within certain limits by linking two extruders to a cascade or by using low throughputs and screw speeds. However, this leads to a production that is not economically viable due to low capacity utilization [17].

One advantage of reactive extrusion is the absence of solvents as a reaction medium, as melts with different viscosities can be processed in a twin-screw extruder [7]. The absence of solvents in a conventional reactor technology causes additional issues regarding the removal of excessive heat, as polymer melts show a low thermal conductivity. In comparison, the reactive extrusion enables efficient heat removal due to the large reactor surface in relation to the reactor volume. The use of twin-screw extruders results in a high flexibility, as twin-screw extruders possess segmented barrels, which enable individual heating and cooling of single segments [7, 12]. In addition, the material is heated by the shear energy of the tightly intermeshing screw elements during conveyance and mixing. The introduced shear heating and the external heating provides the required heat for chemical reactions. The screws of a twin-screw extruder also consist of several segments. The geometry of the screw elements may differ depending on the depth between screw flights, the flight thickness, and the direction and degree of flight pitch [7]. Furthermore, mixing elements can be included for different mixing effects. Therefore, both dispersive and distributive mixing are possible. The design of the screw is key for the resulting mixing effect and the shear energy input.

A typical reactive extrusion process involves the following procedure: The reactants are fed through the main hopper into the twin-screw extruder, where the reactants are heated up to start the chemical reaction. The temperature profile, the use of activators and catalysts, and the residence time of the material in the extruder determine the reaction speed and the reaction progress. Consequently, the throughput, the length of the extruder, and the screw speed limit the degree of polymerization. Otherwise, the polymerization can be stopped specifically by the deactivation of reactive end groups to achieve a low viscosity. The addition of solid, liquid, or gaseous reactants is possible at almost any position at the extruder. For example, heat-sensitive additives can be added at the end of the extruder to minimize the residence time, or an inert gas can be added at the extruder feet to protect the process from degradation by atmospheric oxygen. Low molecular components, reaction byproducts and moisture can be removed by vacuum degassing during the reactive extrusion process. The degassing capability is dependent on the residence time, the melt temperature (which influences the viscosity), the diffusion rate, the attainable vacuum, the number of the degassing zones, and the renewal rates of the surface [11]. In addition, the adding of fillers and additives during the reactive extrusion is possible. The excellent mixing effect of the twin-screw extruder enables a homogenous incorporation of fillers. At the end of the extruder the melt is pressed through a die to either form strings that are pelletized, or to extrude directly sheets, films, or profiles. Thus,

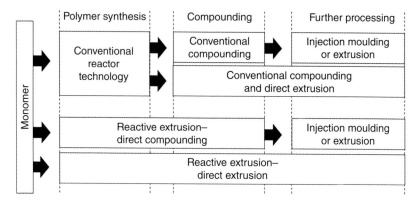

Figure 1.1 Simplified process chain from monomer to the final plastics product [11].

the application of the reactive extrusion shortens the processing chain compared to standard plastic processing (Figure 1.1).

Numerous research projects have demonstrated the possibilities regarding the types of reactions that can be carried out in an extruder [2]. These include the freeradical, anionic, cationic, condensation, and coordination polymerization of monomers or oligomers to high molecular weight polymers. Furthermore, controlled degradation and crosslinking of polymers by means of a free radical initiator for the purpose of producing a product with controlled molecular weight distribution and a higher concentration of reactive sites for grafting are possible. Commodity polymers can be functionalized for the purpose of producing materials to be used in grafting applications. In addition, various properties of the starting material can be improved by grafting monomers or mixture of monomers onto the backbone of existing polymers. Free radical initiators and ionizing radiation can cause the initiation of grafting reactions. Moreover, the formation of interchain copolymers is conceivable. This type of reaction involves combination of reactive groups from several polymers to form a graft copolymer. A final example are coupling reactions that involve reactions of a homopolymer with either a polyfunctional coupling agent or a condensing agent in order to increase the molecular weight by chain extension or branching. Table 1.1 gives an overview about typical types of reactive extrusion [2].

The research work of *Brown* and *Xanthos* gives a detailed overview about reactions that can be carried out in the reactive extrusion [3, 17]. As described earlier, in addition to the reactive extrusion for the synthesis of polymers, the twin-screw extruder provides the possibility of reactive modification as a reaction step after the polymer synthesis in a conventional reactor technique. Existing polymers, for example, apolar polymers, can be functionalized in reactive extrusion to improve compatibility with and adhesion to polar polymers. An overview about the reactive modification for blending immiscible polymers is given in [1, 9]. New functionalities can be introduced to polymers by reactive extrusion with radical polymerization, polycondensation or polyaddition. Furthermore, the viscosity of the product can be adjusted to the requirements of the application by initiated degradation or crosslinking reactions. In the following text, some

Table 1.1 Possibilities of reactive extrusion.

Reaction type	Reactants	Products
Polyaddition	Polyol + diisocyanate + aromatic diamine	Polyurethane
Polycondensation	Bishydroxybutyl terephthalate	Polyethylene terephthalate
	Precondensate	Polyamide
Free radical copolymerization	Styrene + acrylonitrile prepolymer	SAN
Grafting	Polystyrene + maleic anhydride	Polysytrene maleic anhydride adduct
Ionic polymerization	Caprolactam	Polyamide 6
Anionic copolymerization	1,3-Diene + aromatic vinyl compound	Block copolymers of 1,3-diene and aromatic vinyl compounds
Hydrolysis	Polyurethane scrap	Polyol, amines

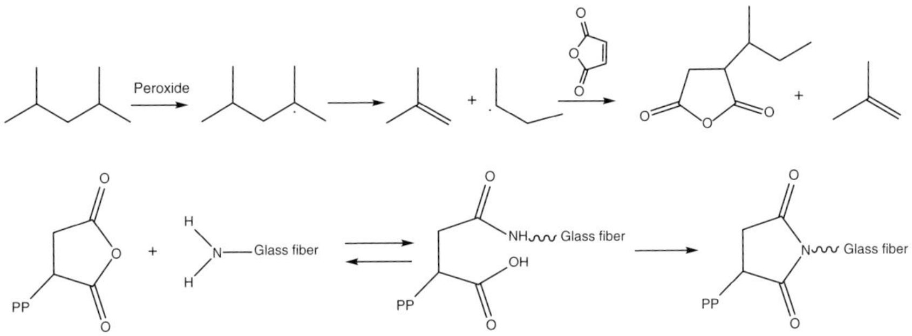

Figure 1.2 Modification of PP with MAH and glass fibers [4].

selected examples of polymer modification are presented that play important roles in industrial practice.

Radical polymerization is the most frequently used type of polymerization for the functionalization of polyolefins. Chain shortening, extension, branching or grafting can be achieved by the formation of radicals. Peroxides are usually used as initiators. This method is used for polyolefins, as the lack of functional groups in polyolefins prevents its wide use in various applications. The grafting of polyolefin (e.g., PP) with maleic anhydride (MAH) is one of the most performed polyolefin modifications [13]. Different formulations of MAH-grafted PP are available in the plastics market. MAH-grafted PP is able to achieve good compatibility with glass fibere that are coated with aminosilane (Figure 1.2). Detailed information on the structure and synthesis of modified polyolefins is given in [14].

Polycondensation reactions can also be carried out in reactive extrusion. In a polycondensation reaction, functionalized monomers or low molecular weight compounds react with polymers by release of smaller molecules in a step-growth reaction. Monomers, low molecular weight compounds, or reactive polymers have functional groups such as carboxyl, hydroxyl, epoxy, or amino groups.

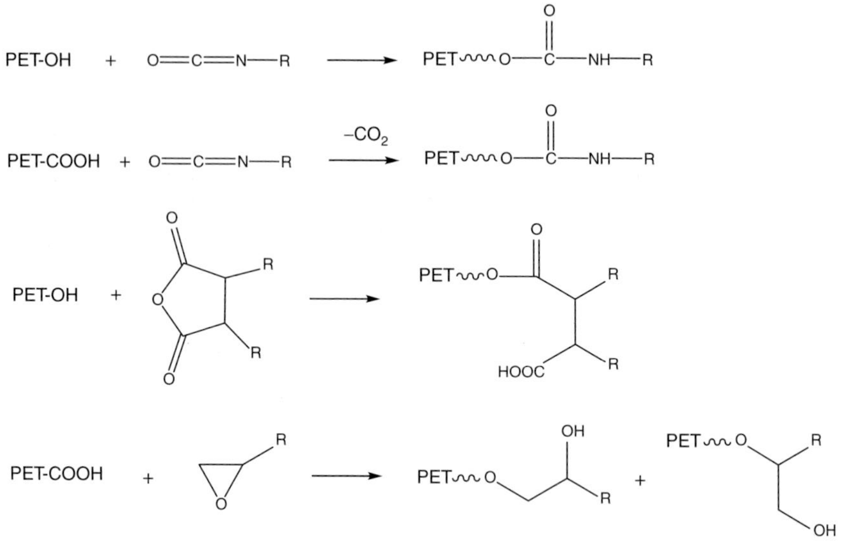

Figure 1.3 Reactions to extend PET chains [4].

Polyethyleneterephthalate (PET) is one of the most important polycondensation products especially in bottle production. The recycling process of PET is very interesting from an economic and ecologic point of view. However, recycled PET often has too low a molecular weight for reuse. Therefore, the industry has developed a discontinuous post-condensation process for increasing the molecular weight of PET. However, the material has a long residence time in the reactor and high temperatures occur. Alternatively, reactive extrusion has been considered. There has been extensive research on the extension of PET chains by functionalized molecules ("chain extenders") in reactive extrusion [4]. The chain extender can be added during the melting of PET in the extruder. Typical chain extenders are isocyanates, anhydrides, and epoxides that react with the hydroxyl and carboxyl groups of PET [15, 16]. Figure 1.3 shows an example of reactions to extend PET chains.

However, the method of chain extension has also disadvantages. The chain extenders are often toxic substances and therefore the products are not suitable for food applications. Despite the shorter reaction times in the reactive extrusion, the process is more expensive to realize established post-condensation processes to increase the molecular weight of the polymer [4].

To summarize, the reactive extrusion offers many possibilities for flexible and economic production of materials with specific characteristics. In addition, the continuous process, the narrow residence time distribution, and the high flexibility also enable the economic production of small amounts of specialty polymers. The reaction enthalpy, which is set free during many polymerization reactions, can be used to heat up the melt. This allows energetically favorable processes. In spite of the advantages, the reactive extrusion from monomer to polymer plays a tangential role in industrial practice. This is exemplified on the basis of the number of existing twin-screw extruders, which are designed for the use in the field of reactive extrusion: In the past 50 years, one of the leading suppliers of twin-screw extruders has sold only 87 optimized twin-screw extruders for reactive extrusion. All these plants are designed for the production of thermoplastic polyurethane by using reactive extrusion. These include 43 twin-screw extruders with throughputs between 600–3000 kg h^{-1} (ZSK 83–133) and 21 twin-screw extruders with throughputs 200–600 kg h^{-1} (ZSK 58–70) [5]. Therefore, theoretically, the production of 250.000 t a^{-1} of thermoplastic polyurethanes is possible by a utilization rate of 80%. In 2012, 430.000 t of thermoplastic polyurethane was produced [6]. The reactive extrusion could potentially satisfy nearly 60% of the global market of thermoplastic polyurethane. However, the production of thermoplastic polyurethane is the only example where reactive extrusion is used for polymerization on an industrial scale. It is, of course, possible, that other standard twin-screw extruders are also used for reactive extrusion or that other polymers are synthesized by the use of reactive extrusion in isolated cases. But the present situation indicates that synthesis of polymers with monomers using reactive extrusion is not given much importance in industrial practice. Reasons for this might be the following disadvantages [4]. The limited residence time allows only fast reactions at high throughputs. Nonideal process conditions could cause local high melt temperatures that can trigger uncontrolled side reactions. A further aspect that should be considered is that the use of reactive extrusion demands a comprehensive knowledge about chemical reactions as well as the extrusion processes. The knowledge about these two complex areas is rarely available within just one company. In addition, the company has the responsibility for both the production and the processing of the material, which is a big challenge and not always appreciated by the customer since it limits the exchangeability with the supplier.

However, the addition of compatibilizers during compounding is a common strategy that is adopted to improve the compatibility and adhesion between different polymers or polymers and fillers that is already widespread today. These compatibilizers can initiate chemical reactions. Therefore, reactive extrusion is already employed to a minor extent in many compounding processes. Consequently, reactive extrusion is an essential part of the compounding process and will gain importance in the future, as the potential of blends or fillers depends on the interactions between components, which are in turn influenced by the reactive extrusion process.

References

1 Baker, W., Scott, C., and Hu, G.H. (2001) *Reactive Polymer Blending*, Carl Hanser Verlag, München.
2 Brown, S.B. and Orlando, C.M. (1988) *Encyclopedia of polymer Science and Engineering*, 2nd edn Vol. 14, Wiley-Interscience, New York, pp. 169–189.
3 Brown, S.B. and Zhu, S. (2002) *A survey of chemical reactions of monomers and polymer during extrusion processing*, in *Reactive Extrusion: Principles and Practice* (ed. M. Xanthos), Hanser-Publishers, New York.

4 Cassagnau, P., Bounor-Legare, V., and Fenouillot, F. (2007) Reactive processing of thermoplastic polymers: a review of the fundamental aspects. *Int. Polym. Process.*, **22** (3), 218–258.

5 Elsaesser, R. (2016) *Personal Communication*, Coperion GmbH, Stuttgart.

6 Galbraith, C. (2013) *A Global Overview of the Thermoplastic Polyurethane (TPU) Market*, IAL Consultants, Press Release.

7 Grefenstein, A. (1996) *Reaktive Extrusion und Aufbereitung: Maschinen-technik und Verfahren*, Carl Hanser Verlag, München, Wien.

8 Hamielec, A.E. and Tobita, E. (2001) *Polymerization Prozess, Industrial Polymers Handbook*, Wiley-VHC Verlag, Weinheim.

9 Harrats, C., Thomas, S., and Groeninckx, G. (2006) *Micro- and nanostructured Multiphase Polymer Blend Systems*, CRC Press LLC, Boca Raton.

10 Keim, W. (2006) *Kunststoffe: Synthese, Herstellungsverfahren, Apparaturen*, Weinheim, Wiley-VCH.

11 Klünker, E. (2015) Restmonomerentgasung und Direktcompoundierung bei der reaktiven Extrusion von Polyamid 6. PhD thesis. RWTH University, ISBN 978-3-95886-031-5.

12 Kohlgrüber, K. (2007) *Der gleichläufige Doppelschneckenextruder*, Carl Hanser Verlag, München.

13 Kowalski, R.C. (1992) *Fit the reactor to the chemistry*, in *Reactive Extrusion: Principles and Practice* (ed. M. Xanthos), Hanser Publishers, New-York.

14 Moad, G. (1999) The synthesis of polyolefin graft copolymers by reactive extrusion. *Prog. Polym. Sci.*, **24**, 81–142.

15 Scheirs, J. (2003) *Additives for the modification of poly(ethyleneterphthalate) to produce engineering-grade polymers*, in *Modern Polyesters, Chemistry and Technology of Polyesters and Copolyesters* (eds. Scheirs, J.; Long, T.), Wiley & Sons, New-York.

16 Xanthos, M. (2002) *Reactive modification/compatibilization of polyesters*, in *Handbook of Thermoplastics Polyesters, Homopolymers, Copolymers, Blends and Composites* (ed. S. Fakirov), Wiley-VCH, Weinheim.

17 Xanthos, M. (1992) *Reactive Extrusion – Principles and Practice*, Carl Hanser Verlag, München, Wien, New York, Barcelona.

Part II

Introduction to Twin-Screw Extruder for Reactive Extrusion

2

The Co-rotating Twin-Screw Extruder for Reactive Extrusion
Frank Lechner

Coperion GmbH, Process Technology, Theodorstrasse 10, D-70469 Stuttgart, Germany

2.1 Introduction

In 1953, Coperion acquired the license to use the Erdmenger patents for co-rotating, closely intermeshing, multi-screw machines from Bayer AG and in 1957 delivered the first twin screw compounder, the ZSK.

Co-rotating twin-screw extruders are particularly suitable for the reactive extrusion process due to several reasons. The reactive extrusion process with twin-screw extruders of the ZSK type applies to mass polymerization, that is, starting from monomers or prepolymers, as well as to modification of polymers by grafting, crosslinking, or degradation. The main advantages of continuous compounding machines as compared to more basic batch type units are that (i) they can cover the whole range of viscosity values from monomers to polymers in one reactor and (ii) they have excellent mixing capabilities. This allows producers to use them for cost-effective mass polymerization processes, instead of having to rely on complex methods in solution, suspension, or emulsion. Conducting several separate operations within one machine, like melting, mixing, and degassing is very typical. Table 2.1 lists some of the reactive extrusion processes, some of which have not been commercialized. However, some of them have been commercialized successfully and are nowadays state of the art.

Being a compact mass device, with relatively small amounts of several raw material being fed into the extruder continuously at the same time, distances for diffusion and convection inside the process part become very short. These continuous compounding machines are characterized by high flexibility, that is, small lot sizes and varying formulations are possible. Additionally, different size extruders for throughput rates from a laboratory scale to production capacities of several ten thousand kilograms per hour are available.

Reactive Extrusion: Principles and Applications, First Edition. Edited by Günter Beyer and Christian Hopmann.
© 2018 Wiley-VCH Verlag GmbH & Co. KGaA. Published 2018 by Wiley-VCH Verlag GmbH & Co. KGaA.

Table 2.1 Overview of reactive extrusion processes on the co-rotating twin-screw compounder.

Product	Feed product	Type of reaction
TPE-V	PP, EPDM, oil, x-linking agent	Free radical
TPU	Polyols, isocyanate	Polyaddition
PP-g-MA	PP, MA + peroxide	Free radical
Partial glycolysis PET	R-PET, glycol	Glycolysis, degradation
HT-PA	Precondensate	Polycondensation
SMAC	Styrene, maleic anhydride	Polyaddition
PA6	Caprolactam	Ionic polymerization
POM	Trioxane, comonomer	Ionic polymerization
PMMA	Methacrylic ester	Radical polymerization
Polyarylate	Bisphenol A, phthalic acid	Polycondensation

Source: Reproduced with permission of Coperion GmbH.

2.2 Development and Key Figures of the Co-rotating Twin-Screw Extruder

While several initial concepts for co-rotating twin screw devices were patented in the early 1900s by Wuensche and Easton [1] the co-rotating design used as the basis for essentially all twin screw compounding systems that are marketed today is based on the self-wiping element profile known as the Erdmenger profile. The initial design and development of this self-wiping element profile is described in German Patent 862,668 granted to W. Meskat and R. Erdmenger in 1952 with a priority date of 1944 (no US Patent filed). The objective of the design at that time was mixing high viscosity liquids already in the fluid state, such as post-polymerization reaction products.

The above noted patent along with the numerous related patents that followed (all issued to Erdmenger or one of his colleagues at Bayer) defined the base design parameters of the ZSK twin-screw extruder, as well as the many copies introduced during the intervening 50 plus years for the eventual development and commercialization by Werner and Pfleiderer in the late 1950s. The key feature of the design is the self-wiping characteristic of one screw with respect to the other. This eliminates stagnation and eventual degradation of material as it is transported along the length of the compounding extruder. Co-rotating and fully intermeshing twin-screw extruders, with one screw wiping the other, and both wiping the eight-shaped barrel inside, are defined by three characteristic dimensions: diameter ratio D_o/D_i specific torque M_d/a^3 (M_d = torque, a = centerline distance) and screw speed RPM, Figure 2.1.

The flight depth ratio D_o/D_i is the characteristic dimension for the free volume whereas the specific torque M_d/a^3 is the key figure for power transmission (gearbox as well as screw shaft design and material of construction). The maximum screw speed of the ZSK MEGA compounder is 1200 rpm. The actual rpm used depends on the product and overall process requirements. The key to the success

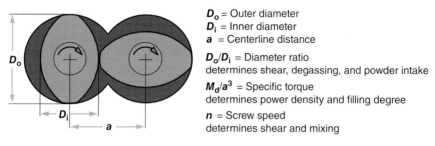

Figure 2.1 Character dimensions of twin-screw extruder. (Reproduced with permission of Coperion GmbH.)

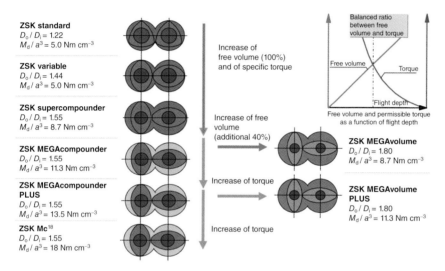

Figure 2.2 Development of the ZSK design parameters diameter ratio and specific torque. (Reproduced with permission of Coperion GmbH.)

of this technology is the increase in the power (torque) transmission capacity in combination with increased screw rpm. A system that simply runs at higher rpm will at some juncture impart enough additional energy to the material being processed to cause degradation. Thus, while maintaining the same centerline distance for torque-limited products, the throughput increased by 7% per year over the entire 50-year life of this kneading system. Seven ZSK generations have been developed to date (see Figure 2.2).

The large-volume ZSK MEGAvolumePLUS, with deeply cut screws (Figure 2.2), is designed for speeds up to 1800 rpm. It is used for products that require a large-volume process chamber, to give better air flow back from fluidizing powders, to create residence time for reactions, or because a temperature or product quality limit occurs long before the torque limit is reached.

Co-rotating twin screws are built using a modular design, see Figure 2.3. The advantage of this modular principle is the individual adaptation of the process section to every application – so that optimum product quality and maximum

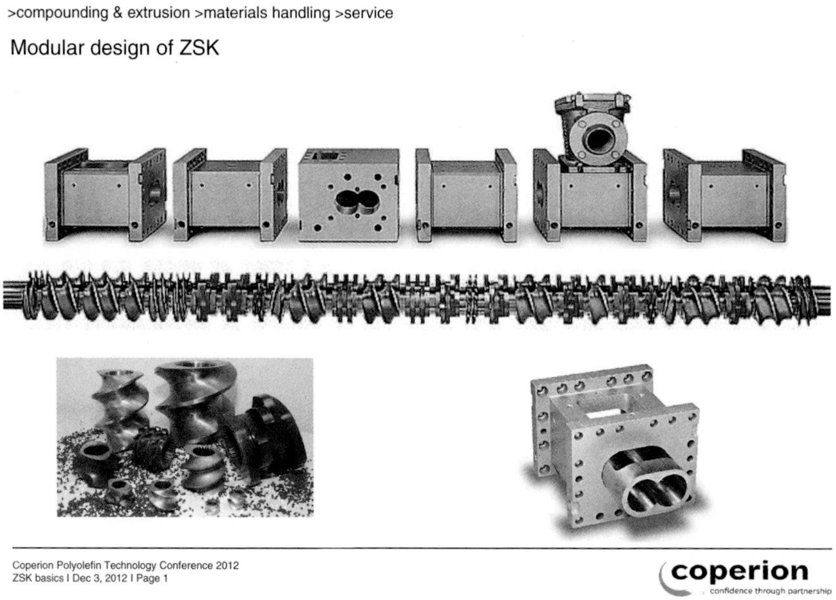

Figure 2.3 Modular design for ZSK barrels and screw elements. (Reproduced with permission of Coperion GmbH.)

throughput are achieved. The entire processing system is constructed from individual barrel sections and individual screw elements (modular design). This will ensure that the processing length and also the screw configuration can be tailored to each process task in order to achieve optimum performance. The barrel sections are manufactured as closed barrels for melting, mixing, and pumping, and as open barrels for feeding and venting/degassing.

Every barrel can be individually temperature controlled. Heating is typically done electrically with heater shells or heater cartridges and cooling by water. The temperature of the barrels can also be controlled with liquid or steam heat media for special applications.

2.3 Screw Elements

As shown in Figure 2.4, there are three types of screw elements for conveying (forward and reverse), kneading and distributive mixing. The wiping between the two screws can be seen from the narrow and constant gap between the two screws. Conveying elements transport material from one section of the extruder, such as the feed section, to another, such as the melting section. The degree of fill is increased as the screw pitch of the forward conveying element is reduced. Reverse pitch screws are 100% full. These elements also create a dynamic seal between different pressurized or vacuum zones in the extruder.

Kneading blocks are used to melt, disperse (break into smaller particles), and homogenize the material being processed. They are normally completely

Figure 2.4 Types of screw elements. (Reproduced with permission of Coperion GmbH.)

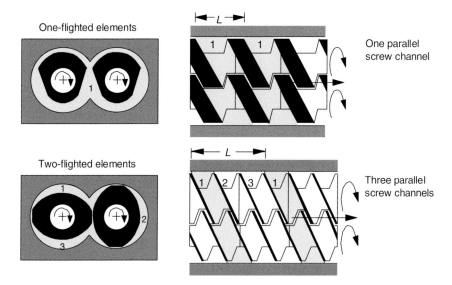

Figure 2.5 One- and two-flighted conveying screws. (Reproduced with permission of Coperion GmbH.)

self-wiping and vary in disk width and disk to disk rotation angle and direction. Distributive mixing elements have a lot of stream splitting channels but introduce minimal energy into the material. They are generally not mechanically self-wiping but kept clean by the shear stresses of the high viscosity product flow. Figure 2.5 shows the typical design of one- or two-flighted conveying elements; however, three-flighted elements are also still used for plastification tasks. The designation 1, 2, or 3 refers to the number of independent helical flow channels along the axis of the element. Today, two-flighted screws with open channels and small crests are the most commonly used design. Single-flighted elements with broad crests and a slightly smaller free volume have advantages in the feed and discharge zone due to their higher conveying efficiency and in the discharge zone due to better bearing surface characteristics in the viscous material.

The working principle of kneading blocks is shown in Figure 2.6. Staggering the kneading disk in the conveying direction creates low shear distributive mixing and material transport (T) in the flow direction. However, staggering the disks in

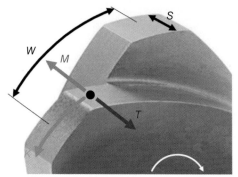

There are the following relations: W↑ --> M↑
--> T↓
S↑ --> M↓
--> K↑

W ... Staggering angle
S ... Width of the disk
T ... Transporting effect
K ... Dispersive mixing (kneading, shearing)
M ... Distribuitive mixing

Figure 2.6 Working principle of kneading blocks. (Reproduced with permission of Coperion GmbH.)

the reverse direction results in high shear dispersive mixing (M) in filled channels. The small percentage of particles flowing in the circumferential direction (K) are extensively kneaded and sheared. The higher the staggering angle (W) the better is the distributive mixing effect, the broader the kneading disck (S), and the better the dispersive mixing.

The increased diversity of materials being utilized in current product formulations, as well as the breadth of required mixing tasks has led to the need for specialized kneading blocks. Some of the issues that have had to be addressed are:

- How to keep soft compactable materials, such as some organic pigments, from being compressed into large agglomerates prior to entering the dispersive mixing section of the screw design.
- How to keep the temperature of rubber compounds from rising almost uncontrollably during processing.
- How to disperse high molecular weight particles that are surrounded by a low molecular weight matrix, that is, bimodal materials.
- How to maximize the incorporation of large amounts of filler in the shortest axial distance along the extruder.
- How to quickly mix reactive ingredients.

To address the issues noted above, several non-typical kneading blocks have been designed. These include three-lobe kneading blocks for the two-lobe geometry, modified disk tip width kneading blocks, tapered kneading blocks, reduced diameter reverse flight cross-cut mixing elements, as well as several other design types.

Three-lobe eccentric kneading blocks for two-lobe extruder systems were designed for more uniform energy input on the two-lobe system (Figure 2.7),

Figure 2.7 Three-lobed kneading blocks. (Reproduced with permission of Coperion GmbH.)

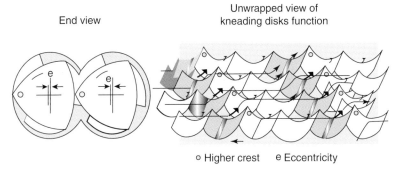

Figure 2.8 Operating principle of three-lobed kneading blocks. (Reproduced with permission of Coperion GmbH.)

particularly in the melting zone. As shown in Figure 2.8, the elements are offset so that one tip wipes the barrel wall while the other two have a significantly greater clearance (three to five times), which permits polymer melt to flow more easily over these tips. This sets up a more circumferential rather than a down channel material flow in the element. The result is a more energy-efficient as well as a more homogeneous polymer melting. The material exits the melting zone with few (if any) remaining unmelted particles.

In addition to providing more uniform melting, three-lobe kneading blocks also are less prone to compacting pigments, fillers, or additives during the melting process. The three-lobe kneading blocks have a lower pressure peak at the apex than do two-lobe kneading blocks. Figure 2.9 illustrates the pressure rise in the apex region for the two-lobe geometry as the two tips of the kneading disk come together. The lower apex pressure for eccentric kneading blocks is primarily the result of the increased clearance of two tips, but three-lobe kneading blocks also have a smaller crest angle (Figure 2.10).

The reduced thickness disk tip (SAM) kneading block was designed for lower shear and material compression (Figure 2.11) in the apex region. By removing a portion of the disk tip, there is less pressure generated at the apex.

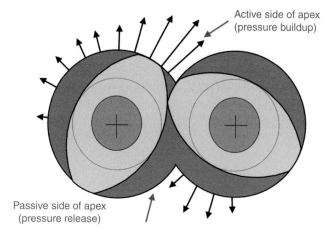

Figure 2.9 Schematic of radial pressure profile in two-lobed kneading disks. (Reproduced with permission of Coperion GmbH.)

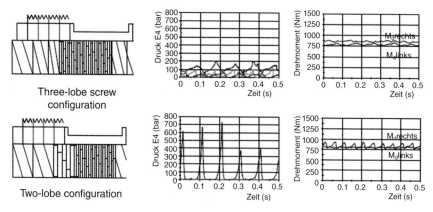

Figure 2.10 Radial pressure profile comparison between two- and three-lobed kneading disks. (Reproduced with permission of Coperion GmbH.)

Tapered kneading blocks were also designed for reduced compression in the apex region (Figure 2.12), but specifically for elastomeric materials that dissipate significant amounts of heat energy with each compression cycle.

Turbine-and-gear mixing elements are defined by the number of teeth around the circumference as well as the tooth angle. The former contributes to stream splitting for generation of interfacial surface and the latter for conveying capacity. Figure 2.13 shows these design considerations for the full barrel self-wiping and partial barrel self-wiping toothed elements, respectively. The main function of both elements is to provide the maximum amount of distributive mixing (little if any dispersive mixing) with minimal energy input. The self-wiping version is an evolution of the design to meet ever increasing quality standards. The elements, which do not wipe the entire barrel wall, provide a low velocity zone for material stagnation and potential degradation for more heat and shear-sensitive materials.

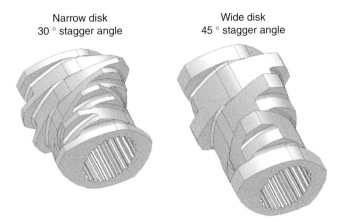

Figure 2.11 Low shear (SAM) kneading blocks. (Reproduced with permission of Coperion GmbH.)

Figure 2.12 Tapered kneading blocks (TKB). (Reproduced with permission of Coperion GmbH.)

Figure 2.13 ZME and TME mixing elements. (Reproduced with permission of Coperion GmbH.)

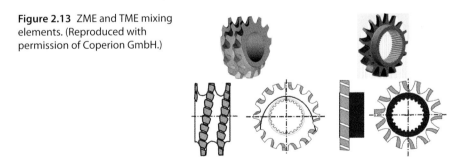

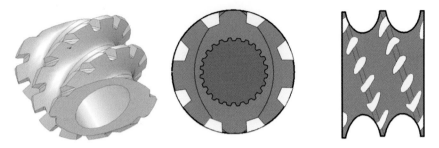

Figure 2.14 SME mixing element. (Reproduced with permission of Coperion GmbH.)

Another type of distributive mixing element (Figure 2.14 is the screw mixing element (SME). This element is based on a standard screw bushing profile but with notches cut across the flight crest. This permits communication in the axial direction between screw channels through increased leak flow. As a consequence of the increased back flow, these elements provide a melt homogenizing effect. They also have a decreased conveying capacity compared to standard screw bushings with the same pitch. This results in a higher fill factor and increased material residence time in the element.

2.4 Co-rotating Twin-Screw Extruder – Unit Operations

The modular building block concept of the co-rotating twin-screw provides design flexibility to configure the various available barrel sections and screw elements in a specific series of unit operations to successfully process a wide variety of mixing tasks from incorporating filler into polymer to blending low viscosity reactants. Figure 2.15 illustrates the most common sequence of unit operations, which will be highlighted in the following sections.

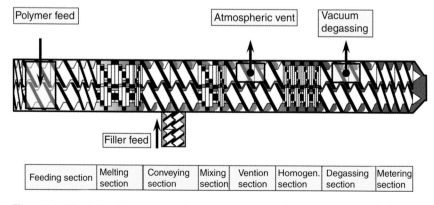

Figure 2.15 Typical unit operations. (Reproduced with permission of Coperion GmbH.)

2.4.1 Feeding

Co-rotating twin-screw extruders are designed to be starve-fed. Therefore, throughput is independent of screw speed. This permits the processor to control residence time, degree of fill, and specific energy input ($kWh\,kg^{-1}$).

Feeding locations are not limited to the first barrel section. Depending on the feedstock or recipe, feed streams, either solid or liquid, may be introduced at a number of locations along the process section.

2.4.2 Upstream Feeding

In compounding operations, the feedstock introduced at the first barrel section is usually a polymeric solid. The form is typically pellets, flake, or powder. Most pellets and flakes are easy to feed since they have a high bulk density, their individual dimensions are sufficiently small with respect to extruder channel depth and they do not lubricate the screw and thus impede the drag flow mechanism. Standard profile screw elements with a pitch of 1D to 2D are appropriate.

In situations where the particle size is large with respect to the channel depth (>1/2 h), the material has a lower bulk density, or absolute maximum feed capacity is needed, special elements with increased volume are useful. The self-wiping profile of the pushing flight (and sometimes the trailing flight) has been transformed into a square channel profile. This modification accomplishes two functions. First, it directs more force acting on the material to be in the down-channel direction. Second, it creates up to 40% additional free volume in the element.

Powders that tend to fluidize, especially those such as silica, which have a very low initial bulk density, are significantly more difficult to feed than pellets or flakes. The first step to successful feeding depends upon eliminating (or at least minimizing) fluidization, and controlling the separation of already entrained air, prior to material entering the extruder. In general, the vertical drop should be as short as possible. Also, the feed has to be directed to the down-turning section of the screw. This is the barrel wall for the co-rotating extruder. Ideally, the screw configuration should be designed to allow air to travel down a channel for removal through a vent section rather than be trapped and forced to flow countercurrent and back out the feed throat. However, in order to be effective, a melting zone typically must be backed up by a restrictive element. This element then blocks the air from traveling downstream. In this case, special elements such as increased volume (undercut) or single flight (SF) element play an important role in feed introduction. The choice of either increased volume or SF elements depends upon the amount and degree of air separation required. For a relatively low amount of air, the undercut element again provides a greater free volume and a more open path for air to flow backward. However, as more air is forced back to the feed throat, the material fluidizes and the undercut elements lose their effectiveness. Under these circumstances the SF should be considered. Unlike the undercut element, which has greater free volume, the SF has approximately 15% lower free volume per unit length than a standard element. The SF elements do not allow air to flow easily in a countercurrent direction and therefore the air is forced to flow past the restriction in the plastification section. These elements function in this manner as a result of the severe flow restrictions through the apex caused by the wide crest. These crests create a positive displacement flow greater than in any other co-rotating twin screw element.

2.4.3 Downstream Feeding

In many compounding operations it is necessary to split the feed streams. This may be required in order to (i) achieve disperse phase size of an impact modifier, (ii) retain aspect ratio of reinforcing fiber filler, and (iii) obtain high level of loading for either low bulk density filler or incompatible low viscosity additives.

The most efficient way to add solids in a downstream location is to use a twin screw side feeder. The twin screw side feeder is normally a co-rotating device with high D_o/D_i ratio (~2.0). It has several advantages over single screw side or top feeders. First, the twin screw has better solids conveying characteristics since as a twin screw device it does not rely totally on drag flow. The twin screw also has a wider longitudinal as well as a larger total cross-sectional discharge opening than a single screw side feeder, and therefore provides lower pressure and more uniform feed introduction. Typically there is a vent above the side feed opening in the barrel. This permits entrained air to escape. For very low bulk density material containing significant air, the vent opening is moved one barrel upstream to avoid material particles getting entrained in the air as it exits the extruder. By having the air travel upstream before discharge, the material particles have to travel a path around the screw channel. This disrupts the air flow and causes the solid to disengage.

Liquid additives are mostly added downstream of the plastification section because they tend to lubricate the pellets or cause powders to agglomerate in the feed throat. If a significant amount of liquid is to be incorporated, it can be added at several locations. The most effective method for low viscosity liquid incorporation is to inject it into a fully filled distributive mixing section. This requires a pressure injection valve and positive displacement pump. For small amounts of compatible liquid, non-pressurized injection into a low degree of fill area of the screw configuration may also be acceptable.

2.4.4 Melting Mechanisms

Melting of polymeric material requires energy to be transferred from an outside source into the material. In the twin-screw extruder this energy transfer occurs through both mechanical and thermal mechanisms. However, as the extruder gets larger, the surface to volume ratio decreases significantly. Therefore mechanical energy transfer is the dominant mechanism for plastification.

2.4.5 Thermal Energy Transfer

Normally, thermal energy is introduced through electric heaters surrounding the barrel or heat transfer medium that is pumped through barrel bores. As mentioned previously, smaller extruders can introduce a greater percentage of energy through heat transfer. Conversely, they can also lose a higher percentage of heat through heat transfer. If too much heat is lost in this manner, then on scale-up to a production unit, the same percentage of heat cannot be removed. This will result in a higher discharge temperature. It is therefore very important that lab or development extruder run as close to adiabatic conditions as possible.

2.4.6 Mechanical Energy Transfer

In most cases, the majority of energy required for plastification comes from the mechanical input of the screw configuration. A kneading block is the primary tool used to accomplish this task. The amount of mechanical energy introduced depends not only on the number of kneading blocks but also on the configuration within the plastification zone, the screw RPM, and the throughput rate. Specific mechanical energy (SME), the energy introduced per pound or kilo of product, increases with RPM, especially when the plastification section is backed-up with a restrictive screw element. A second restrictive element further increases SME.

2.4.7 Mixing Mechanisms

Mixing requirements for polymer compounding can be divided into two basic disciplines: dispersive and distributive. Dispersive mixing breaks down a particle into smaller units, while distributive mixing homogenizes the spatial relationship of the particles (whether dispersed or not).

The basic building blocks for mixing in the co-rotating twin screw are kneading blocks and special mixing bushings. These special elements include toothed mixing units and blister rings, both standard geometry and self-wiping.

It is not surprising that the mixing element selection for distributive and dispersive blending processes is different. The distributive mixing profile uses narrow kneading blocks to maximize the number of flow divisions per machine length.

In dispersive mixing, wide-disk kneading blocks are used to increase shear stress applied to the material. These elements are typically backed up with a restrictive kneading block or screw bushing to increase the degree of fill as well as residence time. The number of restrictive elements that can be used is limited due to the resulting temperature buildup and potential material degradation. Therefore, in order to transmit increased stress to the material, deploying a number of shorter mixing sections (1D–2 D long) is often more efficient than one long mixing section because this permits elastic materials to relax in the conveying sections between the mixing areas.

For very low viscosity products, sufficient energy for dispersive or distributive mixing is introduced by adjusting the feed sequence such that only a small portion of the solvent or dilution oil is added in the upstream part of the twin-screw extruder and therefore high mechanical stress transfer can be achieved. The remainder of the diluent is introduced at the end of the machine.

2.4.8 Devolatilization/Degassing

In compounding lines, devolatilization typically involves removal of entrained air, moisture contained in the incoming feedstock, or byproducts from any reaction between recipe components. The amounts are generally less than 1% or 2% and typically less than 1%. Therefore, the staged vacuum setups and stripping techniques used in devolatilization processes where 10–60% volatiles must be removed, are not necessary. However, even in compounding, devolatilization is a critical step, because while it is necessary to remove these volatiles, it serves as a location for potential contamination. The screw configuration in the vent area typically comprises screw bushing with a pitch from 1D to 2 D. This design permits sufficient residence time at a lower degree of fill to remove volatiles while maintaining the polymer within the bounds of the screw flight.

2.4.9 Discharge

Material is discharged from most twin screw compounding operations as pellets or specialized forms, such as sheets, tubes, ropes, or other more complicated profiles. In order to push material through these dies, the machine must generate the appropriate pressure. This can range from 20 bar or less for strand dies, up to 300 bar or so for sheet and other profile dies. When the discharge pressure exceeds 120 bar, or precise product gauge control is critical, a single-screw extruder or melt gear pump is typically used to generate the pressure in the co-rotating twin screw. In large scale polyolefin pelletizing operations, rates can be dramatically increased by integrating a gear pump into the system. Also, fiber spinning and sheet extrusion operations typically use a gear pump at the end of the machine.

2.5 Suitability of Twin-Screw Extruders for Chemical Reactions

The main advantages of continuous compounding machines compared to simpler design batch units, are that (i) they cover the whole range of feed/product viscosity and (ii) they have a self-wiping screw geometry. A typical chemical reaction starts with a blend of lower viscosity liquids and then, through reaction, produces a higher viscosity material. Stirred vessels can handle initial low-viscosity feedstock material, but will stall out when the polymerizing material reaches higher levels of viscosity. Thus the stirred vessel technology is reaching its limits with increasing viscosity of the reactants or of the final product. One method to handle this issue is to run the reaction in a low viscosity solvent. A more environment friendly method is to run the process in the twin screw compounder.

Theoretically, all processes could be executed continuously in a twin-screw extruder. But because of the comparatively unprofitable price/volume ratio (compared to a stirred vessel) it only makes sense to process high-viscosity materials. There is no exact limit, but tentatively the raw materials or the final product should be on a level higher than approximately 5–10 Pa s [2].

Reaction processes in stirred vessels are typically executed in minutes or even hours. This long reaction time often is required due to poor ratio of heat exchange surface/volume, degassing surface/volume, and low mixing intensity compared to a twin-screw extruder.

To achieve an economical relationship of machine size to throughput, the reaction time in the ZSK should not be longer than 2–3 min. Due to the substantially superior mixing, surface renewal and heat exchange, a much lower reaction

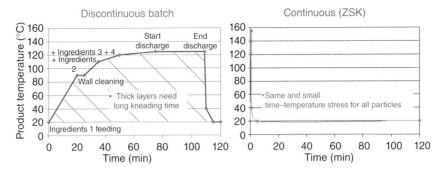

Figure 2.16 Time–temperature curves for discontinuous and continuous kneading. (Reproduced with permission of Coperion GmbH.)

time can be achieved in a twin-screw extruder than in a stirred vessel. TPU (Thermoplastic Polyurethane) formerly was chiefly synthesized with a reaction time of 30–60 min in a stirred vessel. Today, the ZSK twin screw process with a residence time of 1–2 min is state of the art. Even though there is intensive heat transfer from the barrels to the product inside of the process section, the reaction processes can be restricted by the specific heat of reaction. Depending on the heat of reaction, the process can face scale-up limitations because the heat exchange surface/volume ratio is getting smaller and smaller as the machine size (diameter) is increased.

The difference between the batch and the continuous system is shown schematically in Figure 2.7. The batch kneader is stopped several times to feed the different ingredients. These are then mixed into the matrix in thick layers over a long period of time. Conversely, the continuous process meters all ingredients along the process section, mostly with gravimetric feeders. These ingredients are incorporated in very thin layers in an extremely short time. Figure 2.16 shows typical time–temperature curves for both types of processes for hot melt pressure sensitive adhesives. The batch process including the single-screw discharge at the end of the kneading process requires 2 h. The continuous process is completed in 2 min (or less). While the peak temperature is slightly higher, the integral time–temperature history is significantly shorter with a very homogeneous treatment of all particles.

2.6 Processing of TPE-V

Thermoplastic elastomers (TPE) are distinguished both by their elastomeric properties and their thermoplastic processing properties. They are divided into TPE-S, TPE-O, TPE-U, TPE-E, TPE-A, and TPE-V based on their chemical–morphological structure.

TPE-O in this case describes elastomer alloys that consist of different strengths of thermoplastics and elastomers, usually SBS/SEBS.

TPE-V describes TPE with a crosslinked rubber component. The crosslinking and dispersion of the EPDM and the mixing of, in some instances, large quantities

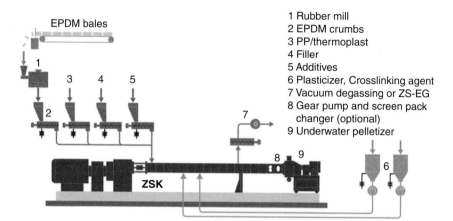

Figure 2.17 Typical setup for TPE-V. (Reproduced with permission of Coperion GmbH.)

of plasticisers (>40%) demand twin-screw extruders with process lengths of an average $L/D = 56$. The configuration of the process section specifically is adapted to the process steps to be executed. The exact setup depends upon the raw materials and the crosslink system. Figure 2.17 illustrates a typical setup for EPDM/PP-based TPE-V.

The PP, EPDM, and any small amount of fillers (e.g., talc, chalk, etc.) are fed at the beginning of the extruder. The PP and EPDM are melt mixed in the first few barrels. Depending on the ratio of the components, the EPDM is a continuous phase or the EPDM/PP form an interpenetrating network (IPN) of two co-continuous phases. Once the crosslinking agent is introduced downstream, it is mixed into the EPDM/PP blend and starts to crosslink the EPDM. As the EPDM continues to be crosslinked while undergoing high shear "dynamic" processing, the system transforms from one where the EPDM is the continuous or co-continuous phase to one where the PP is the continuous matrix surrounding the crosslinked EPDM.

One of the key productivity advantages of the reactive extrusion process for TPE-V is that it does not produce just one product but a family of products. The ratio of EPDM to PP in the formulation is easily modified to influence the product elasticity and hardness. Additionally, heat distortion temperature (HDT) and hardness of the product are influenced by the amount of crosslinking agent added and amount of plasticizing oil introduced with the crosslinking agent. If the EPDM is too lightly crosslinked, the rubber will maintain more elasticity and therefore have a lower HDT.

Another productivity advantage for TPE-V reactive extrusion is the advent of higher torque extruders. When compared to non-crosslinked TPE-O materials, the TPE-V process requires significantly more SME. As example, a typical range of SME for TPE-O recipes would be between $0.10\,\text{kWh}\,\text{kg}^{-1}$ and $0.25\,\text{kWh}\,\text{kg}^{-1}$. The lower SME would be for higher oil- and filler-content formulations. On the other hand, a typical range of SME for TPE-V formulations would be $0.16-0.4\,\text{kWh}\,\text{kg}^{-1}$. In general, they require a higher SME than TPE-O materials due to the longer and more intense mixing process length required for

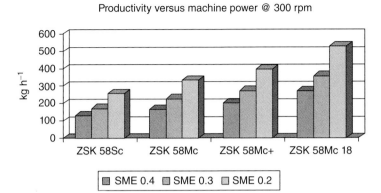

Figure 2.18 Productivity improvement with increased machine torque capacity. (Reproduced with permission of Coperion GmbH.)

crosslinking and dispersing the EPDM. For example, a 90 Shore A formulation may require between 0.33 and 0.40 kWh kg^{-1} while one 64 Shore A hardness TPE-V might require an SME of between 0.28 and 0.34 kWh kg^{-1} and a different formulation with a 67 Shore A product might require an SME of only 0.18 and 0.20 kWh kg^{-1}. The higher SME for similar Shore A formulations would correspond with recipes with less oil and higher viscosity (molecular weight) EPDM and PP.

Typically, an SME of 0.4 kWh kg^{-1} will limit the productivity of the manufacturing to such an extent that it will not be economical to produce. However, with the latest high torque extruder designs this can be significantly mitigated. Figure 2.18 shows that potential productivity has been more than doubled from 130 kg h^{-1} (286 lb h^{-1}) for the ZSK 58 Supercompounder (Sc) generation of the 1980s to 270 kg h^{-1} (594 lb h^{-1}) for the ZSK 58 Mc[18] currently deployed [3].

Decorative films for car interiors for example can be produced in the single-stage process by Coperion twin-screw extruders. The intermediate pelletizing typical for plastics and other products is omitted in inline compounding; the investment, operating costs, and the energy requirements of the production process drop drastically as a result.

2.7 Polymerization of Thermoplastic Polyurethane (TPU)

The most common reaction process that is commercially practiced on ZSKs is the polyaddition of isocyanates (MDI) with polyester- or polyether-diols to produce thermoplasic polyurethane (TPU).

More than 80 machines have been sold during the past 40 years and virtually all renowned TPU producers worldwide are using ZSK machines for their production. The production of TPU with the ZSK has become the preferred technology in the chemical industry (see Figure 2.19). ZSK permits continuous production of an extremely broad product spectrum of polyurethanes, from soft polyurethane

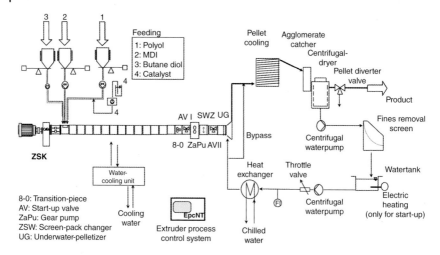

Figure 2.19 Typical setup for polymerization of TPU. (Reproduced with permission of Coperion GmbH.)

Table 2.2 Advantages of polymerization of TPU with ZSK.

Lower production costs	Better product quality
• Lower space requirement • Less manpower requirement • Less energy demand	• Excellent mixing and homogenizing during the total reaction phase

Source: Reproduced with permission of Coperion GmbH.

adhesives to the hardest thermoplastic polyurethane elastomers with Shore hardnesses of D60.

A look back to the historical development of the TPU process provides an excellent roadmap to illustrate the advantages of the reactive extrusion. The original process was a batch polymerization: polymerization in a kettle, grinding, melting, and pelletizing. The process took almost an hour from start to finish. A comparison of the key advantages with the historical process is shown in Table 2.2.

Again, the main advantage of these continuous compounding machines compared to simpler batch units, is that they cover the whole range of viscosity values from monomers to polymers in one reactor.

An important point is easily addressed with a well-designed liquid loss in the weight feeding system. Narrow residence time, good homogenization of reactants, and removal of heat can be accomplished with screw design and efficient barrel tempering. An underwater pelletizer will cut sticky and elastic materials and a long tempered water transfer line will provide post-reaction finishing. Table 2.3 shows typical throughput rates on the ZSK MEGAvolumePlusserie.

TPU products are customized during polymerization. To control stiffness, the methylene diphenyl diisocyanate (MDI) to polyol ratio is modified. Table 2.4 shows three variations of TPU from very soft adhesive grades to rigid grades. For each increase in stiffness the relative amount of MDI is increased

Table 2.3 Typical throughput ranges for TPU on ZSK MvPLUS.

ZSK 54 MvPlus	100–250 kg h^{-1}
ZSK 62 MvPlus	150–400 kg h^{-1}
ZSK 76 MvPlus	300–700 kg h^{-1}
ZSK 98 MvPlus	500–1500 kg h^{-1}
ZSK 125 MvPlus	1500–3500 kg h^{-1}

Source: Reproduced with permission of Coperion GmbH.

Table 2.4 TPU formulation variations.

	Polyole (%)	Methyl isocyanate (%)	Butanediol (%)
Adhesives	85–90	5–10	0–2
Soft-Grades TPU Sh 75–95 A	50–75	20–45	3–12
Rigid-Grades TPU Sh 50–75 D	35–55	35–50	10–15

Source: Reproduced with permission of Coperion GmbH.

with respect to the polyol. For UV resistance formulators the aliphatic hard segment, toluene diisocyanate (TDI), is used rather than the more typical aromatic MDI. For hydrocarbon resistance the polyol is typically polyester based while for a wet environment a polyether polyol is chosen.

2.8 Grafting of Maleic Anhydride on Polyolefines

The grafting of maleic anhydride (MAH) has been commercialized for many years. The reaction is generally performed to improve adhesion of polyolefines to metals, glass fibers, and other polymers. Furthermore, the reaction is limited by diffusion, which indicates that good mixing performance is required. The MAH can be fed into a liquid or solid form similar to the peroxide.

As the polymer is mainly present as pellets (LLDPE, PP, EVA), the process is torque limited. The process requires high SME and will be less economic on a lower powered system. For example, a 92-mm high torque extruder operating at 600 rpm will produce between 2400 and 3400 kg h^{-1} depending upon polymer type and viscosity. The typical setup is illustrated in Figure 2.20. Overall residence time needs to be less than one minute and narrow to avoid gel formation and to be economically feasible. Efficient mixing of a small amount of reactant to assure uniform grafting onto the matrix molecules is essential as the reaction is limited by diffusion [4].

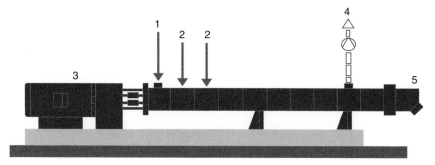

1. Polymer (PP, EVA, LLDPE) + MAH / peroxide; **2.** MAH / peroxid injection (alternative); **3.** ZSK Twin screw extruder; **4.** Vacuum pump; and **5.** Die head

Figure 2.20 Schematic of process configuration for grafting MAH onto a polyolefin polymer. (Reproduced with permission of Coperion GmbH.)

2.9 Partial Glycolysis of PET

PET bottle flakes are depolymerized by a partial glycolysis process to low molecular weight PET using a ZSK twin-screw extruder. The low viscous melt, achieved with that process, is filtered with fine filter mesh, separating impurities in a very efficient way. The final product is good for reuse, "nearly as virgin" for further processing. Washed and undried PET bottle flake is fed into a ZSK twin screw compounder, see Figure 2.21. After melting and degassing, ethylene glycol is fed into the machine, which effects a partial glycolysis during the further passage through the extruder. Special mixing and kneading sections in the compounder support the degradation process. The partially depolymerized, low viscous melt has an intrinsic viscosity (IV) of approx. $0.3-0.4\,dl\,g^{-1}$. Downstream of the ZSK, it is passed through a comparably fine filter (e.g., 20–30 μm, or even less). For filtration, fully continuous screen changers are used. The ZSK125MvPLUS in operation produces depolymerized PET by using this technology at rates up to $4500\,kg\,h^{-1}$.

There are diverse options on how to further process the decontaminated melt discharged from the ZSK. Without solidifying/pelletizing, the melt can be fed

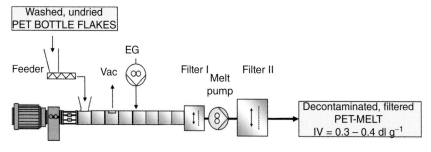

Figure 2.21 Schematic of process configuration for partial glycolysis on ZSK. (Reproduced with permission of Coperion GmbH.)

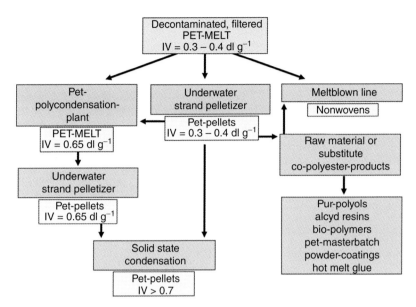

Figure 2.22 Diverse product range after partial glycolysis of PET. (Reproduced with permission of Coperion GmbH.)

directly, via pipeline, from the discharge of the fine filter into a PET polycondensation production line, for example, the inlet of the finisher or in an early stage of a PET polymerization plant as a substitute for PTA and EG, see Figure 2.22.

2.10 Peroxide Break-Down of Polypropylene

Co-rotating twin-screw extruders with screw diameter ranging from 250 to 420 mm are widely applied in compounding/pelletizing of polyolefins, whether it is polypropylene or polyethylene (HD-, LD-, or LLDPE). They achieve throughput capacities of up to $100\,t\,h^{-1}$ and, for those capacities, drive motors with 18–20 MW are required while the die plates of the pelletizers for these huge units may hold close to 12 000 holes for extrusion and cutting the melt strands into single pellets.

Besides the reactor, extruders are the second most important piece of equipment in a polyolefin producing plant. While in the reactor the polymer is polymerized to the desired molecular weight, branching of the molecule chains, and incorporation of co-monomers, the extruder will incorporate additives into the molten polymer, will perform viscosity breaking to adjust the molecular weight, and will shape it to a uniform size and shape pellet for easy handling in subsequent processing steps. Such an installation requires meticulous selection and design phase way upfront of a specific project and even more so during a project involving substantial investment.

Typically, the molecular weight distribution is relatively broad directly after the polymerization of polypropylene. Using a controlled viscosity breaking by feeding peroxide during the compounding process, polymer chains with high molecular

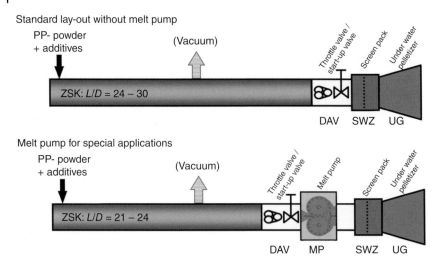

Figure 2.23 ZSK extruder layout: Compounding of PP. (Reproduced with permission of Coperion GmbH.)

weight are cut into shorter polymer chains resulting in a more narrow molecular weight distribution and higher melt flow index. In addition, this leads to better flow characteristics of the melt with favorable properties for subsequent injection molding or fiber spinning. Figure 2.23 shows the process setup for compounding polypropylene with and without the use of a melt pump.

The reactive additive, peroxide, has to be mixed in with an excellent homogeneity to avoid unreacted or only partially reacted polypropylene. Those particles would render a subsequent film blowing or spinning process prone to interruptions due to its inhomogeneity. Only an elongation of the process section will provide the required additional residence time. In some cases the necessity to remove volatiles from the process section, either at atmospheric pressure or under vacuum, requires a vent port.

Visc-breaking is mainly performed by using 2,5-dimethyl-2,5-di(t-butylperoxy) hexane (DHBP) or di-tertiary butyl peroxide (DTBP). Differences in boiling point and volatiles have to considered with relevance to safety. The amount of peroxide depends on the degradation rate and the reactivity of the peroxide. Typically, the reaction starts at 200 °C, above the melting point of the crystalline parts of the polypropylene. Typically, the initial melt flow index (MFI) starts between 0.5 and 40 g/10 min (at 216 kg and 230 °C) with a target MFI of 35–200 g/10 min; in some individual cases, even higher.

ZSK twin-screw extruders can generate sufficient pressure to force the polymer melt flow through the downstream equipment. However, a wide product portfolio, a high capacity, and a fine mesh of the screen pack can necessitate installation of a gear pump for pressurization. In such case the process section of the twin screw is shortened to an extent.

2.11 Summary

The modular co-rotating Twin-screw extruder provides flexibility for reactive extrusion. It provides good heat transfer, efficient mixing, and, with the right screw design, either a narrow residence time distribution or a broader one for improved back-mixing and homogenization. The high torque (power) systems are appropriate for materials such as TPE-V and MAH grafted LLDPE that require a significant amount of SME. High free volume systems are appropriate for materials such as TPU that require longer residence time for completion of the reaction. The two current ZSK machine series ZSK Mc[18] and ZSK MEGAvolumePLUS allow a choice depending on the application between an especially high-torque machine (ZSK Mc[18]: $D_o/D_i = 1.55$) and an especially high free volume extruder (ZSK MEGAvolumePLUS: $D_o/D_i = 1.8$).

References

1 Wunsche, A. (1901) German Patent 131,392. Easton, R.W. (1916) British Patent 109,663. Easton, R.W. (1923), US Patent 1,468,379.
2 Gärtner, S. (2004) *Chemische Reaktionen Zweischneckenknetern Typ ZSK*, VDI, Kunststofftechnik Düsseldorf. ISBN 3-18-234263-0.
3 Andersen, P. (2014) Reactive extrusion: opportunity for improved performance products and manufacturing productivity. *Rubber World*.
4 Janssen, N.N. (1990) The grafting of maleic anhydride on high density polyethylene: a two component reaction. *Polym. Eng. Sci.*, **30**, 1529.

Part III

Simulation and Modeling

3

Modeling of Twin Screw Reactive Extrusion: Challenges and Applications

Françoise Berzin[1] and Bruno Vergnes[2]

[1] *Université de Reims Champagne-Ardenne, UMR 0614 Fractionnement des AgroRessources et Environnement, 2 esplanade Roland Garros, BP 224, 51686 Reims Cedex, France*
[2] *MINES ParisTech., Centre de Mise en Forme des Matériaux (CEMEF), UMR CNRS 7635, CS 10207, 06904 Sophia Antipolis Cedex, France*

3.1 Introduction

3.1.1 Presentation of the Reactive Extrusion Process

Reactive extrusion, which consists in using an extruder as a continuous chemical reactor, is not a recent process; it has rapidly developed during the past 30 years and is used extensively today for the chemical modification of existing polymers [1, 2]. Besides the classical functions of a screw extruder (solid conveying, melting, mixing, pumping), additional functions are desired, principally for the enhancement and control of a chemical reaction. Compared to a classical batch process in solution, reactive extrusion provides the following advantages [3]:

- The reaction is conducted in the melt, in the absence of any kind of solvent, which feature is interesting in terms of environmental issues.
- An extruder is able to work with highly viscous products, which is not always the case for batch reactors.
- Processing conditions are much more flexible in an extruder (mixing conditions, high temperatures, modular geometry, and so on.).

Compared to a batch process in the melt (e.g., by using a Banbury mixer), reactive extrusion is a continuous process, which makes it much more interesting from an economic perspective. However, some drawbacks also need to be mentioned.

- The main drawback is the residence time, which is very short in an extruder (typically, of the order of a few seconds to a few minutes). Consequently, only fast reactions can be considered in reactive extrusion, even though the temperature and concentration of reagents are very often more important in molten state than in solution.
- The cooling capacity of an extruder is limited and therefore it could be difficult to manage high exothermic reactions in reactive extrusion.

Reactive Extrusion: Principles and Applications, First Edition. Edited by Günter Beyer and Christian Hopmann.
© 2018 Wiley-VCH Verlag GmbH & Co. KGaA. Published 2018 by Wiley-VCH Verlag GmbH & Co. KGaA.

Among the existing extrusion systems (single-screw extruders, counter- and co-rotating twin-screw extruders, co-kneaders), the co-rotating twin-screw extruders are the most widely used in reactive extrusion, on account of the many advantages they offer [1–3]:

- Twin-screw extruders have modular screw and barrel geometries, which allows one to adapt the screw profile to the reaction to be carried out. The profile can be divided into successive independent sections, with specific functions: feeding and melting of the polymer, injection of the reagents, mixing, reaction development, devolatilization, pumping, and shaping.
- Twin-screw extruders are starve-fed. The filling ratio of the screws is only partial, and therefore it is easy to introduce or remove different ingredients along the barrel, either in liquid or solid form.
- Mixing capabilities are generally more important in a twin-screw extruder, and the mixing conditions (distributive or dispersive) can be easily controlled according to the geometry of the kneading disk sections (number of disks, number of tips, disk thickness, staggering angle) [4].

3.1.2 Examples of Industrial Applications

As explained earlier, all the chemical modifications conducted in batch reactors cannot be transposed to a reactive extrusion process. However, a number of applications have been developed and used industrially.

- Radical polymerization of styrene: polymerization is initiated in a pre-polymerizer (batch reactor) until reaction rates of 20–40% are achieved. The pre-polymerized compound is subsequently introduced in an extruder to achieve polymerization and for the elimination of residual monomers [5]. Other polymers such as polyesters [6], polyamides [7], or polyacrylates [8] can also be produced by reactive extrusion.
- Chemical modifications: the objective is to modify the chemical or physico-chemical properties of existing polymers, without altering the other properties. The most popular example is the grafting of maleic anhydride onto polypropylene [9, 10] and polyethylene. Exchange reactions or modification of functional groups can also be conducted in extruders [11]. Two examples will be discussed in Sections 3.3.1 and 3.3.4.
- Rheological modifications: the controlled degradation of polypropylene has been used for many years for adjusting, after polymerization, the molecular weight distribution of these resins, and thus their rheological properties. This is achieved by the use of peroxide and will also be detailed in Section 3.3.2.
- Reactive blending: polymer blends have been largely studied with relevance to producing new materials with enhanced properties. As many polymers are incompatible, it is necessary to add a compatibilizer to improve interfacial adhesion and to obtain satisfactory properties. In some cases, this compatibilizer can be created *in situ* by an interfacial reaction conducted during the extrusion process [12].
- Dynamic vulcanization: thermoplastic elastomers are obtained by the dispersion of a vulcanized elastomer in a thermoplastic matrix. A phase inversion between the elastomer and the thermoplastic takes place during the

vulcanization process. All these products are traditionally produced by using reactive extrusion [13, 14].

3.1.3 Interest of Reactive Extrusion Process Modeling

As can be seen from the previous examples, reactive extrusion is an important process in the modification and development of new polymeric systems. However, it is also a complex process that involves many aspects, and the development and the optimization of such a system addresses a lot of questions that arise due to the high number of operating variables and the numerous interactions that occur during the process [15]. For example, flow conditions (residence time, temperature, mixing conditions) will govern reaction development, which will modify the polymer properties (viscosity change, exothermicity), leading to changes in flow conditions. To illustrate, Figure 3.1 shows the effects of changing a single parameter (e.g., screw speed) on all other parameters [16]: due to all the possible interactions, it is almost impossible for the operator to predict what would be the impact of such a change. Therefore, a theoretical model will be an effective aid to understand the process.

Consequently, it is of primary importance to develop dedicated theoretical models that are able to provide predictive tools in order to define the best conditions for conducting reactive extrusion.

3.2 Principles and Challenges of the Modeling

Some authors have attempted to model reactive extrusion using a chemical engineering approach, that is, a process essentially based on residence time distribution and the description of the flow conditions using combinations of ideal reactors [17–20]. However, this approach often necessitates the use of various adjustable parameters and is not easy to generalize. In our case, an approach based on continuum mechanics is preferred, because it generally does not present these drawbacks and provides more predictive results.

For modeling a reactive extrusion process using such an approach, the following three modules are necessary:

- A module for the calculation of the flow conditions in the considered extrusion process. In this chapter, we will focus on the co-rotating twin-screw extruder.
- A module for the calculation of the extent of the chemical reaction as a function of local values of time, temperature, mixing conditions, concentration of reagents, and so on.
- A module for the calculation of the physical and rheological changes induced in the polymer during the reaction development. This last module is only necessary when the studied reaction significantly modifies the rheological behavior. For example, during polymerization, the viscosity changes from that of a monomer ($\approx 10^{-3}$ Pa s) to that of a polymer ($\approx 10^3$ Pa s) over a short length of time. This is also the case, as we will see in Section 3.3.2, for controlled degradation of polypropylene. On the contrary, a grafting operation or an exchange reaction, which only slightly modifies the structure of the chain, will not require the use of such a module, as we will see in Sections 3.3.1 and 3.3.4.

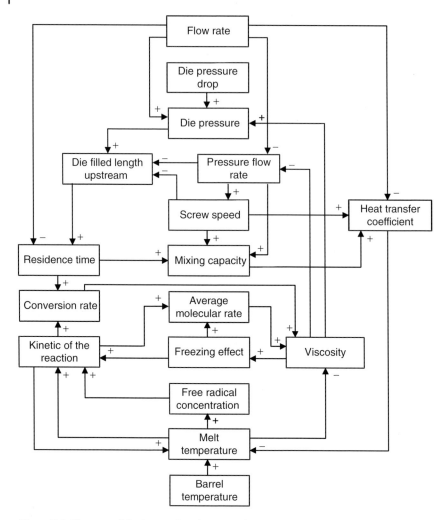

Figure 3.1 Diagram of the interactions between the process parameters in the case of the polymerization of PMMA in a twin-screw extruder; the signs + or − indicate an increase or a decrease in the considered parameter. (Jongbloed *et al.* 1995 [16]. Reproduced with permission of Wiley.)

Subsequently, all these modules have to be coupled in order to provide the desired results.

3.2.1 Twin Screw Flow Module

The first ingredient of a reactive extrusion model is a tool for computing the flow conditions along the twin-screw extruder. In such machines, the geometry and the local flow conditions are complex: the flow is unsteady, highly non-isothermal, and three-dimensional. 3D models, based on the finite element method, are able to handle this complexity [21]. However, to obtain a user-friendly software, a lot of simplifications are necessary. Therefore, a few years ago, the Ludovic model was developed in our laboratory [22]. This commercialized software is based on a local one-dimensional (1D) approach. It allows one to calculate, along the screws from the hopper to the die exit, the main thermomechanical parameters of the process (pressure, temperature, residence time, shear rate, filling ratio, etc.).

The screws are divided into sub-elements corresponding to C-shaped chambers. Each C-shaped chamber accommodates volume occupied by the material around one screw. Computation of the various parameters is done separately for each type of element (partially or totally filled right-handed screw elements, left-handed screw elements, and blocks of kneading disks (KDs)) and for the die components. The flow in screw elements is computed using cylindrical co-ordinates in which the channel section is perpendicular to the screw flights. This analysis can be applied to both right- and left-handed screw elements. The flow in a block of KDs is modeled considering only the peripheral flow around a disk. Due to the geometry and the relative barrel velocity, this flow is characterized by a pressure peak located just before the tip of the disk. As the tips of the adjacent disks are staggered, the pressure profiles are also staggered, creating an axial pressure gradient parallel to the screw axis and pushing the material downstream. This axial pressure gradient is determined by staggering the adjacent pressure profiles and adjusting the pressure level to match the imposed flow rate in the axial direction.

These elementary models are linked together to obtain a global description of the flow field along the extruder. We can either consider a specific melting model [23] or more simply assume that the melting is instantaneous and takes place before the first restrictive element of the screw profile. The material is now considered to be fully molten and can fill the screw channel, totally or partially according to local geometry and flow conditions. Twin screw extrusion is a highly non-isothermal process. To calculate the mean temperature $\overline{T}$ over the channel depth (the only choice compatible with a 1D mechanical approach), a thermal balance is written, which takes into account the dissipated power, the heat transfer toward barrel and screws, and the heat of the reaction.

As the screws are starve-fed, the filling ratio of the system is not known. Therefore, the calculation has to start at the die exit and proceed backward. But, as the final product's mass temperature is unknown, an iterative procedure must be used. Starting from an arbitrary value of exit mass temperature, the software computes the successive pressures and temperatures in each element are computed, until the first restrictive element where the melting is assumed to take place. Convergence is achieved when, at this location, the temperature equals the melting temperature of the product. Otherwise, the exit temperature is modified and the computation is restarted.

This software has been validated through comparisons with experiments [24] as well as more developed numerical simulations [25]. It has largely been used for more than 15 years to study specific developments in various extrusion fields, for example, the extrusion-cooking of starchy products [26, 27], the development of gluten-based materials [28], the preparation of polymer blends [29], the compounding of nanocomposites [30–32], and, more recently, the breakage of glass or vegetal fibers during compounding [33, 34]. As Ludovic® is a 1D model, it provides only the average local values of temperature and residence time. But, as can be seen later on, these values are largely sufficient for the prediction of the reaction extent along the screws.

3.2.2 Kinetic Equations

A chemical reaction is usually described by one or many kinetic equations, controlled by kinetic constants. Balance equations written for the different chemical species generally allow one to define a conversion degree or a rate of advancement as a function of time and temperature. Depending on the complexity of the case, it can be obtained either by calculating a simple analytical function or by solving a set of partial differential equations. In the latter case, numerical resolutions are necessary at each step of the reaction development along the screws.

The complexity of the kinetic module depends on the studied reaction. In the case of the polymerization of ε-caprolactone, for example, the conversion C is simply provided by an analytical expression (see Section 3.3.3). On the contrary, for the peroxide degradation of polypropylene, the kinetic scheme takes into account several reactions, like peroxide dissociation, β-scission, intermolecular transfer, and thermal degradation and termination. It is then necessary to write a balance of chemical species (peroxides, free-radicals, polymer macroradicals, and polymer chains), which leads to a complex system of eight partial differential equations that must be solved at each time step, using a numerical method. This will be presented in detail in Section 3.3.2.

For this module, it is most often difficult to define accurately the different kinetic constants involved in the kinetic equations. We have shown that, in some cases, a small error in the values of such constants may induce large discrepancies in the simulation of the whole process [35].

3.2.3 Rheokinetic Model

In some cases (for instance exchange reactions), the modifications induced by the reaction on the rheological behavior of the polymer are negligible. The viscosity of the unmodified polymer can then be used for computation throughout the process. In other cases (controlled degradation, polymerization), these changes are dominant. For example, in a polymerization process, the viscosity increases typically from 10^{-3} Pa s (monomer) to 10^3 Pa s (polymer) between the entry and the exit of the extruder. This has, obviously, to be taken into account and it is necessary to develop rheokinetic models, describing the change in viscosity (and eventually other material parameters) as a function of the reaction extent. Detailed examples will be provided in Section 3.3.2 for the controlled degradation of polypropylene and in Section 3.3.3 for the polymerization of ε-caprolactone. In these cases, the kinetic equations allow for the calculation of the average molecular weight and the parameters of the viscosity law are functions of this molecular weight.

3.2.4 Coupling

Once the above mentioned models have been established, it is necessary to ensure that the coupling that exists between the different modules.

We start with the simplest case, when the rheological behavior does not change with the chemical reaction. In this case, a single calculation with Ludovic® software provides, among other information, the evolutions of local residence time and temperature along the screw profile. From this information, the evolution of the reaction can be calculated using the kinetic module.

If the progress of the reaction modifies the rheological behavior, the calculation is a little bit more complex. Indeed, the reaction develops according to the local values of temperature and residence time, which modifies the viscosity of the polymer, which in turn affects the local flow conditions. To handle these complex couplings, the following procedure is adopted [36]:

- A first calculation is performed with the initial rheological behavior in order to have a first estimate of the filled sections, local residence times, and temperatures. The local viscosities are then defined according to these values. As usual, this calculation is made iteratively, from downstream (die) to upstream (hopper).
- Starting now from upstream (at the introduction of the reactants), according to the previously estimated parameters, the coupling between flow, viscosity, and reaction is calculated step by step:
  - The residence time t_i and the temperature T_i are known in the element i.
  - From the kinetic equation, the conversion increment is calculated in the element i.
  - From the new conversion rate, the corresponding viscosity is calculated using the rheokinetic module.
  - With the new viscosity, the flow and thereby the new values of residence time and temperature are calculated in the element $i + 1$.

The calculation progresses thus, element by element, along the extruder until the die exit.

3.2.5 Open Problems and Remaining Challenges

As explained before, the proposed simulations are based on a number of approximations and assumptions. We will see in Section 3.3 that the results are often satisfactory, but important questions still remain open to discussion:

- Kinetic equations are generally established in static and isothermal conditions, very different from the strong conditions encountered in the extruder (high temperatures, large strains, high stresses). However, it needs to be established if the values of kinetic constants measured in static conditions are still valid for the extrusion simulation.
- The reactive medium is assumed to be homogeneous in the models. The reagents are supposed to be instantaneously and perfectly mixed with the molten polymer, as soon as they are introduced in the extruder. It is clear that, in some cases, local mixing should be evaluated and introduced as a

parameter in the kinetic equations to account for the heterogeneity of the real system. This is, for example, the case when a liquid reagent is injected into a very viscous polymer [37].
- Degradation of reagents, side (or undesired) reactions induced by high temperatures are often difficult to be taken into account, even if they play a role in the real process.
- The examples presented hereafter concern chemical modifications of single polymers. In the case of reactive blends for example, a new difficulty arises from the morphology of the system. Depending on the flow and reaction conditions, the blend morphology (dispersed or co-continuous phase, dimensions of droplets or fibrils) will change during the process. It is thus necessary to take this change into account, which poses important problems for the modeling.

3.3 Examples of Modeling

3.3.1 Esterification of EVA Copolymer

This first example [36, 38] concerns the modification of an ethylene–vinyl acetate copolymer by an aliphatic primary alcohol (octanol-1) through a transesterification reaction. This reaction, catalyzed by an organometallic compound (dibutyltin dilaurate, DBTDL), leads to the substitution of vinyl acetate groups of ethylene vinyl acetate (EVA) by hydroxyl groups of the alcohol (Figure 3.2).

It is an equilibrated reaction, with a maximum of 65% conversion at the equilibrium, whose kinetics is well known and documented [39]. As no important changes are induced in the EVA molecular structure, the reaction does not modify the viscosity and a rheokinetic model is not necessary in this case.

The conversion percentage χ is calculated from the relation:

$$\chi(\%) = \frac{[A]_0 - [A]_t}{[A]_0} \times 100 \tag{3.1}$$

where $[A]_t$ and $[A]_0$ are the EVA concentrations at time t and $t = 0$, respectively.

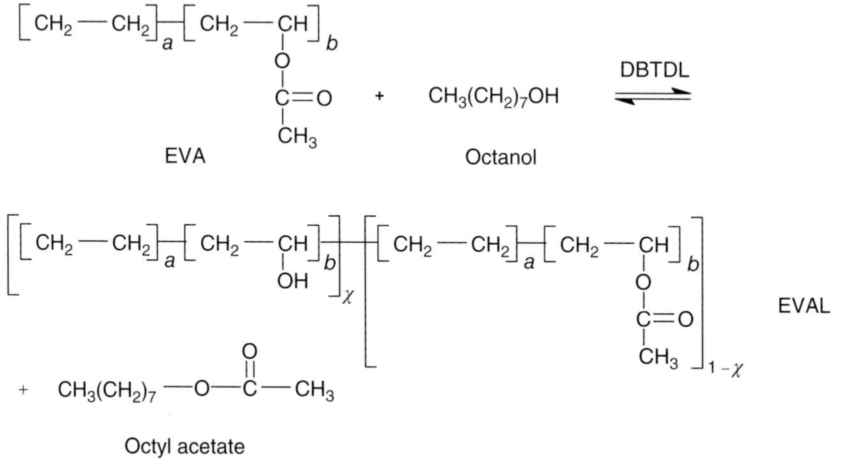

Figure 3.2 Transesterification reaction of EVA copolymer.

The concentration of the copolymer EVA in relation to time and temperature is given by [36]:

$$[A]_t = \frac{(C_2 + C_3)\exp[-C_3 k_1 (1 - 1/K_e)t] + (C_3 - C_2)C_4}{2C_4 - 2\exp[-C_3 k_1 (1 - 1/K_e)t]} \quad (3.2)$$

where

$$C_1 = -\frac{[A]_0^2 + [A]_0[C]_0 + [A]_0[D]_0 + [C]_0[D]_0}{K_e - 1} \quad (3.3)$$

$$C_2 = \frac{K_e}{K_e - 1}\left([B]_0 - [A]_0 + \frac{2[A]_0 + [C]_0 + [D]_0}{K_e}\right) \quad (3.4)$$

$$C_3 = \sqrt{C_2^2 - 4C_1} \quad (3.5)$$

$$C_4 = \frac{2[A]_0 + C_2 + C_3}{2[A]_0 + C_2 - C_3} \quad (3.6)$$

where $[B]_0$, $[C]_0$, $[D]_0$, are the initial concentrations of octanol, modified EVA, and octyl acetate, respectively. k_1 and k_2 are the rate constants of the forward and reverse reactions, depending on temperature through Arrhenius equations. K_e is the ratio of these constants, that is, $K_e = k_1/k_2$.

The extruder used in the experiments (Clextral BC 45, Firminy, France) has the following characteristics: centerline distance 45 mm, screw diameter 50 mm, and length 1200 mm. Figure 3.3 presents the screw profile. It includes two left-handed elements (LHEs) and a block of 12 KDs, staggered at 60°: 6 left-handed preceded by 6 right-handed. EVA pellets are introduced in the hopper and melted in the first LHE. Reagents (mixture of alcohol and catalyst, in liquid form) are injected after the melting, that is, before the second LHE.

Figure 3.4 shows the changes in the local mean temperature and the cumulative residence time along the screws, together with the conversion rate calculated from these two parameters (Eqs (3.1)–(3.6)), for the following conditions: flow rates: EVA 10 kg h^{-1}, alcohol 2.12 kg h^{-1} and catalyst 0.3 kg h^{-1}, screw speed 150 rpm, and barrel temperature 185 °C. We can see that the extent of the reaction closely follows the evolution of the residence time. It increases mainly in the filled sections of the extruder, that is, in the second LHE and in the block of KDs. These zones are also characterized by a long residence time and a local increase in temperature, due to viscous dissipation, which accelerates the reaction. The conversion at the extruder exit is 38% (it is 65% maximum at equilibrium).

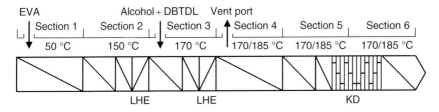

Figure 3.3 Screw profile used for the esterification of EVA.

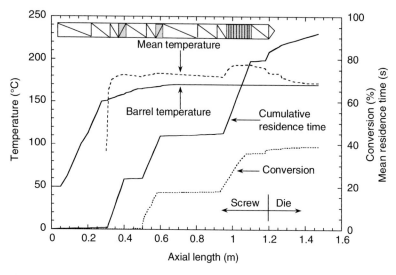

Figure 3.4 Esterification of EVA; change along the screws of the mean temperature, the cumulative residence time, and the conversion. (Adapted from Berzin and Vergnes 1998 [38]. Reproduced with permission of Wiley.)

To test the influence of flow rate (1.9, 4.1, and 6.7 kg h^{-1}), the screw speed was set at 100 rpm, the temperature of the barrel at 170 °C, and the functional ratio of the committed species (molar ratio of alcohol and EVA) was 1. During these tests, we used comparatively low feed rates, so that the residence time of the reactive system in the extruder could be long enough to obtain significant conversion rates at the die exit. At constant screw speed, any decrease in the throughput results in a drop in the filling ratio, an increase in the residence time, and nearly no influence on the temperature along the screws [4]. Consequently, the reaction progress is directly linked to the residence time and therefore decreases when the flow rate is increased (Figure 3.5). However, it can be observed that, even for the lowest throughput, reaction equilibrium is not achieved in one single pass through the extruder. We can see also in Figure 3.5 that the correlation between software predictions and the results obtained from infrared measurements are satisfactory.

Figure 3.6 shows the influence of the screw speed (50, 100, and 150 rpm) for a total flow rate of 4.1 kg h^{-1} (EVA 3.45 kg h^{-1}, alcohol + DBTDL 0.65 kg h^{-1}) and a barrel temperature of 185 °C. An increase in rotation speed induces both a decrease in residence time and an increase in product temperature by viscous heating [4]. Thus, the parameters controlling the reaction vary in opposite directions and therefore it is impossible to intuitively foresee the final influence on reaction extent. The model gives the answer: it can be seen in Figure 3.6 that the reaction is more progressed at a higher screw speed. An increase in screw speed from 50 to 150 rpm increases the conversion from 27% to 38%. Thus, in this case, the temperature increase takes precedence over the shorter time, but this result cannot be considered as a general rule. It can be checked that it is effectively due to a higher product temperature all along the screws, leading to a higher reaction

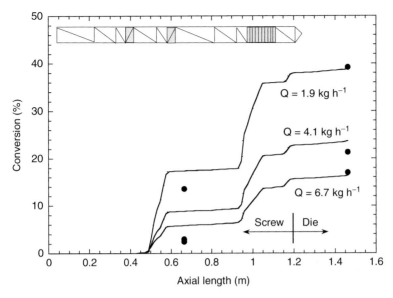

Figure 3.5 Esterification of EVA; evolution of the conversion with the flow rate. Symbols: experimental points, lines: theoretical model. (Adapted from Berzin and Vergnes 1998 [38]. Reproduced with permission of Wiley.)

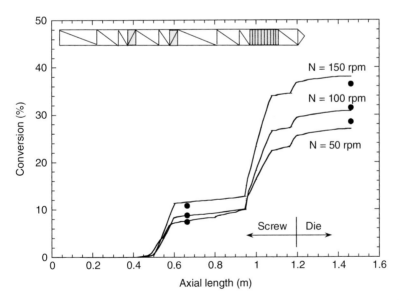

Figure 3.6 Esterification of an EVA; evolution of the conversion with the screw speed. Symbols: experimental points, lines: theoretical model. (Adapted from Berzin and Vergnes 1998 [38]. Reproduced with permission of Wiley.)

extent despite the shorter residence time. It is also confirmed in Figure 3.6 that the agreement with experiments is pretty good, both at the die exit and along the screws.

We also tested the influence of barrel temperature. As expected, a higher temperature leads to a higher conversion, from 21% to 32% when the temperature is increased from 170 to 185 °C.

General validation has been made for various feed rates, barrel temperatures, stoichiometry, and extruder size. In all situations, this model, without any adjustable parameter, is very close to reality.

3.3.2 Controlled Degradation of Polypropylene

As explained previously, peroxide-induced degradation is used to adjust the level of viscosity of polypropylene that is polymerized using Ziegler–Natta catalysis. The principle of this reaction is as follows. The peroxide is decomposed at high temperature, leading to free-radicals. These free-radicals then attack the C–H links of polypropylene, which have the weakest link energy and give rise to tertiary polymer macroradicals. The macromolecular chain of the polypropylene is then fragmented by β-scission, which leads to the creation of polypropylene of reduced mass and of a new polymer radical, but this time, of secondary type [40]. This mechanism has been proposed and explicated in numerous publications [41–43]. The kinetic model that will be used is taken from this literature [44, 45]. It includes peroxide decomposition, hydrogen abstraction, β-scission, intermolecular chain transfer, thermal degradation and termination by disproportionation. Besides the usual kinetic constants, the kinetic model necessitates using an adjustable parameter, which is the efficiency of the peroxide, f. It is the ratio of free-radicals leading to β-scissions to the whole number of free-radicals. The kinetic scheme can be expressed by a set of eight partial differential equations [45]. The solutions for these sets of equations are obtained at each time step through a Runge–Kutta procedure, which provides the values of the various average molecular weights [36].

As the viscosity is directly linked to the molecular weight, we have conducted a specific study on the rheological behavior of these products [46]. We have shown that the viscosity can be described by a Carreau–Yasuda law, whose parameters are a function of the weight average molecular weight:

$$\eta = \eta_0 a_T [1 + (\lambda \dot{\gamma} a_T)^a]^{(n-1)/a} \quad \text{with} \quad a_T = \exp\left[\frac{E_a}{R}\left(\frac{1}{T} - \frac{1}{T_0}\right)\right] \quad (3.7)$$

where η_0 is the zero-shear viscosity, λ is the characteristic time, n is the power-law index, a is the Yasuda parameter, a_T is the shift factor, E_a is the activation energy, R is the gas constant, and T_0 is the reference temperature. In this equation, η_0, λ, n, and a are power or polynomial functions of the weight average molecular weight $\overline{M}_w$, obtained by minimizing the differences between the measured and calculated viscosities. The zero-shear viscosity varies with a 3.9 power of $\overline{M}_w$ whereas, when increasing the molecular weight, characteristic time λ increases while n and a decrease continuously. Finally, the activation energy of the different degraded samples slightly decreases (from 44.9 to 39.3 kJ mol^{-1}) when increasing the molecular weight (from 63 to 301.6 kg mol^{-1}).

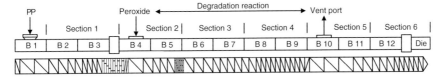

Figure 3.7 Peroxide-controlled degradation of PP; screw profile used in the experiments.

Experiments were carried out on the screw profile presented in Figure 3.7, in various processing conditions: peroxide concentration from 0.025 to 0.4 wt%, feed rate from 2.5 to 20 kg h^{-1}, and screw speed from 150 to 300 rpm [47]. The extruder was a Werner ZSK 30 (centerline distance 26 mm, screw diameter 30.85 mm, length 1160 mm). The peroxide (here, 2,5-dimethyl-2,5-di(*tert*-butylperoxy)hexane, namely DHBP Trigonox 101 from AKZO) was injected with a pump after the polypropylene melting. The DHBP was previously diluted in 1,2,4-Trichlorobenzene (solution at 15%) to have a good precision on the pump flow rate, even for concentrations less than 1000 ppm (i.e., 0.1 wt%). The melting of the polymer was assured by KD blocks followed by left-handed screw elements. The injection of DHBP was housed in an unfilled area, just after the melting zone. The reaction area first included two KD blocks designed to improve the mixing and homogenization between free-radicals and polypropylene. Next, were placed several right-handed conveying screw elements with decreasing pitches and finally a left-handed screw element was placed just before the devolatilization zone (localized at barrel 10). At this level, with the help of a vacuum pump, the degradation reaction of polypropylene was stopped by eliminating all the residual volatile reactants.

Calculations were made according to the methodology presented in Section 3.2.3, with a strong coupling between flow, chemical reactions, and rheological behavior. An example of the result is shown in Figure 3.8, which represents the evolution along the screws of the temperature, the cumulative residence time, and the weight average molecular weight.

As already stated, the calculation of reaction starts at the place where the DHBP has been injected and stops at the level of the vent zone. We can see that the temperature of the material after reaction decreases compared to the temperature for virgin polypropylene. This is due to the fact that the dissipated power decreases if the viscosity decreases. At high concentrations of peroxide, the dissipated power becomes quickly very low and the average temperature depends essentially on the heat transfer. The temperature is then close to the barrel regulation temperature (here, 170 °C). On the contrary, as far as the residence time is concerned, no significant variation has been observed, because residence time is quite independent of polymer viscosity [4]. Finally, considering molecular weight $\overline{M}_w$, we observe a very fast decrease after the injection point, from 283 to 93 kg mol^{-1}, followed by a much slower variation until the vent port.

As explained previously, the peroxide efficiency f is an unknown parameter of the model. In the literature, values between 0.6 and 1 are often used for peroxide concentration lower than 0.1 wt% [42, 45, 48]. In the present work, we decided to define f as a function of peroxide concentration. Therefore, efficiency values were obtained by fitting the model to obtain, at die exit, the experimental weight

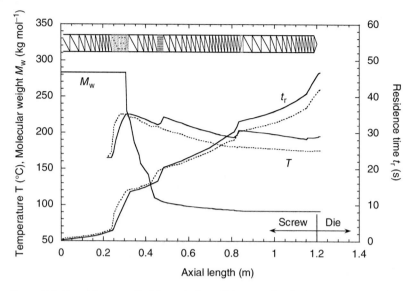

Figure 3.8 Peroxide-controlled degradation of PP; example of calculated results; evolution of weight average molecular weight M_w, cumulative residence time t_r, and temperature T along the screws. Full lines correspond to first calculation and dotted lines to the final result, after coupling between reaction extent and viscosity. (Adapted from Berzin et al. 2000 [47]. Reproduced with permission of Wiley.)

average molecular weight measured by size exclusion chromatography (SEC), in fixed processing conditions (flow rate 10 kg h^{-1}, screw speed 225 rpm, and barrel temperature 170 °C) (Figure 3.9).

The change in efficiency with the peroxide concentration can then be deduced (Figure 3.10). We observe that f decreases with peroxide content (from 0.96 to 0.44 when the peroxide content increases from 280 to 3780 ppm); these values are in agreement with those found in the literature. We can thus now study the influence of processing conditions, by keeping f constant and just defined according to the chosen amount of peroxide.

Figure 3.11 shows the influence of feed rate (2.5, 10, and 20 kg h^{-1}) at a constant screw speed of 225 rpm for a peroxide content of 1000 ppm (0.1 wt%). We observe that the degradation is more important at low feed rate, which is directly explained by the increase in residence time. The agreement with the experimental points at the die exit is fairly satisfactory.

As seen in the case of EVA transesterification, the influence of screw speed at constant feed rate is more difficult to foresee, because screw speed has an opposite effect on the two parameters controlling the reaction. As observed in Figure 3.12, the polypropylene is more degraded at high rotation speed despite a reduction of residence time from 65 to 40 s when the screw speed is increased from 150 to 300 rpm. This is because all along the screws the temperatures are higher at higher screw speed. Once again, the predictions of the model are confirmed by the SEC measurements made at the die exit.

Other experimentations have been carried out to check the validity of the theoretical model, using a larger extruder (Werner & Pfleiderer ZSK 58, with 58 mm

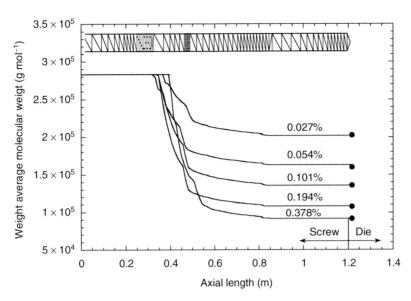

Figure 3.9 Peroxide degradation of polypropylene; changes in the average molecular weight along the screws for various peroxide contents. Symbols are experimental points, lines are the results of the model (Adapted from Berzin et al. 2000 [47]. Reproduced with permission of Wiley.)

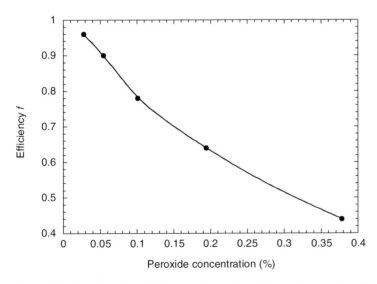

Figure 3.10 Peroxide degradation of polypropylene; changes in peroxide efficiency with peroxide concentration. (Adapted from Berzin et al. 2000 [47]. Reproduced with permission of Wiley.)

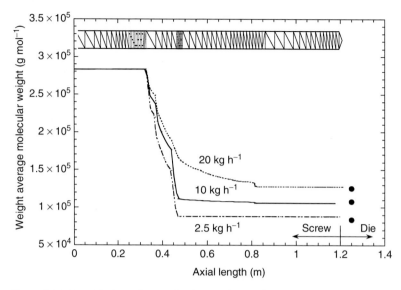

Figure 3.11 Peroxide degradation of polypropylene. Influence of feed rate on average molecular weight at constant screw speed (225 rpm, 170 °C, 0.1 wt%). Symbols are experimental points, lines are the results of the model. (Adapted from Berzin et al. 2000 [47]. Reproduced with permission of Wiley.)

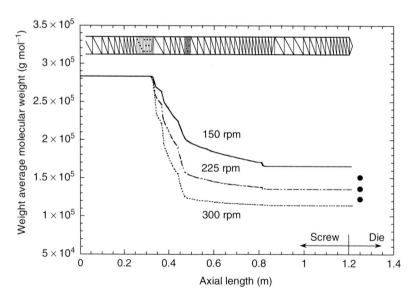

Figure 3.12 Peroxide degradation of polypropylene. Influence of screw speed on average molecular weight at constant feed rate (10 kg h^{-1}, 170 °C, 0.1 wt%). Symbols are experimental points, lines are the results of the model. (Adapted from Berzin et al. 2000 [47]. Reproduced with permission of Wiley.)

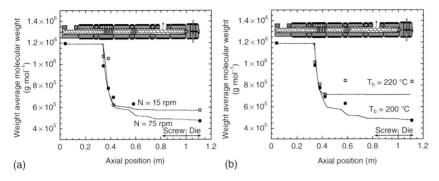

Figure 3.13 Peroxide degradation of polypropylene; change in average molecular weight along the screws with screw speed (a) and barrel temperature (b). Symbols are experimental points, lines are the results of the model. (Berzin et al. 2006 [51]. Reproduced with permission of Wiley.)

screw diameter and an L/D of 44) and another polymer (polypropylene copolymer). In all cases, there was agreement between computed and measured molecular weights [36, 49]. It has also been confirmed that the results remain valid for samples collected along the screws. For that purpose, we used a laboratory scale co-rotating extruder (Leistritz LSM 30–34, with 34 mm screw diameter and an L/D of 29) from the university of Minho (Portugal), which was equipped with specific sampling devices [50]. Samples were thus quickly collected along the extruder axis and immediately quenched in liquid nitrogen, in order to avoid further reaction. It can be seen in Figure 3.13 that the strong decrease provided by the model after the injection of peroxide is also observed in all the experiments [51]. The agreement noticed when varying parameters (screw speed, barrel temperature, screw profile, extrusion size) has also proven to be satisfactory for the study.

3.3.3 Polymerization of ε-Caprolactone

Polycaprolactone is a synthetic polymer that is both biodegradable and biocompatible. It is mainly used in the medical domain for applications such as development of artificial skin and resorbable prostheses and sustained release systems for drugs. It can be obtained from the polymerization of ε-caprolactone, initiated by organometallic compounds, such as metal alkoxides and carboxylates. In this study, tetrapropoxy-titanium was used as initiator, and a two-step coordination–insertion process was assumed for the polymerization [52, 53]. It consists in monomer coordination onto the initiator, followed by monomer insertion into the titanium-alkoxy bond (Figure 3.14). The polymer chain grows from the titanium atom by successive insertion of the monomer. This polymerization continues as long as the reactive sites are not destroyed.

The bulk polymerization was carried out in a laboratory scale co-rotating twin-screw extruder (Leistritz LSM 30–34, screw diameter: 34 mm, barrel length: 1.2 m, L/D: 35), equipped with a slit die (length: 44 mm, width: 50 mm, thickness: 2 mm) [54]. The screw profile was composed of right-handed screw elements and three blocks of KDs, staggered at 30°, which are used to improve the mixing between ε-caprolactone monomers and tetrapropoxy-titanium initiator.

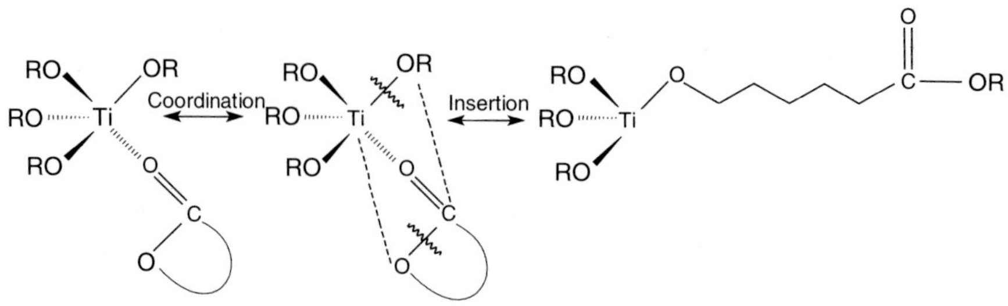

Figure 3.14 Chemical mechanism of ε-caprolactone polymerization initiated by tetrapropoxy-titanium.

The reactive mixture (monomer and initiator) was injected with a pump into the barrel, just before the first block of KDs. Two other kneading blocks downstream allow mixing and homogenization of the polymerizing material.

The global kinetic equation was determined by Gimenez et al. [53, 54]. The monomer conversion $C(t)$ is defined as:

$$C(t) = \frac{[M]_0 - [M]_t}{[M]_0} \tag{3.8}$$

where $[M]_0$ is the initial monomer concentration and $[M]_t$ the concentration at time t. The variation of the monomer conversion with time in isothermal conditions can be written as:

$$C(t) = 1 - \exp(-Kt) \tag{3.9}$$

where the kinetic constant K is expressed as:

$$K = k[I]_0^{\alpha_c} \exp\left(-\frac{E_a}{RT}\right) \tag{3.10}$$

where k is a constant, $[I]_0$ is the initial concentration of initiator, α_c is the partial order related to the initiator, E_a is the activation energy, and R is the gas constant. The conversion rate is thus a function of the time and temperature, the partial order related to the initiator, and the $[M]_0/[I]_0$ ratio.

From the conversion rate, the weight average molecular weight can be determined [53] as:

$$M_w(t) = (0.39[M]_0/[I]_0 + 79)C(t) + \overline{M}_0 \tag{3.11}$$

Obviously, the polymerization reaction induces a huge variation in viscosity. Consequently, the definition of an evolutionary viscosity is necessary. Gimenez et al. [55] showed that it is necessary to consider two regimes, with a dilution effect:

- Below a critical mass, there is a Rouse regime, essentially with a Newtonian behavior:

$$\eta = K_0 a_T M_w^{1.2} C \tag{3.12}$$

where K_0 is a parameter, a_T is the temperature shift factor, and C is the conversion defined by Eq. (3.9).

- Above this critical mass, there is an entangled regime with a viscosity described by a Carreau–Yasuda law with zero-shear viscosity η_0 and characteristic time λ depending on the average molar mass:

$$\eta = \eta_0 [1 + (\lambda \dot{\gamma})^a]^{(m-1)/a} \tag{3.13}$$

$$\eta_0 = A \exp\left[\frac{E_v}{R}\left(\frac{1}{T} - \frac{1}{T_0}\right)\right] M_w^{\alpha_v} C^4 \alpha_c \tag{3.14}$$

$$\lambda = B \exp\left[\frac{E_v}{R}\left(\frac{1}{T} - \frac{1}{T_0}\right)\right] M_w^{\alpha_t} C^{1.75} \alpha_c \tag{3.15}$$

where A, B, E_v, α_v, and α_t are constants that have been experimentally defined by Gimenez et al. [55].

These different models have been implemented in Ludovic® software to calculate the change in monomer conversion along the screws [35]. An example of the results is shown in Figure 3.15. It corresponds to the following processing conditions: feed rate 4 kg h^{-1}, screw speed 100 rpm, and barrel temperature 160 °C, $[M]_0/[I]_0 = 1000$. It is observed that the polymerization starts at the injection point, with a very low viscosity (10^{-4} Pa s), and increases mainly in the blocks of KDs, because of the long residence times. The polymerization is achieved just before the die exit, where the viscosity reaches a value of 900 Pa s.

In Figure 3.16, for a lower barrel temperature (105 °C), we can see the influence of the feed rate at constant screw speed. In these conditions, the polymerization is only partial (maximum 65%) and is lower at high flow rate, because of the shorter residence time. The experimental measurements of conversion rate by ^{1}H NMR confirm these results [35].

Various conditions of screw speed, barrel temperature, and $[M]_0/[I]_0$ ratio have been tested [35]. A general comparison for all processing conditions is proposed in Figure 3.17. Irrespective of the conditions, the agreement is quite good and

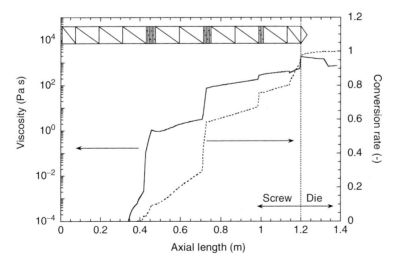

Figure 3.15 Polymerization of ε-caprolactone; evolution of conversion rate (CR) (--) and viscosity (—) along the screws. (Adapted from Poulesquen et al. 2001 [35]. Reproduced with permission of Carl Hanser Verlag GmbH & Co. KG.)

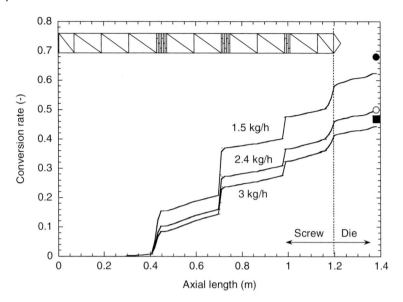

Figure 3.16 Polymerization of ε-caprolactone; influence of feed rate on CR at constant screw speed. Symbols (●: 1.5 kg h^{-1}; ○: 2.4 kg h^{-1}; ■: 3 kg h^{-1}) represent experimental measurements by ^{1}H NMR. (Adapted from Poulesquen *et al.* 2001 [35]. Reproduced with permission of Carl Hanser Verlag GmbH & Co. KG.)

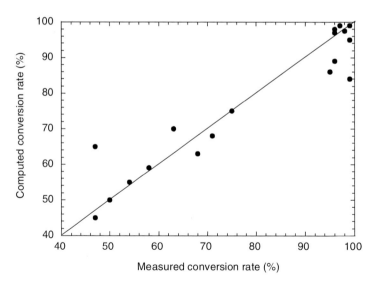

Figure 3.17 General comparison between computed and measured CRs. (Adapted from Poulesquen *et al.* 2001 [35]. Reproduced with permission of Carl Hanser Verlag GmbH & Co. KG.)

validates both the simulation and the kinetic data. However, it must be said that the quality of the prediction strongly depends on the accuracy of the kinetic data. For the cases at low barrel temperature, the calculation with a partial order α_c of 3, experimentally measured, led to an underestimation of the conversion by a factor 2. Correct results were only obtained by changing to a value of 2.74, indicating the high sensitivity of the model to this parameter.

3.3.4 Starch Cationization

The last example concerns the chemical modification of a natural polymer. Starch, obtained from natural and renewable resources, is more widely used for industrial applications instead of synthetic polymers, for example in the packaging industry. Starch can be chemically modified to obtain products with specific properties [56]. For example, in the papermaking industry, cationic starches can increase the strength, filler and fines retention, and drainage rate of the pulp. They can also lower biological oxygen demand of the white water when used as wet-end additives. These cationic starches are produced from starch with reagents containing amino, ammonium, sulfonium, or phosphonium groups, which are able to carry a positive charge [57]. Starch cationization consists thus in substituting the hydroxyl groups of the glycosyl units by one of these functional groups. In our case, a wheat starch and two reagents were used: 2,3-epoxypropyl-trimethylammonium chloride (Quab 151®, from Degussa) and 3-chloro 2-hydroxypropyl-trimethylammonium (Quab 188®, from Degussa). As shown in Figure 3.18, the reaction of cationization involves two stages with the Quab 188®: the first stage for transforming the reagent (in alkaline medium) into an active epoxy form, and the second stage for operating the substitution on starch backbone. With the Quab 151®, being initially under epoxy form, the first stage is eliminated.

The degree of substitution (*DS*) indicates the average number of sites per anhydroglucose unit on which there are substituent groups. Thus, if one hydroxyl group on those of the anhydroglucose units has been cationized, *DS* is equal to 1.

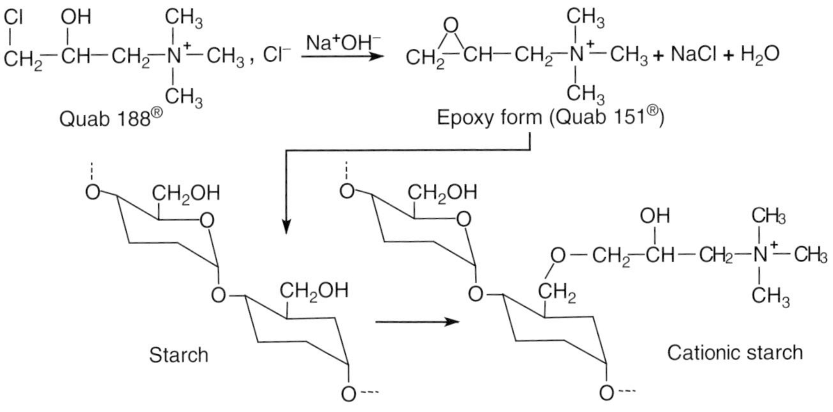

Figure 3.18 Reaction scheme of starch cationization.

If all three hydroxyl groups have been cationized, DS is maximum and equal to 3. Cationic starches used in industry usually have low DS, in the range 0.02–0.10.

Cationic starches are usually prepared using conventional batch reaction procedures. However, the cationization in a batch reactor presents some drawbacks, like discontinuous process with low yield, residual reactive agent elimination, and environment pollution. Thus, to overcome these drawbacks, some attempts have been made in the 1990s to develop starch cationization using reactive extrusion process [58, 59]. However, without modeling, it was difficult to define optimal conditions for real industrial applications.

In our experiments [60], a laboratory scale co-rotating twin-screw extruder was used (Clextral BC 21, Firminy, France) (centerline distance: 21 mm, screw diameter: 25 mm, L/D: 36). The screw profile was composed of two-flighted conveying screw elements of various pitches, and two blocks of KDs, negatively staggered ($-45°$). The barrel was made of nine sections of 100 mm each, regulated at a fixed temperature (50 °C for the first one and 80 or 90 °C for the others). Starch was plasticized with 40% water. It was fed in the first barrel element, when water and reagent were mixed together and injected downstream using a pump. After extrusion, the nitrogen content (%N, value between 0 and 100) of each sample of cationic starch was determined by Kjeldahl method (Kjeldatherm, Gerhardt GmbH., Oberdollendorf, Germany). DS was then calculated by:

$$DS = \frac{M_S \%N}{100 M_N - M_R \%N} \tag{3.16}$$

where M_S, M_N, and M_R are the molar masses of starch anhydroglucose monomer (162 g mol^{-1}), nitrogen (14 g mol^{-1}), and reagent once fixed on glycosyl unit (152.5 g mol^{-1}), respectively. The theoretical degree of substitution, noted DS_{th}, is the theoretical value that would be obtained if the reaction efficiency was 100%. It corresponds to the molar ratio between the reagent and anhydroglucose monomer. In the experiments, it defines the target and allows to adjust the values of the starch and reagent flow rates, according to:

$$DS_{th} = \frac{\rho_r Q_r}{Q_S} \frac{M_S}{M_{fr}} \frac{I_{pr}}{I_{ps}} \tag{3.17}$$

where Q_r (L h^{-1}) and Q_S (kg h^{-1}) are the flow rates of reagent and starch, as fed in the extruder. M_{fr} is the molar mass of free reagent, ρ_r the reagent density, and I_{pr} and I_{ps} the purity indices of reagent and starch. The reaction efficiency RE is logically defined as:

$$RE = \frac{DS}{DS_{th}} \tag{3.18}$$

As the reaction is an exchange reaction, its influence on native starch viscosity is low. Therefore, we will neglect this aspect. A study of kinetic data allowed defining an overall kinetic law, permitting to describe, for each reagent, the change in DS as a function of time and temperature [61]:

$$DS = \frac{DS_{th}}{[B]_0}([A]_0 - [A]_t) \tag{3.19}$$

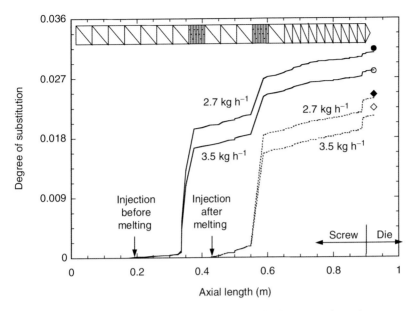

Figure 3.19 Starch cationization; variation of degree of substitution along the screws with feed rate at constant screw speed (400 rpm, 80 °C, $DS_{th} = 0.04$). (Adapted from Berzin et al. 2007 [62]. Reproduced with permission of Wiley.)

where $[A]_0$ and $[B]_0$ are the initial concentrations in starch and reagent, respectively. The starch concentration at time t is given by:

$$[A]_t = \frac{\frac{[A]_0}{[B]_0}([A]_0 - [B]_0)\exp[k([A]_0 - [B]_0)t]}{\frac{[A]_0}{[B]_0}\exp[k([A]_0 - [B]_0)t] - 1} \quad (3.20)$$

where k is the kinetic constant, depending on the temperature.

After implementation into Ludovic® software, the evolution of the DS along the screws was calculated [62]. Figure 3.19 shows an example of results for various flow rates at 400 rpm and 80 °C. We observe that the reaction starts once the reagents have been injected. The increase in DS is important in the blocks of KDs, where the residence time is long. As seen in the experiments, an increase in feed rate tends to reduce the reaction extent. Moreover, it is clear that the injection of the reagents after the melting zone reduces the residence time and thus leads to a lower DS.

Systematic comparisons have been made between experiments and computed results. It can be seen in Figure 3.20 that, irrespective of the reagent, its amount or the processing conditions (feed rate, screw speed, barrel temperature), the agreement is good on a wide range of values (efficiency between 30% and 90%).

3.3.5 Optimization and Scale-up

In such complex problems as those we have just presented, modeling is a great help both to understand what happens during the process and to try to improve it.

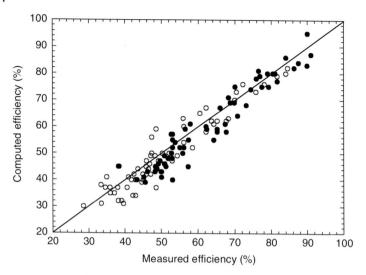

Figure 3.20 Starch cationization; comparison between computed and experimental efficiency (● Quab 151®; ○ Quab 188®). (Adapted from Berzin et al. 2007 [62]. Reproduced with permission of Wiley.)

Another recurring problem in industrial applications is the scale-up, that is, the passage from a laboratory test (flow rate of a few kilogram per hours) to an industrial one (several tons per hours, on a machine with a diameter 10 times larger). This can be done on the basis of trial and error, but this is very expensive in terms of time, immobilization of machines and personnel, and consumption of raw material. Again, the use of a modeling approach can be largely beneficial. In this last section, without going into details, some examples of the use of a software like Ludovic® to solve this type of problem will be presented.

Of course, as we will see later on, a computer code can be directly used to optimize a process by trial and error, just as it could do experimentally (but faster and cheaper!). However, it is also possible, if the numerical solution is fast enough, to implement more sophisticated optimization techniques. Genetic Algorithms were used to optimize extrusion conditions in the case of a fixed screw profile [63]. These algorithms mimic the natural evolution processes in which an initial population evolves over successive generations with the aid of different operations (selection, crossover, mutation, etc.). The optimization procedure then comprises the following three modules:

- an objective function that quantifies the performance of each individual (i.e., each case of extrusion calculated) with respect to the target performance;
- a calculation module, here Ludovic®, which calculates the value of the objective function for each individual;
- an optimization algorithm, which controls the evolution of the population toward the goal. The initial population is randomly chosen (e.g., by varying screw speed, flow rate, and barrel temperature). With each new generation, the algorithm selects the best individuals and generates new ones.

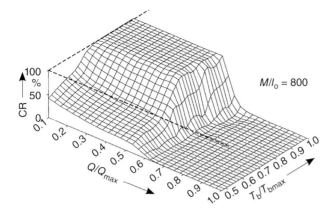

Figure 3.21 Polymerization of ε-caprolactone; changes in the CR with flow rate and barrel temperature. (Adapted from Gaspar-Cunha et al. 2002 [63]. Reproduced with permission of Carl Hanser Verlag GmbH & Co. KG.)

A global objective function can be defined, comprising several different targets, possibly conflicting, with associated weights:

$$F_o = \sum_i w_i f_i \tag{3.21}$$

where w_i are the weights ($\sum_i w_i = 1$) and f_i specific functions defined to maximize, minimize, or keep a parameter in a fixed interval. If the case of the polymerization of ε-caprolactone introduced in Section 3.3.3 is considered, the following objectives can be defined: maximum flow rate, complete conversion, maximum material temperature along the screws lower than 220 °C, and consumed power less than 9.2 kW. Figure 3.21 shows the conversion rate obtained at 150 rpm, depending on the values of Q/Q_{max} and T_b/T_{bmax} (maximum flow rate Q_{max}: 30 kg h^{-1} and maximum barrel temperature T_{bmax}: 210 °C).

There is a 100% conversion plateau, which indicates that several possible combinations of flow rate and temperature lead to the same result. The best value of the objective function corresponds to 200 rpm and 26 kg h^{-1}, giving a consumption of 6.4 kW and a residence time of 32 s.

With more advanced optimization techniques (multiobjective evolutionary algorithms), the problem of screw profile optimization was then studied [64–66]. The optimizations performed often lead to many solutions that are more or less equivalent, among which it is often difficult to discern the best. It is possible that Ludovic® software is sometimes too simplified to get the best of these techniques. In this case, we can go back to a "manual" optimization, which is illustrated with respect to starch cationization, presented in Section 3.3.4. We start by the optimization of the screw profile of a laboratory extruder (Clextral BC 21, 25 mm diameter). We first define various screw profiles, the severity of which increases from profile 1 to profile 5, by incorporating more and more kneading blocks [67]. Then the reaction progress is calculated for various feed rates between 2.5 and 40 kg h^{-1}, at a screw speed of 400 rpm. Figure 3.22 shows the change in the efficiency of the reaction for the different profiles and flow rates.

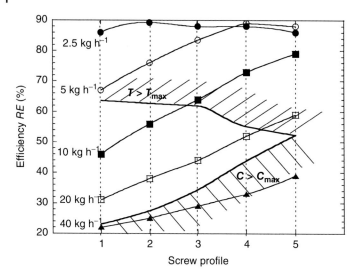

Figure 3.22 Laboratory machine; variation of reaction efficiency with screw profile and flow rate (○ 2.5 kg h^{-1}, ● 5 kg h^{-1}, □ 10 kg h^{-1}, ■ 20 kg h^{-1}, □ 40 kg h^{-1}). (Adapted from Berzin et al. 2007 [67]. Reproduced with permission of Wiley.)

Efficiency is higher when the flow rate is lower (because of longer residence time) and the profile severe (because of higher temperature). If one wants to keep the temperature below 165 °C to avoid degradation and the torque to be lower than C_{max} = 50 mN, we see that the processability window is reduced. Depending on whether the efficiency or the throughput is privileged, various points may be chosen along the upper boundary: 64% efficiency at 5.5 kg h^{-1} with profile 1, or 63% at 8.5 kg h^{-1} with profile 2, or 62% at 11 kg h^{-1} with profile 3, or 55% at 18 kg h^{-1} with profile 4, or ultimately 52% at 25 kg h^{-1} with the profile 5. Here, it is up to the user to make a choice according to his/her priorities.

Suppose now that we want to predict the processing conditions of the cationization reaction on a larger machine (Clextral BC 72, 82 mm diameter). First the screw profiles must be defined that are similar to those of the laboratory extruder, that is to say with the same number of restrictive elements and identical relative positions and length ratios. Then, the flows in these new profiles are simulated for a range of flow rates that are suited to the size of the industrial machine, at the maximum speed, which in this case is 360 rpm. The results are shown in Figure 3.23, with a torque limit C_{max} of 1500 mN.

We observe the same type of results as for the laboratory scale extruder. As seen earlier, the restrictions on temperature and torque lead to a minimized window, which allows to define the optimum conditions: 60% efficiency at 155 kg h^{-1} with profile 1, or 58% at 225 kg h^{-1} with profile 2, or 58% at 310 kg h^{-1} with profile 3, or 49% at 560 kg h^{-1} with profile 4. In contrast with the small extruder, no conditions can be satisfied with profile 5 and the processing window of profile 4 is largely reduced. Again, the final choice is left to the user. However, by providing a realistic estimate of the available flow rates the calculation helps to predict the potential profitability of the industrial process.

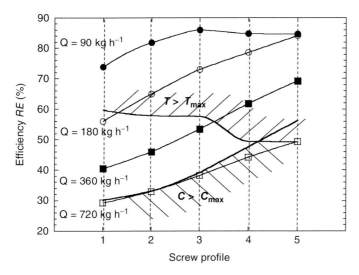

Figure 3.23 Production machine; variation of reaction efficiency with screw profile and flow rate (●: 90 kg h⁻¹, ○: 180 kg h⁻¹, ■: 360 kg h⁻¹, □: 720 kg h⁻¹).)(Adapted from Berzin et al. 2007 [67]. Reproduced with permission of Wiley.)

If we compare the two extruders in the case of profile 3, we obtain a similar efficiency (≈60%) with a flow rate ratio Q_l/Q_s of 28.2, where Q_l and Q_s are the flow rates of the large and the small extruder, respectively. If the classical rule of thumb for estimating this ratio from the ratio of screw diameters is considered ($Q_l/Q_s = (D_l/D_s)^3$), a value of 35.3 is obtained. Even though the order of magnitude is the same, the numerical modeling allows a better and more accurate estimation of the extrapolated values.

3.4 Conclusion

It has been shown in this chapter that it is possible to model the operations of reactive extrusion with good accuracy. The various examples presented as illustrations (on different systems, different types of reaction, and different machines) clearly show the versatility of the method. For that purpose, it is necessary to have accurate kinetic data available and, if required, rheokinetic laws. Depending on the required reaction, predictive models without any adjustable parameters can be derived. They are very useful to better understand the conditions of the process and to optimize it, for example by modifying screw profile and/or processing conditions. Moreover, the difficult scale-up problems can be easily solved using such models. In other cases, adjustable parameters remain in the kinetic equations, which have to be estimated through experiments before any application. Nevertheless, although these models may present some imperfections, they appear as efficient tools to address such complex systems, compared to systematic trial and error procedures.

Besides the examples we have just presented, this modeling approach was successfully applied to other reactions, such as the extension of polyamide

12 chains with a coupling agent [68] and the synthesis of nanoparticles using the sol–gel method [69]. By considering the breakage and erosion mechanisms of solid fillers as kinetic equations, it was possible to even address the dispersive mixing of particles during compounding, using these reactive extrusion models [70].

References

1 Xanthos, M. (1992) *Reactive Extrusion: Principles and Practice*, Hanser, Munich.
2 Janssen, L.P.B.M. (2004) *Reactive Extrusion Systems*, CRC Press, Boca Raton.
3 Berzin, F. and Hu, G.H. (2003) Procédés d'extrusion réactive, Techniques de l'Ingénieur, AM3654, 1–16.
4 Lafleur, P.G. and Vergnes, B. (2014) *Polymer Extrusion*, ISTE-Wiley, London.
5 Kelley, J.M. (1993) Styrene polymerization process. US Patent 5,274,029.
6 Gouinlock, E.V., Marciniak, H.W., Shatz, M.H., Quinn, E.J., and Hindersinn, R.R. (1968) Preparation and properties of copolyesters polymerized in a vented extruder. *J. Appl. Polym. Sci.*, **12**, 2403–2413.
7 Bartilla, T., Kirch, D., Nordmeier, J., Proemper, E., and Strauch, T. (1986) Physical and chemical changes during the extrusion process. *Adv. Polym. Technol.*, **9**, 339–387.
8 Stuber, N.P. and Tirrell, M. (1985) Continuous polymerization studies in a twin-screw extruder. *Polym. Proc. Eng.*, **3**, 71–83.
9 Hu, G.H., Flat, J.J., and Lambla, M. (1997) Free radical grafting of monomers onto polymers by reactive extrusion: principles and applications, in *Reactive Modifiers for Polymers* (ed. S. Al-Malaika), Thomson Science & Professional, London pp. 1–83.
10 Berzin, F., Flat, J.J., and Vergnes, B. (2013) Grafting of maleic anhydride on polypropylene by reactive extrusion: effect of maleic anhydride and peroxide concentrations on reaction yield and products characteristics. *J. Polym. Eng.*, **33**, 673–682.
11 Hu, G.H. and Lambla, M. (1994) Catalysis and reactivity of the transesterification of ethylene and alkyl acrylate copolymers in solution and in the melt. *Polymer*, **35**, 3082–3090.
12 Baker, W., Scott, C., and Hu, G.H. (2001) *Reactive Polymer Blending*, Hanser, Munich.
13 Abdou-Sabet, S. and Shen, K.S. (1986) Process for the preparation of thermoplastic elastomers US Patent 4,594,390.
14 Naskar, K. (2007) Thermoplastic elastomers based on PP/EPDM blends by dynamic vulcanization. *Rubber Chem. Technol.*, **80**, 504–519.
15 Michaeli, W. and Grefenstein, A. (1995) Engineering analysis and design of twin-screw extruders for reactive extrusion. *Adv. Polym. Technol.*, **14**, 263–276.
16 Jongbloed, H.A., Kiewiet, J.A., Van Dijk, J.H., and Janssen, L.P.B.M. (1995) Self-wiping co-rotating twin-screw extruder as a polymerization reactor for methacrylates. *Polym. Eng. Sci.*, **35**, 1569–1579.

17 Michaeli, W., Grefenstein, A., and Berghaus, U. (1995) Twin-screw extruders for reactive extrusion. *Polym. Eng. Sci.*, **35**, 1485–1504.
18 De Graaf, R.A., Rohde, M., and Janssen, L.P.B.M. (1997) A novel model predicting the residence-time distribution during reactive extrusion. *Chem. Eng. Sci.*, **52**, 4345–4356.
19 Choulak, S., Couenne, F., Le Gorrec, Y., Jallut, C., Cassagnau, P., and Michel, A. (2004) Generic dynamic model for simulation and control of reactive extrusion. *Ind. Eng. Chem. Res.*, **43**, 7373–7382.
20 Puaux, J.P., Cassagnau, P., Bozga, G., and Nagy, I. (2006) Modeling of polyurethane synthesis by reactive extrusion. *Chem. Eng. Proc.*, **45**, 481–487.
21 Valette, R., Coupez, T., David, C., and Vergnes, B. (2009) A direct 3D numerical simulation code for extrusion and mixing processes. *Int. Polym. Proc.*, **24**, 141–147.
22 Vergnes, B., Della Valle, G., and Delamare, L. (1998) A global computer software for polymer flows in co-rotating twin screw extruders. *Polym. Eng. Sci.*, **38**, 1781–1792.
23 Vergnes, B., Souveton, G., Delacour, M.L., and Ainser, A. (2001) Experimental and theoretical study of polymer melting in a co-rotating twin screw extruder. *Int. Polym. Proc.*, **16**, 351–362.
24 Carneiro, O.S., Covas, J.A., and Vergnes, B. (2000) Experimental and theoretical study of the twin screw extrusion of polypropylene. *J. Appl. Polym. Sci.*, **78**, 1419–1430.
25 Durin, A., De Micheli, P., Nguyen, H.C., David, C., Valette, R., and Vergnes, B. (2014) Comparison between 1D and 3D approaches for twin-screw extrusion simulation. *Int. Polym. Proc.*, **29**, 641–648.
26 Della Valle, G., Barrès, C., Plewa, J., Tayeb, J., and Vergnes, B. (1993) Computer simulation of starchy products transformation by twin-screw extrusion. *J. Food Eng.*, **19**, 1–31.
27 Della Valle, G., Berzin, F., and Vergnes, B. (2011) Modelling of twin screw extrusion process for food products design and process optimization, in *Advances in Food Extrusion Technology* (eds M. Maskan and A. Altan), CRC Press, Boca Raton, pp. 327–354.
28 Redl, A., Morel, M.H., Bonicel, J., Vergnes, B., and Guilbert, S. (1999) Extrusion of wheat gluten plasticized with glycerol: influence of process conditions on flow behaviour, rheological properties and molecular size distribution. *Cereal Chem.*, **76**, 361–370.
29 Delamare, L. and Vergnes, B. (1996) Computation of the morphological changes of a polymer blend along a twin screw extruder. *Polym. Eng. Sci.*, **36**, 1685–1693.
30 Lertwimolnun, W. and Vergnes, B. (2007) Influence of screw profile and extrusion conditions on the microstructure of polypropylene/organoclay nanocomposites. *Polym. Eng. Sci.*, **47**, 2100–2109.
31 Domenech, T., Peuvrel-Disdier, E., and Vergnes, B. (2012) Influence of twin-screw processing conditions on structure and properties of polypropylene – organoclay nanocomposites. *Int. Polym. Proc.*, **27**, 517–526.

32 Domenech, T., Peuvrel-Disdier, E., and Vergnes, B. (2013) The importance of specific mechanical energy during twin screw extrusion of organoclay based polypropylene nanocomposites. *Compos. Sci. Technol.*, **75**, 7–14.

33 Ville, J., Inceoglu, F., Ghamri, N., Pradel, J.L., Durin, A., Valette, R., and Vergnes, B. (2013) Influence of extrusion conditions on fiber breakage along the screw profile during twin screw compounding of glass fiber-reinforced PA. *Int. Polym. Proc.*, **28**, 49–57.

34 Berzin, F., Vergnes, B., and Beaugrand, J. (2014) Evolution of lignocellulosic fibre length along the screw profile during twin screw compounding with polycaprolactone. *Comp. Part A*, **59**, 30–36.

35 Poulesquen, A., Vergnes, B., Cassagnau, P., Gimenez, J., and Michel, A. (2001) Polymerization of ε-caprolactone in a twin screw extruder: experimental study and modelling. *Int. Polym. Proc.*, **16**, 31–38.

36 Berzin, F. (1998) Etude expérimentale et modélisation d'une opération d'extrusion réactive. PhD dissertation. Ecole des Mines de Paris.

37 Cassagnau, P., Courmont, M., Melis, F., and Puaux, J.P. (2005) Study of mixing of liquid/polymer in twin screw extruder by residence time distribution. *Polym. Eng. Sci.*, **45**, 926–934.

38 Berzin, F. and Vergnes, B. (1998) Transesterification of ethylene acetate copolymer in a twin screw extruder. *Int. Polym. Proc.*, **13**, 13–22.

39 Bouilloux, A. (1985) Etude de la transestérification à l'état fondu des copolymères d'éthylène et d'acétate de vinyle. PhD dissertation. Université Louis Pasteur, Strasbourg.

40 Fritz, H.G. and Stöhrer, B. (1986) Polymer compounding process for controlled peroxide-degradation of PP. *Int. Polym. Proc.*, **1**, 31–41.

41 Balke, S.T., Suwanda, D., and Lew, R. (1987) A kinetic model for the degradation of polypropylene. *J. Polym. Sci. C*, **25**, 313–320.

42 Tzoganakis, C., Vlachopoulos, J., and Hamielec, A.E. (1988) Production of controlled-rheology polypropylene resins by peroxide promoted degradation during extrusion. *Polym. Eng. Sci.*, **28**, 170–180.

43 Triacca, V.J., Gloor, P.E., Zhu, S., Hrymak, A.N., and Hamielec, A.E. (1993) Free radical degradation of polypropylene: random chain scission. *Polym. Eng. Sci.*, **33**, 445–454.

44 Krell, M.J., Brandolin, A., and Valles, E.M. (1994) Controlled rheology polypropylenes – an improved model with experimental validation for the single-screw extruder process. *Polym. React. Eng.*, **2**, 389–408.

45 Tzoganakis, C., Vlachopoulos, J., and Hamielec, A.E. (1988) Modeling of the peroxide degradation of polypropylene. *Int. Polym. Proc.*, **3**, 141–150.

46 Berzin, F., Vergnes, B., and Delamare, L. (2001) Rheological behavior of controlled-rheology polypropylenes obtained by peroxide promoted degradation during extrusion: comparison between homopolymer and copolymer. *J. Appl. Polym. Sci.*, **80**, 1243–1252.

47 Berzin, F., Vergnes, B., Dufossé, P., and Delamare, L. (2000) Modelling of peroxide initiated controlled degradation of polypropylene in a twin screw extruder. *Polym. Eng. Sci.*, **40**, 344–356.

48 Ryu, R.H., Gogos, C.G., and Xanthos, X. (1992) Parameters affecting process efficiency of peroxide-initiated controlled degradation of polypropylene. *Adv. Polym. Technol.*, **11**, 121–131.

49 Vergnes, B. and Berzin, F. (2000) Peroxide-controled degradation of poly(propylene): rheological behavior and process modelling, in *Rheology of Polymer Systems* (ed. J. Kahovec), Wiley-VCH, Weinheim pp. 77–90.

50 Machado, A.V., Covas, J.A., and van Duin, M. (1999) Evolution of morphology and of chemical conversion along the screw in a corotating twin-screw extruder. *J. Appl. Polym. Sci.*, **71**, 135–141.

51 Berzin, F., Vergnes, B., Canevarolo, S.V., Machado, A.V., and Covas, J.A. (2006) Evolution of peroxide-induced degradation of polypropylene along a twin-screw extruder: experimental data and theoretical predictions. *J. Appl. Polym. Sci.*, **99**, 2082–2090.

52 Dubois, P., Jacobs, C., Jérôme, R., and Teyssié, P. (1991) Macromolecular engineering of polylactones and polylactides. 4. Mechanism and kinetics of lactide homopolymerization by aluminum isopropoxide. *Macromolecules*, **24**, 2266–2270.

53 Gimenez, J., Boudris, M., Cassagnau, P., and Michel, A. (2000) Control of bulk ε-caprolactone polymerization in a twin screw extruder. *Polym. React. Eng.*, **8**, 135–157.

54 Gimenez, J., Boudris, M., Cassagnau, P., and Michel, A. (2000) Bulk polymerization of ε-caprolactone in twin screw extruder: a step toward the process control. *Int. Polym. Proc.*, **15**, 20–27.

55 Gimenez, J., Cassagnau, P., and Michel, A. (2000) Bulk polymerization of ε-caprolactone: rheological predictive laws. *J. Rheol.*, **44**, 527–547.

56 Moad, G. (2011) Chemical modification of starch by reactive extrusion. *Prog. Polym. Sci.*, **36**, 218–237.

57 Rutenberg, M.W. and Solareck, D. (1984) Starch derivatives: production and uses, in *Starch: Chemistry and Technology* (eds R.L. Whistler, J.N. BeMiller, and E.F. Paschall), Academic Press, Orlando pp. 311–388.

58 Della Valle, G., Colonna, P., Tayeb, J., and Vergnes, B. (1993) Use of extrusion processes for enzymic and chemical modifications of starch, in *Plant Polymeric Carbohydrates* (eds F. Meuser, D.J. Manners, and W. Seibel), Royal Society of Chemistry, Cambridge pp. 240–251.

59 Carr, M.E. (1994) Preparation of cationic starch containing quaternary ammonium substituents by reactive twin-screw extrusion processing. *J. Appl. Polym. Sci.*, **54**, 1855–1861.

60 Tara, A., Berzin, F., Tighzert, L., and Vergnes, B. (2004) Preparation of cationic wheat starch by twin-screw reactive extrusion. *J. Appl. Polym. Sci.*, **93**, 201–208.

61 Ayoub, A., Berzin, F., Tighzert, L., and Bliard, C. (2004) Study of the thermoplastic wheat starch cationisation reaction under molten condition. *Starch/Stärke*, **56**, 513–519.

62 Berzin, F., Tara, A., Tighzert, L., and Vergnes, B. (2007) Computation of starch cationization performances by twin screw extrusion. *Polym. Eng. Sci.*, **47**, 112–119.

63 Gaspar-Cunha, A., Poulesquen, A., Vergnes, B., and Covas, J.A. (2002) Optimization of processing conditions for polymer twin-screw extrusion. *Int. Polym. Proc.*, **17**, 201–213.

64 Gaspar-Cunha, A., Vergnes, B., and Covas, J.A. (2005) Defining the configuration of co-rotating twin-screw extruders with multi-objective evolutionary algorithms. *Polym. Eng. Sci.*, **45**, 1169–1173.

65 Gaspar-Cunha, A., Covas, J.A., Vergnes, B., and Berzin, F. (2011) Reactive extrusion – optimization of representative processes, in *Optimization of Polymer Processing* (eds A. Gaspar-Cunha and J.A. Covas), Nova Science Publishers, New York, pp. 115–143.

66 Teixeira, C., Covas, J.A., Berzin, F., Vergnes, B., and Gaspar-Cunha, A. (2011) Applications of evolutionary algorithms for the definition of optimal twin-screw extruder configuration for starch cationisation. *Polym. Eng. Sci.*, **51**, 330–340.

67 Berzin, F., Tara, A., and Vergnes, B. (2007) Optimization and scale-up of thermoplastic wheat starch cationization in a twin screw extruder. *Polym. Eng. Sci.*, **47**, 814–823.

68 Chalamet, Y., Taha, M., Berzin, F., and Vergnes, B. (2002) Carboxyl terminated polyamide 12 chain extension by reactive extrusion using a dioxazoline coupling agent. Part II: effects of extrusion conditions. *Polym. Eng. Sci.*, **42**, 2317–2327.

69 Bahloul, W., Oddes, O., Bounor-Legaré, V., Mélis, F., Cassagnau, P., and Vergnes, B. (2011) Reactive extrusion processing of polypropylene/TiO_2 nanocomposites by in-situ synthesis of the nanofillers. *AIChE J.*, **57**, 2174–2184.

70 Berzin, F., Vergnes, B., Lafleur, P.G., and Grmela, M. (2002) A theoretical approach of solid filler dispersion in a twin screw extruder. *Polym. Eng. Sci.*, **42**, 473–481.

4

Measurement and Modeling of Local Residence Time Distributions in a Twin-Screw Extruder

Xian-Ming Zhang[1], Lian-Fang Feng[2], and Guo-Hua Hu[3]

[1] Key Laboratory of Advanced Textile Materials and Manufacturing Technology, Ministry of Education, Zhejiang Sci-Tech University, Hangzhou 310018, China
[2] State Key Laboratory of Chemical Engineering, College of Chemical and Biochemical Engineering, Zhejiang University, Hangzhou 310027, China
[3] Université de Lorraine–CNRS, Laboratoire Réactions et Génie des Procédés (UMR7274), 1 rue Grandville–BP 20451, 54001 Nancy, France

4.1 Introduction

Screw extruders are the most important components of a polymer processing equipment. They are divided into two main types: single- and twin-screw extruders (TSEs). The key advantages of TSEs are more functionalities, lower energy consumption, and more flexibility. As such they are used for mixing various types of rheologically complex systems such as polymer materials. They are also often used as continuous reactors for chemically modifying polymers. During an extrusion process, each material element may undergo different temporal, thermal, and/or mechanical changes. As a result, its final properties may be different from those of other material elements [1]. The residence time of a material element is the time it spends in the extruder. Since all material elements do not necessarily spend the same amount of time in the extruder, residence times have a distribution, called residence time distribution (RTD). The residence time and RTD have the following applications [2].

1) Residence time provides information on the time a fraction of the material stays inside the extruder.
2) The RTD characterizes, to some extent, the transport capacity, flow pattern, and axial mixing performance of the extruder. Therefore, the RTD helps to select appropriate screw elements and mixing devices.
3) The residence time of a flow element can be a measure of the extent of thermal, mechanical, and/or chemical changes it undergoes. Thus, the RTD reflects, to some extent, the changes of different fractions of the material, and can help understand the effects of extrusion conditions on the extrusion process and the extruded product.
4) For reactive extrusion processes, whose reaction time is the same order of magnitude as the average residence time, the RTD may significantly influence

Reactive Extrusion: Principles and Applications, First Edition. Edited by Günter Beyer and Christian Hopmann.
© 2018 Wiley-VCH Verlag GmbH & Co. KGaA. Published 2018 by Wiley-VCH Verlag GmbH & Co. KGaA.

the extent of the reaction and affect the product quality. When a reaction is of the first order, the reaction yield is solely determined by the RTD.
5) The RTD helps to diagnose abnormalities in flow. If the average residence time is lower than the expected value (the ratio of fill volume to throughput), there may be stagnant zones in the extruder. A longer average residence time is more likely to be caused by experimental errors.
6) A local RTD provides information about the above five aspects for the corresponding target zone in the extruder.

Despite the fact that local RTDs are very important for understanding and controlling the working of (reactive) extrusion processes and are necessary for validating flow and mixing models reported in the literature, there is lack of such data due to experimental difficulties. This chapter describes an in-line instrument capable of measuring in real time both the global and local RTD in an extruder, on the one hand; and 3D finite elements based simulations for flow and distributive mixing in kneading disk (KD) zones of co-rotating TSEs, on the other hand. The local RTDs are used to validate the simulations. Only when the simulations are validated by the local RTD data can they be used to predict distributive mixing that characterizes interfacial area generation and is much more difficult to measure experimentally.

4.2 Measurement of the Global and Local RTD

4.2.1 Theory of RTD

Typically, the RTD is measured when the extrusion process has reached a steady state. A tracer is injected to the flow stream (usually at the hopper) as a pulse, the time interval being so short that the input can be considered only as a Dirac function. The relationship between the variation of the tracer concentration and time at a chosen test point is the RTD density function, $E(t)$, between the tracer injection point and the test point. $E(t)$ can be determined using the following equation [3]:

$$E(t) = \frac{c(t)}{\int_0^\infty c \, dt} = \frac{c(t)}{\sum_{i=1}^\infty c(t) \Delta t} \tag{4.1}$$

where c is the tracer concentration at time t. Another function closely related to $E(t)$ is the cumulative RTD function $F(t)$:

$$F(t) = \int_0^t E(t) dt = \sum_0^t E(t) \Delta t = \frac{\sum_0^t c \Delta t}{\sum_0^\infty c \Delta t} \tag{4.2}$$

The first moment of $E(t)$ is the mean residence time, $\bar{t}$, and is defined as:

$$\bar{t} = \int_0^\infty t E(t) = \frac{\int_0^\infty tc \, dt}{\int_0^\infty c \, dt} = \frac{\sum_0^\infty tc \Delta t}{\sum_0^\infty c \Delta t} \tag{4.3}$$

The second moment of $E(t)$, σ_t^2, called the variance, is a measure of the spread of the RTD about the mean residence time and is defined as:

$$\sigma_t^2 = \int_0^\infty (t - \bar{t})^2 E(t) \mathrm{d}t = \sum_0^\infty (t - \bar{t})^2 E(t) \Delta t \tag{4.4}$$

The RTD can be measured off-line, on-line, or in-line. Off-line methods are relatively simple to use. However, measurements have to be done in a discrete manner. Thus, they are often very time-consuming and generally cannot be performed automatically. Moreover, the resulting data points may not be numerous enough and/or may lack accuracy. By contrast, in-line methods continuously analyze the flow material in real time and have no need for sampling like off-line methods do. Thus, they have many advantages over off-line ones. On-line methods slightly differ from in-line ones in that samples are collected through a bypass and then analyzed. Thus the flow stream is more or less perturbed. However, both sample collection and analysis are done in a continuous manner.

On-line and in-line methods for RTD measurement are developed. They use different tracers and detection devices. Detection principles are based on properties such as radioactivity (MnO_2, La_2O_3, etc.) [4], ultrasound reflectivity (filling, carbon black, etc.) [5], optical reflectivity (TiO_2) [6], electrical conductivity (KNO_3, $NaNO_3$, KCl, etc.) [7], light transmission (carbon black) [8], near infrared (NIR) spectroscopy [9], and fluorescence emission (anthracene) [10–12]. Hu *et al.* developed an in-line RTD measuring device based on fluorescence emission and applied it to TSEs [2, 13]. However, the device, as it was reported in Ref. [2], could only measure the RTD at a single test point, the die exit, because the optical probe was made of poly(methyl methacrylate). Later, they developed a new device based on the principle described in Ref. [2]. The new in-line RTD measurement device possessed many advantages over the one described in Ref. [2], as will be described below.

4.2.2 In-line RTD Measuring System

The in-line RTD measuring system described in this chapter is mainly composed of the following three parts: fluorescent light generation, fluorescent light detection, and signal processing (see Figure 4.1). The fluorescent light source is an ultraviolet high-pressure mercury lamp (125 W) and is split into two beams. Each beam successively passes through a coupler and a bifurcated optical fiber before

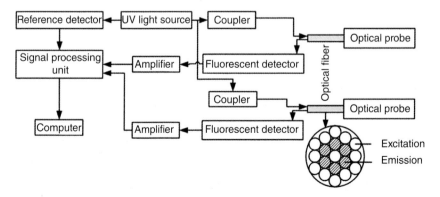

Figure 4.1 Diagram of the in-line RTD measuring system involving three main parts: a fluorescent light generating source, fluorescent light detection and signal processing. (Zhang *et al.* 2006 [14]. Reproduced with permission of Wiley.)

it irradiates the tracer-containing polymer flow stream in the extruder. The light with a specific wavelength emitted from the tracer is subsequently transmitted to a fluorescent detector (a photomultiplier) through the bifurcated optical fiber probe and is then amplified through an amplifier. Finally, the amplified optical signals coming from the two fluorescent detectors reach the signal processing unit. The latter converts them to two analog ones. They are then collected by the computer system and are displayed in real time on the screen. The sensitivity of the measuring system with respect to the tracer concentration can be regulated by a knob. Signals are collected once every second.

It should be noted that although the above in-line RTD measuring instrument follows exactly the same basic principle as the one developed by Hu et al. [2] there are several important differences as well as improvements. First, a reference detector has been added to the system to improve its resistance to both external and internal disturbances. As a result, baseline noises are smoothed out and the quality of the signals is enhanced. Second, it has two light paths or two optical probes instead of one in [2]. Thus, it allows for simultaneous measurement of the RTDs at two different locations. Assessment of local RTDs then becomes possible. Third, the material used for the optical fiber is silica instead of poly(methyl methacrylate) as used in [2]. Silica withstands much higher temperatures than poly(methyl methacrylate). Therefore, the probe can be installed anywhere along the extruder. Finally, the design of the probe is very different from that of the probe in [2], as shown in Figure 4.2. Its front end is a quartz window. The latter prevents the probe from polymer deposition. Moreover, part of the probe body is a ribbed radiator capable of rapidly dissipating heat, which is very important for improving the robustness of the probe because temperatures to which it is subjected could be very high (150–350 °C). As for the light filter, it serves to ensure that only the desired monochromatic light passes through.

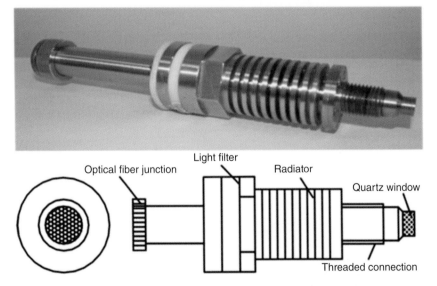

Figure 4.2 A picture and a schematic diagram of the optical probe. (Zhang et al. 2006 [14]. Reproduced with permission of Wiley.)

4.2.3 Extruder and Screw Configurations

Experiments are carried out on a co-rotating TSE (TSE-35A, Nanjing Rui Ya, China) with a diameter of 35 mm and a length-to-diameter ratio of 48. Figure 4.3a shows the locations of the RTD probes and three screw configurations used in the study. The latter differ only in the kneading zone in terms of the geometry of the KDs (Figure 4.3b). The kneading zone is composed of four identical KDs whose geometry is either of type 30/7/32, 45/5/32, or of type 90/5/32 (Figure 4.3c). The head of the extruder is equipped with a strip die, which is 30 mm in length and 1.2 mm in height.

There are three RTD test points, as marked by probes 1, 2, and 3, respectively. Probe 1 is placed at the front end and 2 at the the rear end of the kneading zone,

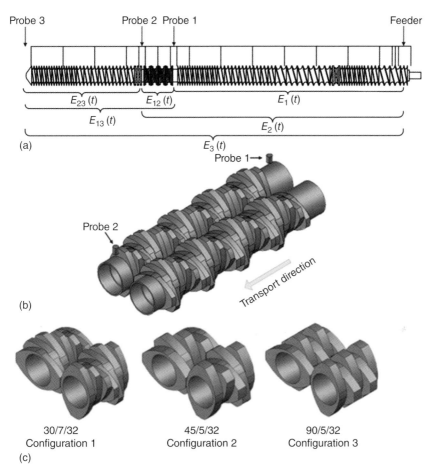

Figure 4.3 (a) Three locations (three test points) where the RTD probes are placed; (b) details of the screw profile of the kneading zone between probes 1 and 2; (c) details of the three different types of kneading disks used for the kneading zone. A kneading block $x/y/z$ is one which has a length of z mm and y disks. The latter are assembled x degrees one with respect to the other. The processing temperature is set at 220 °C. (Zhang et al. 2006 [14]. Reproduced with permission of Wiley.)

and they measure the partial RTD density functions $E_1(t)$ and $E_2(t)$, respectively. Probe 3 is placed in the die and measures the overall RTD density function of the extruder between the feeder and the die, $E_3(t)$. The local RTDs, $E_{12}(t)$, $E_{23}(t)$, and $E_{13}(t)$, corresponding to the three local zones of the extruder, can be obtained by the deconvolution of experimental RTDs $E_1(t)$, $E_2(t)$, and $E_3(t)$, as will be further discussed.

4.2.4 Performance of the In-line RTD Measuring System

Polystyrene, PS (158K, YANGZI - BASF, China) is used as the flow material. Figure 4.4 shows the variation of its viscosity as a function of shear rate at 220 °C. The tracer is in the form of a masterbatch containing PS and anthracene. The concentration of anthracene in the masterbatch is 1%, 3%, 5%, or 10% by mass. These masterbatches are prepared by blending the PS with anthracene in a Haake batch mixer (HBI SYSTEM 90) and then extruded into pellets whose dimensions are similar to those of the PS.

The in-line RTD measuring instrument is very different from the one reported in the literature [2], as outlined in Section 4.2.2. Its performance is evaluated in terms of the reproducibility, on the one hand, and the linearity between the amplitude of the response signal and the amount of the tracer, on the other hand. The former is verified by repeating experiments. Figure 4.5 shows the raw analog signals collected from probes 1 and 2 for three repeated experiments. The relative voltage versus time curves are very smooth and the reproducibility is very good for each of the two probes. However, it is noted that the baselines are not at zero. Nevertheless, they are all virtually at the same and small value.

Figure 4.6 shows the effect of the concentration of the tracer (anthracene) in the masterbatch on the raw analog signal (relative voltage) versus time detected by probe 3. As the concentration of the tracer in the masterbatch increases, the relative voltage versus time curve is shifted upward, as expected. The curves in Figure 4.6a should all fall on two single and separate master curves if they are normalized with respect to the integrated areas and if the relative voltage is linearly proportional to the concentration of the tracer in the masterbatch. This is basically the case, as shown in Figure 4.6b. However, the baselines are more scattered when the concentration of the tracer in the masterbatch is lower, say, 1 or 3 wt%.

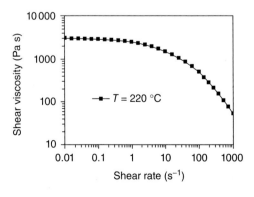

Figure 4.4 Variation of the viscosity of the polystyrene as a function of shear rate at 220 °C. (Zhang et al. 2006 [14]. Reproduced with permission of Wiley.)

Figure 4.5 Raw analog signal (relative voltage) versus time curves for three repeated experiments carried out under the following conditions: screw configuration 3; screw speed = 90 rpm, feed rate = 8 kg h^{-1}; tracer = masterbatch containing 5% anthracene by mass; amount of the tracer = three pellets. (Zhang et al. 2006 [14]. Reproduced with permission of Wiley.)

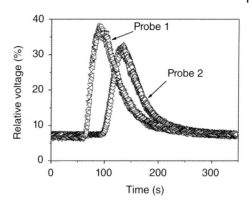

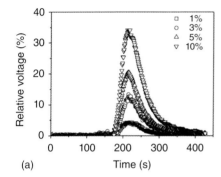

(a)

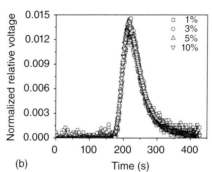

(b)

Figure 4.6 (a) Effect of the concentration of the tracer (anthracene) in the masterbatch on the raw analog signal at probe 3; (b) Normalized relative voltage versus time curves based on Figure 4.6a. Screw configuration 1; screw speed = 60 rpm; feed rate = 10.7 kg h^{-1}; amount of the tracer masterbatch = 0.1 g (four pellets); concentration of the tracer in the masterbatch varying from 1% to 10% by mass. The baselines of the raw analog signal are scaled to zero. (Zhang et al. 2006 [14]. Reproduced with permission of Wiley.)

Based on these results, the masterbatch containing 5 wt% anthracene is selected for subsequent RTD experiments.

Figure 4.7a shows the effect of the amount of the masterbach (number of pellets) on the relative voltage measured at probe 3. The relative voltage versus time curve is shifted upward when the amount of the masterbatch increases to four pellets. Figure 4.7b shows the normalized relative voltage versus time for the results in Figure 4.7a. A single master curve is obtained. Based on the above results obtained with the different amounts of tracer, three pellets of the masterbatch containing 5 wt% anthracene were used for subsequent experiments. Since under these conditions the relative voltage is linearly proportional to the tracer concentration, a normalized relative voltage versus time curve obtained under these conditions can be taken as the RTD density function $E(t)$.

4.2.5 Effects of Screw Speed and Feed Rate on RTD

A TSE often works under starved conditions, namely, feed rate ($\dot{Q}$) is smaller than the maximum pumping capacity of the extruder. Therefore, screw speed (N)

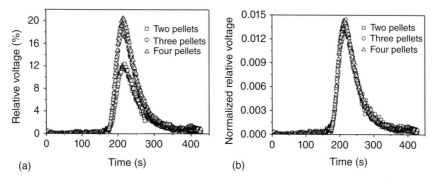

Figure 4.7 (a) Effect of the amount of the masterbatch (number of pellets) on the raw analog signal at probe 3; (b) Normalized relative voltage versus time curves based on Figure 4.7a. Screw configuration 1; screw speed = 60 rpm; feed rate = 10.7 kg h^{-1}. The baselines of the raw analog signal are scaled to zero. (Zhang et al. 2006 [14]. Reproduced with permission of Wiley.)

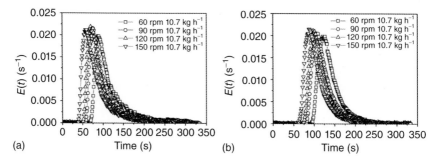

Figure 4.8 Effect of screw speed on the RTD. (a) screw configuration 3 and probe 1; (b) screw configuration 3 and probe 2. (Zhang et al. 2006 [14]. Reproduced with permission of Wiley.)

and feed rate ($\dot{Q}$) are two independent operating parameters. Increasing screw speed shifts the RTD curve toward the shorter time domain, irrespective of the probe location. Meanwhile, their width is slightly affected. Increasing feed rate also shifts the RTD curve toward the shorter time domain, irrespective of the probe location, its effect being similar to the effect of increasing screw speed. However, the RTD curves have significantly narrowed, which is different from the effect of increasing screw speed. The effects of screw speed and feed rate on RTD are shown in Figures 4.8 and 4.9, respectively.

The finding that increasing screw speed or feed rate shifts the RTD curve to the shorter time domain is in agreement with all the works reported in the literature [2, 8, 15–17]. As for the finding that the shape of the RTD curve is more affected by feed rate than screw speed is in line with only some works reported in the literature [16, 17] and is in contradiction to some others [2, 8, 15]. Understanding the exact reasons for the contradiction is of practical significance because it will help conduct (reactive) extrusion processes under optimal conditions. However, thus far, they remain unclear and are likely related to differences that exist among those works in terms of the type, the size and the screw profile of the extruder, the material, and the operating conditions.

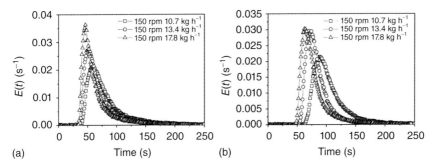

Figure 4.9 Effect of feed rate on the RTD. (a) screw configuration 3 and probe 1; (b) screw configuration 3 and probe 2. (Zhang et al. 2006 [14]. Reproduced with permission of Wiley.)

4.2.6 Assessment of the Local RTD in the Kneading Disk Zone

The flow and mixing conditions in a kneading zone are expected to meet those under which the statistical theory for the RTD would apply. Detailed discussion on the validity and applicability of this theory can be found elsewhere [18–21]. This theory stipulates that for a closed system composed of two statistically independent elements, A and B, the overall RTD density function $E(t)$ is related to those of the two elements $E_A(t)$ and $E_B(t)$ by the following equation:

$$E(t) = \int_0^t E_A(t) \cdot E_B(t - t') dt' \tag{4.5}$$

where t' is a dummy integral variable of time. This equation shows that knowing any two of the three RTD density functions allows to calculate the third one either by convolution or by deconvolution. Assessment of the local RTDs calls upon deconvolution. Details on the deconvolution procedure are given in Ref. [14].

Figure 4.10 shows the effects of the staggering angle of the KDs on the partial and overall RTDs, $E_1(t)$, $E_2(t)$, and $E_3(t)$, measured by probes 1, 2, and 3, respectively. While all these RTD curves are very similar for the 30° and 45° KDs, they are significantly different from those obtained with the 90° disks. The width of the partial RTD curves with the 90° KDs is larger, indicating that the 90° KDs have a better axial mixing capacity than the 30° or 45° ones. It is also seen from Figure 4.10a that the RTD of the screw zone located upstream with regard to the kneading zone between the feeder and probe 1, $E_1(t)$, could possibly be affected by the screw configuration of the kneading zone itself. Otherwise, $E_1(t)$ should have always been the same, irrespective of the type of the KDs used for the kneading zone. Nevertheless, under the conditions of this work, the effect is relatively small.

Figure 4.11 shows the local RTDs in the kneading zone between probes 1 and 2, $E_{12}(t)$, for the three screw configurations that differ only in the staggering angle of the KDs used for the kneading zone. They are obtained by deconvolution using the RTD curves in Figure 4.10 and following the deconvolution procedure. The effect of the staggering angle of the KDs can now be appreciated in a clearer manner. Both the mean residence time and the axial mixing quality characterized quantitatively by the width of the $E_{12}(t)$ curve followed the order: $30° < 45° \ll 90°$.

Figure 4.12 shows the local RTDs in the last part of the extruder between probes 2 and 3, $E_{23}(t)$. The width of the $E_{23}(t)$ is slightly different among the

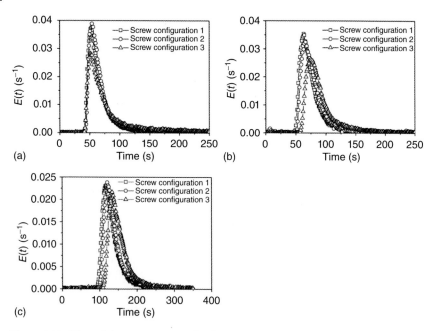

Figure 4.10 Effect of the staggering angle of the kneading disks on the partial RTDs. (a) Probe 1 (b) Probe 2 (c) Probe 3. Screw speed = 120 rpm; feed rate = 15.5 kg h^{-1}. (Zhang et al. 2006 [14]. Reproduced with permission of Wiley.)

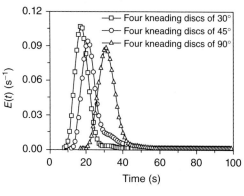

Figure 4.11 Effect of the staggering angle on the RTD over the kneading zone between probes 1 and 2. See Figure 4.10 for the screw configurations and the operating conditions. (Zhang et al. 2006 [14]. Reproduced with permission of Wiley.)

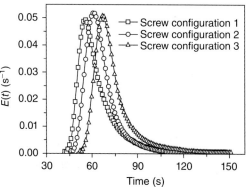

Figure 4.12 Effect of the staggering angle on the RTD between probes 2 and 3. See Figure 4.10 for the operating conditions. (Zhang et al. 2006 [14]. Reproduced with permission of Wiley.)

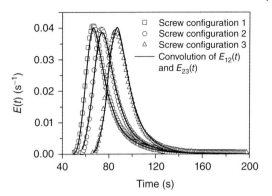

Figure 4.13 Effect of the staggering angle on the RTD between probes 1 and 3. The deconvolution results of $E_3(t)$ and $E_1(t)$ agree with the convolution results of $E_{12}(t)$ and $E_{23}(t)$ corresponding to Figures 4.11 and 4.12 well. See Figure 4.10 for the operating conditions. (Zhang et al. 2006 [14]. Reproduced with permission of Wiley.)

three screw configurations, implying that the mixing performance of the screw zone between probes 2 and 3 is almost unaffected by the screw configuration of the upstream kneading zone. However, obvious differences exist among the three screw configurations in terms of the delay time and mean residence time. The above results imply that the screw profile of the kneading zone affects the downstream velocity more than it does the flow pattern.

Figure 4.13 shows the local RTDs between probes 1 and 3, $E_{13}(t)$, for the three screw configurations. They are obtained by the deconvolution of the experimental data collected at probes 3 and 1 (see the discrete points). They are compared with those obtained by the convolution of $E_{12}(t)$ and $E_{23}(t)$ (see the solid curves). The $E_{13}(t)$ curves obtained by the deconvolution of the experimentally measured $E_3(t)$ and $E_1(t)$ agree well with those obtained by the convolution of the $E_{12}(t)$ and $E_{23}(t)$, showing that both the deconvolution and convolution methods have worked well.

4.3 Residence Time, Residence Revolution, and Residence Volume Distributions

In addition to the distribution in residence time which is characterized by $E(t)$, there are also distributions in residence revolution and residence volume, called residence revolution distribution (RRD) and residence volume distribution (RVD), respectively. The RRD and RVD are obtained when the time coordinate of the RTD is converted to the number of accumulated screw revolutions (n) and the volume of the extrudate (v), respectively. They are given by Eqs (4.6) and (4.7), respectively [16]:

$$F(n) = \frac{c(n/N)}{\int_0^\infty c(n/N)dn} \quad (4.6)$$

$$G(v) = \frac{c(v/\dot{Q})}{\int_0^\infty c(v/\dot{Q})dv} \quad (4.7)$$

where N and $\dot{Q}$ are the screw speed (revolutions per min or rpm) and material volume throughput (l/min), respectively. The integration variables of $F(n)$ and

$G(v)$ are the cumulative screw revolutions (n) and cumulative extrudate volume (v), respectively. Their relationships with time are the following:

$$t = \frac{n}{N} = \frac{v}{\dot{Q}} \tag{4.8}$$

The RVD is a direct measure of the tracer concentration distribution along the screw length, and thus a measure of the degree of the axial mixing, while the RRD may shed light on the transport behavior of the material in the extruder.

4.3.1 Partial RTD, RRD, and RVD

The locations of the RTD probes are the same as those shown in Figure 4.3. Three types of kneading blocks (30/7/32, 60/4/32, and 90/5/32) and one type of gear block are chosen to study the effect of the screw configuration on the overall, partial, and local RTD (Figure 4.14). They differ only in the screw configuration of the test zone. Screw configurations 1, 2, and 3 correspond to the cases where the test zone is composed of four kneading blocks (30/7/32, 60/4/32, and 90/5/32, respectively). In screw configuration 4, the test zone is composed of four gear blocks.

Table 4.1 shows the experiments and their operating conditions. The volume throughput is calculated by dividing the mass throughput by the melt density of the PS, 970 kg m^{-3}.

In order to better understand the process, it would be useful to search for a process parameter whose variation would not change the intensity of mixing but

| 30/7/32 | 60/4/32 | 90/5/32 | Gear block |
| Screw configuration 1 | Screw configuration 2 | Screw configuration 3 | Screw configuration 4 |

Figure 4.14 Geometries of the three different types of kneading blocks and one type of gear blocks used for the test zone. A kneading block x/y/z has a length of z mm and y disks. The latter are assembled x degrees one with respect to the adjacent one. A gear block has two rows of gears along its length of 32 mm. There are 10 gears per row. (Zhang et al. 2008 [22]. Reproduced with permission of Wiley.)

Table 4.1 Experiments and their operating conditions.

Experimental Nb.	1	2	3	4	5	6	7	8	9
Screw speed N (rpm)	60	90	120	150	120	120	150	150	90
Mass throughput (kg h^{-1})	10.7	10.7	10.7	10.7	14.3	15.5	13.4	17.8	8.0
Volume throughput $\dot{Q}$ ($\times 10^{-3}$ liter/min)	183.8	183.8	183.8	183.8	245.7	266.3	230.2	305.8	137.5
Specific throughput $\dot{Q}/N$ ($\times 10^{-3}$ liter/rev.)	3.06	2.04	1.53	1.23	2.04	2.22	1.53	2.04	1.53

The barrel temperature is set at 220 °C.

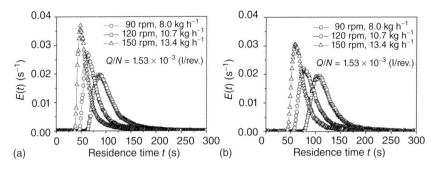

Figure 4.15 Effect of increasing screw speed and throughput on the RTD for a $\dot{Q}/N$ of 1.53×10^{-3} liter/revolution. (a) Probe 1; (b) Probe 2. Screw configuration 1. (Zhang et al. 2008 [22]. Reproduced with permission of Wiley.)

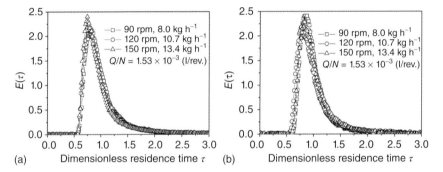

Figure 4.16 Dimensionless residence time distribution density function $E(\tau)$ versus τ curves corresponding to the $E(t)$ versus t curves in Figure 4.15. (a) Probe 1; (b) Probe 2. Screw configuration 1, $\dot{Q}/N = 1.53 \times 10^{-3}$ liter/revolution. Note that all the $E(\tau)$ versus τ curves fall on a single curve. (Zhang et al. 2008 [22]. Reproduced with permission of Wiley.)

only the RTD, or vice versa. Early studies show that specific throughput ($\dot{Q}/N$) could be one [2, 10, 23–26]. Figure 4.15 shows the RTDs of screw configuration 1 measured at probes 1 and 2 for given specific throughputs. An increase in screw speed with a corresponding increase in throughput shifts the RTD curve to the shorter time domain, as expected. Moreover, the RTD curve becomes narrower. Besides these two classical observations, there is no other obvious correlation or relationship among the RTD curves obtained at different screw speeds and throughputs at a given $\dot{Q}/N$. However, when they are normalized with regard to their respective mean residence times, then they all superimpose on single master curves, respectively. These are shown in Figures 4.16 in which the $E(t)$ versus t curves are converted to $E(\tau)$ versus τ curves. $E(\tau)$ is a dimensionless RTD density function and τ is a dimensionless residence time ($\tau = t/\bar{t}$) [2, 13]. This confirms the results of the literature that the dimensionless RTD density function $E(\tau)$ versus τ curve is unique when $\dot{Q}/N$ and screw configuration are fixed, irrespective of $\dot{Q}$ and N.

When the RTD curves shown in Figure 4.15 are converted to the screw revolution and extrudate volume coordinates using Eqs (4.6) and (4.7), respectively,

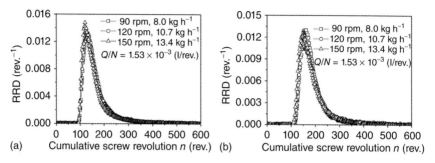

Figure 4.17 RRD corresponding to probes 1 and 2, respectively. (a) Probe 1; (b) probe 2; screw configuration 1; $\dot{Q}/N = 1.53 \times 10^{-3}$ liter/revolution. Note that all the RRD curves overlap. (Zhang et al. 2008 [22]. Reproduced with permission of Wiley.)

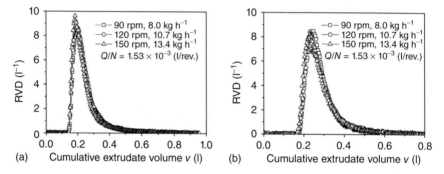

Figure 4.18 RVD corresponding to probes 1 and 2, respectively. (a) Probe 1; (b) probe 2; screw configuration 1; $\dot{Q}/N = 1.53 \times 10^{-3}$ liter/revolution. Note that all the RVD curves overlap. (Zhang et al. 2008 [22]. Reproduced with permission of Wiley.)

one pair of master curves is obtained. They are shown in Figures 4.17 and 4.18, respectively. This indicates that the RRD and RVD are also unique when $\dot{Q}/N$ and screw configuration are fixed, irrespective of $\dot{Q}$ and N.

The key question is then what are the inherent factors that lead to the master curves of the $E(\tau)$ versus τ, RRD versus n and RVD versus v shown in Figures 4.16–4.18, respectively? What are the relationships among the $E(\tau)$ versus τ, RRD versus n and RVD versus v? Actually all of them are derived from the $E(t)$ versus t and their respective abscissas are as follows:

$$\tau = \frac{t}{\bar{t}} \tag{4.9}$$

$$n = tN \tag{4.10}$$

$$v = t\dot{Q} \tag{4.11}$$

According to Eq. (4.8), one has the following equality:

$$\tau = \frac{t}{\bar{t}} = \frac{n}{\bar{t}N} = \frac{v}{\bar{t}\dot{Q}} \tag{4.12}$$

Considering Eq. (4.12) and the fact that when $\dot{Q}/N$ is fixed, $E(\tau)$ versus τ, RRD versus n, and RVD versus v are master curves and are independent of $\dot{Q}$ and N, one arrives at the following relationships for a given $\dot{Q}/N$:

$$N\bar{t} = k_1 = \text{constant 1} \tag{4.13}$$

$$\dot{Q}\bar{t} = k_2 = \text{constant 2} \tag{4.14}$$

where k_1 and k_2 are constants and $k_2/k_1 = \dot{Q}/N$. In other words, when $\dot{Q}/N$ is fixed, $N\bar{t}$ and $\dot{Q}\bar{t}$ should be constants, irrespective of the values of $\dot{Q}$ and N. This will be discussed later.

The mean residence time can be determined by:

$$\bar{t} = \frac{V \times f}{\dot{Q}} \tag{4.15}$$

where V is the free volume in the extruder barrel and f is the mean degree of fill. When Eq. (4.15) is introduced to Eqs (4.13) and (4.14), respectively, the following equation is obtained:

$$f = \frac{k_1}{V} \times \frac{\dot{Q}}{N} = \frac{k_2}{V} = \text{constant 3} \tag{4.16}$$

Equation (4.16) indicates that when $\dot{Q}/N$ is fixed, the mean degree of fill is fixed, regardless of $\dot{Q}$ and N. Mudalamane and Bigio proposed the following equation to describe the relationship between the complete fill length (FL), and $\dot{Q}/N$ [27]:

$$\text{FL} = K_{sc}\left(\frac{\dot{Q}/N}{K_{pump} - \dot{Q}/N}\right) \tag{4.17}$$

where K_{sc} is a constant that only depends on screw geometry and K_{pump} is a function of screw design. This equation shows that for a given screw geometry and screw design, once $\dot{Q}/N$ is fixed, the complete FL is fixed.

According to Eqs (4.13) and (4.14), for a given screw profile and $\dot{Q}/N$, the value of $\bar{t}N$ and that of $\bar{t}\dot{Q}$ should be constant, respectively, regardless of N or $\dot{Q}$. This is confirmed by the data shown in Table 4.2 corresponding to the values of $\bar{t}\dot{Q}$ obtained at probes 1 and 2 for the four different screw configurations in Figure 4.14 and for a $\dot{Q}/N$ of 2.04×10^{-3} liter/revolution with three pairs of $\dot{Q}$ and N values. When $\dot{Q}/N$, screw configuration, and probe location are fixed, the values of $\bar{t}\dot{Q}$ for all the three different values of $\dot{Q}$ are indeed the same within

Table 4.2 Values of $\bar{t}\dot{Q}$ corresponding to a given $\dot{Q}/N$ for four screw configurations in Figure 4.14 at probes 1 and 2.

Probe	4 × 30° kneading blocks			4 × 60° kneading blocks			4 × 90° kneading blocks			4 × gear blocks		
	$\bar{t}_1\dot{Q}_1$ (l)	$\bar{t}_2\dot{Q}_2$ (l)	$\bar{t}_3\dot{Q}_3$ (l)	$\bar{t}_1\dot{Q}_1$ (l)	$\bar{t}_2\dot{Q}_2$ (l)	$\bar{t}_3\dot{Q}_3$ (l)	$\bar{t}_1\dot{Q}_1$ (l)	$\bar{t}_2\dot{Q}_2$ (l)	$\bar{t}_3\dot{Q}_3$ (l)	$\bar{t}_1\dot{Q}_1$ (l)	$\bar{t}_2\dot{Q}_2$ (l)	$\bar{t}_3\dot{Q}_3$ (l)
Probe 1	0.29	0.30	0.29	0.30	0.31	0.31	0.32	0.34	0.32	0.31	0.29	0.33
Probe 2	0.34	0.37	0.32	0.35	0.37	0.36	0.39	0.40	0.40	0.42	0.44	0.42

the experimental and numerical calculation errors. For example, the value of $\bar{t}\dot{Q}$ is between 0.42 and 0.44 for the four gear block configurations and probe 2, regardless of the values of $\dot{Q}$ and N.

4.3.2 Local RTD, RRD, and RVD

The local RTD curves in the test zone between probes 1 and 2, $E_{12}(t)$, are obtained by the deconvolution of the experimental data of $E_1(t)$ and $E_2(t)$. Figure 4.19 shows the effects of screw speed and volume throughput on $E_{12}(t)$ for screw configuration 1. Increasing screw speed with throughput being held constant shifts the local RTD curve to the short time domain. However, its shape does not change much (Figure 4.19a). Increasing throughput with screw speed being held constant also shifts the RTD curve to the short time domain. Moreover, its width becomes narrower. Therefore, the shape of the RTD is much more affected by throughput than by screw speed. The effects of screw speed and volume throughput on local RTD curves $E_{12}(t)$ are similar to those on global RTD curves $E_3(t)$.

Figure 4.20 shows the local RTD, RRD, and RVD curves between probes 1 and 2 for given values of $\dot{Q}/N$. Again, they all fall on a single curve. This indicates that for a given screw profile, $\dot{Q}/N$ not only controls the overall and partial RTD, RRD, and RVD, but also local ones. Moreover, local RRD and RVD do not provide more information on a given screw zone than the corresponding local RTD does.

Figure 4.21 shows the effect of screw configuration on the local RTD, RRD, and RVD curves of the test zone between probes 1 and 2. They are obtained by deconvolution of the data obtained by probes 1 and 2 using Eq. (4.5). It is seen that both the mean residence time and the axial mixing quality characterized by the width of the RTD, follow the order: $30° < 60° \ll 90° <$ gear block.

The RRD and RVD curves follow the same trend as the RTD, as expected. When the operating conditions are fixed, an increase in the staggering angle of the kneading block from 30° to 90° shifts the local RRD and RVD to higher screw revolution and larger extrudate volume domains, respectively. The shift with the gear blocks is the most significant, indicating that conveying a given amount of tracer from probe 1 to probe 2 needs more screw revolutions or more extrudate volume when the gear blocks are used.

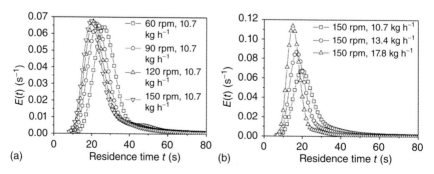

Figure 4.19 Effects of the screw speed (a) and throughput (b) on the local RTD of the test zone of screw configuration 1. (Zhang et al. 2008 [22]. Reproduced with permission of Wiley.)

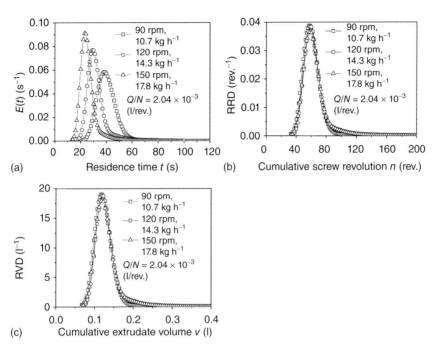

Figure 4.20 Local RTD (a), RRD (b), and RVD (c) curves between probes 1 and 2 for screw configuration 3 at a given $\dot{Q}/N$ (2.04×10^{-3} liter/revolution). They are obtained by deconvolution. Note that all the local RTD, RRD, and RVD curves fall on single master curves, respectively. (Zhang et al. 2008 [22]. Reproduced with permission of Wiley.)

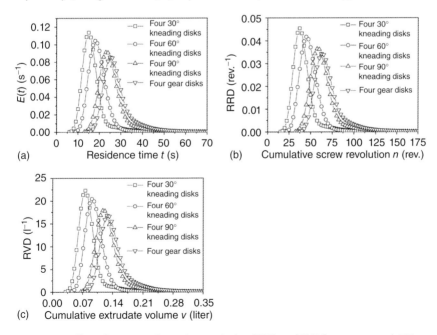

Figure 4.21 Effect of screw configuration on the local RRD and RVD for screw speed: 150 rpm, feed rate: 17.8 kg h^{-1}. (a) RTD, (b) RRD, and (c) RVD. (Zhang et al. 2008 [22]. Reproduced with permission of Wiley.)

4.4 Modeling of Local Residence Time Distributions

Kneading blocks are the main mixing elements for a co-rotating TSE. However, it remains empirical when it comes to designing KDs and optimizing extrusion conditions for a specific application due to the very large number of geometrical, rheological, and operating parameters that may influence the flow and mixing [28, 29]. Currently, simplified 1-D models are often used in predicting the temperature, pressure profile, and RTD in extrusion processing [16, 30]. However, they only provide the average values of different parameters along the extruder, which are insufficient for describing flow pattern and mixing [31, 32]. In recent years, much attention has been paid to 3-D simulation of the flow field in different regions of a co-rotating TSE [33–36].

It is not easy to simulate flow patterns in complex geometries such as KDs. Time-dependent boundaries as KDs rotate make numerical simulations even more difficult. Experimental validation of simulated results is another big challenge. There have been studies on numerical simulations of pressure, temperature, and velocity profiles and their comparison with experiments. Although numerical simulations and experimental validations of the local RTD in mixing sections of a TSE are crucial, they are rare due to experimental difficulties.

4.4.1 Kinematic Modeling of Distributive Mixing

For a viscous fluid, the distributive mixing mechanism is the stretching and reorientation of fluid elements. Ottino *et al.* developed a lamellar model to quantify the capacity of a flow to deform matter and to generate an interface between two components in a mixture, which gives a kinematic approach to modeling distributive mixing by tracking the amount of deformation experienced by a fluid element [37–39]. In the initial fluid domain, when an infinitesimal material area $|dA|$ with normal direction $\hat{N}$ is transformed into an area $|da|$ with normal direction $\hat{n}$ at time t, the area stretch ratio of material surface η is defined as [38]:

$$\eta = \frac{|da|}{|dA|} \tag{4.18}$$

A good distributive mixing quality requires a high value of η over space and time. The arithmetic mean of logarithm of the area stretch ratio $\overline{\log \eta}$ can be used to compare the distributive mixing performance of different mixers, and is defined as:

$$\overline{\log \eta} = \frac{\log \eta_1 + \log \eta_2 + \cdots + \log \eta_n}{n} \tag{4.19}$$

An instantaneous stretching efficiency e_η based on stretching of material surfaces is defined as [39]:

$$e_\eta = \frac{\dot{\eta}/\eta}{(D:D)^{1/2}} = \frac{d(\log \eta)/dt}{(D:D)^{1/2}} = \frac{-D:\hat{m}\hat{m}}{(D:D)^{1/2}} \tag{4.20}$$

where $\dot{\eta}$ is a material time derivative of η, D is a rate of deformation tensor and $\hat{m}$ is a local normal unit vector of current surface. e_η characterizes the fraction of the dissipated energy used for stretching the area and falls in the range $[-1, 1]$. A positive (negative) value indicates that the dissipated energy is used for stretching (shrinking) the surface. A time-averaged efficiency $\langle e_\eta \rangle$ is defined as [39]:

$$\langle e_\eta \rangle = \frac{1}{t} \int_0^t e_\eta \, dt \qquad (4.21)$$

The above mixing parameters can be calculated when the velocity field is known. Distributive mixing efficiency can be characterized by time-average efficiency $\langle e_\eta \rangle$. Some researchers defined a distribution function F_α associated with the field α. The quantity $F_\alpha(\alpha_p, t)$ is given as [40, 41]:

$$F_\alpha(\alpha_p, t) = p \qquad (4.22)$$

where α_p is a critical value, which indicates that $p\%$ of marker particles have a value of α lower than or equal to α_p at time t. The field α may be η, e_η, or $\langle e_\eta \rangle$.

The velocity field in kneading zones is simulated using a commercial CFD code: POLYFLOW 3.10.0® of ANSYS Inc. The local RTD of the same domains is then calculated based on the velocity field. Simulated results are compared with experimental ones obtained by an in-line measuring instrument as described earlier. The distributive mixing parameters such as the area stretch ratio of material surface η, instantaneous stretching efficiency e_η, and time-averaged efficiency $\langle e_\eta \rangle$ are calculated using the interface tracking techniques. These parameters are then used to compare the distributive mixing performance and efficiency of different types of KDs.

4.4.2 Numerical Simulation

Figure 4.22 shows the geometries of eight different KDs simulated in this work. They all have the same axial length, that is, 64 mm, and differ in stagger angle, disk width, and/or disk gap. Table 4.3 gives the description of KDs (KD1 to KD8) and flow channel in Figure 4.22. The first three KDs, KD1 to KD3 are used for experimentally validating simulated results. KD4 to KD6 have the same disk width and gap, but differ in stagger angle. They are chosen to analyze the effect of the stagger angle on the local RTD as well as the distributive mixing parameters. The effects of the disk width and gap on the axial and distributive mixing are studied by KD6, KD7, and KD8. There is no gap between two adjacent disks of KD7 and KD8. The flow channel includes three parts: inlet section, kneading section, and outlet section. The kneading section accommodates a KD. Sufficiently long inlet and outlet sections are added to the computational domains so that the flows can be assumed to be fully developed. Figure 4.23 shows the disk gaps of a KD.

Meshing is carried out using the commercial pre-processor software GAMBIT (Fluent Inc.). The meshes of each KD are chosen after a preliminary mesh-discretization study. A convergence analysis compares the velocity and pressure solutions of different mesh numbers along the axial direction. The number of meshes is increased till the solutions are no more sensitive to it.

The flow is assumed to be a time-dependent, isothermal, incompressible, and generalized Newtonian fluid with no body forces. Meanwhile, it is assumed

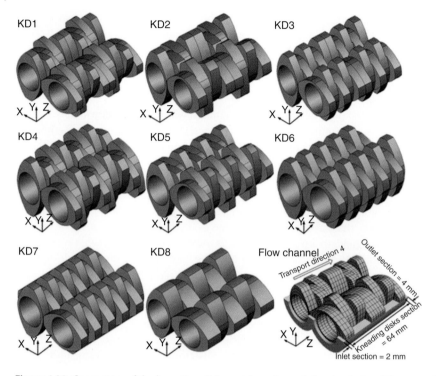

Figure 4.22 Geometries of the kneading disks and flow channels for simulation. (Zhang *et al.* 2009 [42]. Reproduced with permission of Wiley.)

Table 4.3 The description of kneading disks KD1 to KD8 and flow channel in Figure 4.22.

	The description of kneading disks
KD1	Two pairs of intermeshing 45/5/32 kneading disks
KD2	Two pairs of intermeshing 60/4/32 kneading disks
KD3	Two pairs of intermeshing 90/5/32 kneading disks
KD4	One pair of intermeshing 45/10/64 kneading disks
KD5	One pair of intermeshing 60/10/64 kneading disks
KD6	One pair of intermeshing 90/10/64 kneading disks
KD7	One pair of intermeshing 90/10/64 kneading disks
KD8	One pair of intermeshing 90/5/64 kneading disks
Flow channel	Inlet section (2 mm) + kneading section (64 mm) + outlet section (4 mm)

that the free volume of KD is filled with melt. The governing equations are the momentum and continuity equations and can be written as:

$$\Delta \cdot \tau - \Delta P = 0 \tag{4.23}$$

$$\nabla \cdot u = 0 \tag{4.24}$$

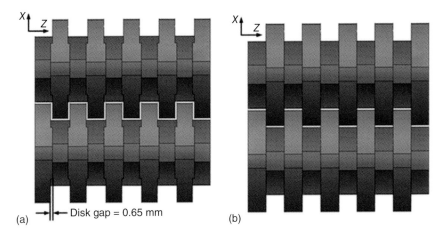

Figure 4.23 Top view of (a) KD6 with gaps and (b) KD7 without gap. (Zhang et al. 2009 [42]. Reproduced with permission of Wiley.)

where τ is the extra stress tensor, P is the pressure, and u is the velocity vector. τ is given by

$$\tau = 2\eta D \tag{4.25}$$

where D is the rate of deformation tensor and η is the kinematic viscosity that depends on the shear rate $\dot{\gamma}$. The constitutive equation is represented by the Carreau model:

$$\eta(\dot{\gamma}) = \eta_0[1 + (\lambda\dot{\gamma})^2]^{(n-1)/2} \tag{4.26}$$

where η_0 is the zero shear rate viscosity, λ is the natural time, and n is the power law index. The parameters for the Carreau model are: $\eta_0 = 2814$ Pa s, $\lambda = 0.148$ s, and $n = 0.278$.

No-slip boundary conditions are imposed on the surfaces of the KDs and barrel bore walls. A volume flow rate (PS melt) is set to the inlet section and a normal stress boundary condition is applied to the outlet section. Table 4.4 lists the operating conditions for the numerical simulations and experiments. The volume throughput is calculated by dividing the mass throughput by the melt density of the PS, 970 kg m^{-3}.

To calculate the local RTD provided by each of the above eight types of KDs, virtual particles are launched at the same time in the flow domain. It is assumed that the marker particles are massless and non-interacting with each other. Under

Table 4.4 Operating conditions for the numerical simulations and experiments carried out in this work (barrel temperature: 220 °C).

Experimental Nb.	1	2	3	4
Screw speed (rpm)	120	150	150	150
Mass throughput (kg h^{-1})	10.7	10.7	13.4	17.8
Volume throughput $\dot{Q}$ (×10^{-6} m^3 s^{-1})	3.06	3.06	3.84	5.10

these assumptions, the particles can be located by integrating the velocity vectors with respect to time. Initially, these particles are randomly distributed in an inlet vertical plane and their trajectories between the inlet and outlet are calculated from the velocity profiles. Along each trajectory, the residence time is computed. The RTD is then obtained based on the residence time of each of these particles. Two thousand particles with random directions are used to calculate mixing parameters.

4.4.3 Experimental Validation

A better understanding of the flow occurring in the KDs of the co-rotating TSE can be obtained from Figure 4.24 that shows the streamlines of two particles initially located at different places of the inlet. They have followed different streamlines to the exit, implying that they have different residence times. Particle one proceeds around the screws in a "figure-eight" mode, which is in agreement with the experimental results of previous researchers [43, 44]. For particle two, its motion first follows a "figure-eight" mode and then circles around one screw. It finally returns to a "figure-eight" mode. The above motion modes show that as time evolves, the fluid particles undergo different degrees of shear and stress. The complex flow in the KDs is indicative of good mixing.

Figure 4.25 compares the numerical local RTD results with the experimental ones for KD2. Their overall trends are in good agreement. However, the former are wider than the latter. Reasons for these differences are as follows. The simulation assumes that the flow channel be fully filled and isothermal. In practice these assumptions are not fully met, which is the main reason for these differences. The melt density is a function of pressure and temperature, not a constant, which may also contribute to these differences. Meanwhile, the subtle effects of the rheology on the velocity profile could impart significant differences to mixing [45]. Figure 4.26 compares the simulated and experimental RTD for different screw configurations. Again, the 90° KDs have the highest axial mixing performance, while the 45° and 60° KDs have similar axial mixing performance.

The KD4, KD5, and KD6 have the same disk width and gap, but different stagger angles. Figure 4.27 shows the effect of stagger angle on the local RTD. The delay

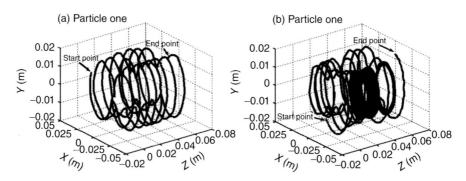

Figure 4.24 Streamlines of two particles for KD3. Screw speed = 150 rpm; feed rate = 17.8 kg h^{-1}. (Zhang et al. 2009 [42]. Reproduced with permission of Wiley.)

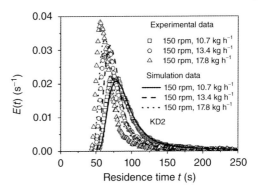

Figure 4.25 Comparison of the local RTD between the numerical and experimental results for different feed rates. (Zhang et al. 2009 [42]. Reproduced with permission of Wiley.)

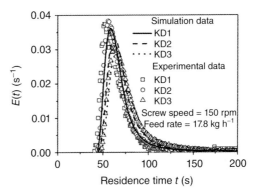

Figure 4.26 Comparison of the local RTD between numerical and experimental results for different screw configurations. (Zhang et al. 2009 [42]. Reproduced with permission of Wiley.)

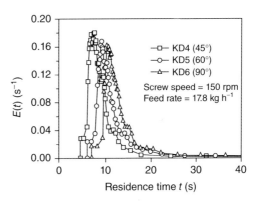

Figure 4.27 Effect of stagger angle on the local RTD. (Zhang et al. 2009 [42]. Reproduced with permission of Wiley.)

time and mean residence time increase with an increase in the stagger angle. Also, the 90° KD provides the highest axial mixing quality.

Figure 4.28 shows the effects of the disk gap and width on the local RTD. The three local RTD curves are similar in delay time and shape, indicating that the KD6, KD7, and KD8 have similar axial mixing quality.

4.4.4 Distributive Mixing Performance and Efficiency

The area stretch is calculated to compare the distributive mixing performance of different types of KDs. A KD having a high distributive mixing performance

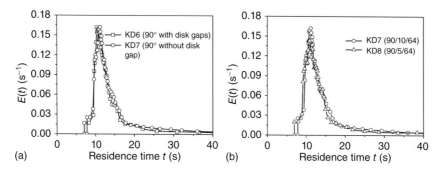

Figure 4.28 Effects of the disk gap (a) and disk width (b) on the local RTD. (Zhang et al. 2009 [42]. Reproduced with permission of Wiley.)

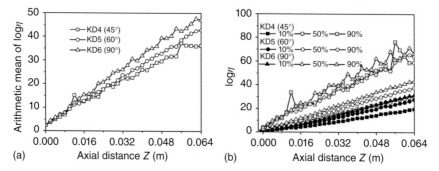

Figure 4.29 Axial evolution of log η: (a) arithmetic mean of log η, (b) critical value of log η for given percentiles of marker particles. Screw speed: 150 rpm, feed rate: 17.8 kg h^{-1}. (Zhang et al. 2009 [42]. Reproduced with permission of Wiley.)

requires high values of the logarithm of the area stretch ratio over the axial distance. Distributive mixing efficiency is characterized by time-average efficiency. Figure 4.29a shows the arithmetic mean of log η (Eq. (4.19)) from the entrance to the exit. For KD4, KD5, and KD6, it increases more or less linearly with increasing axial distance Z, indicating that η increases exponentially with Z. Among the three screw configurations, for a given axial distance the arithmetic mean of log η follows the order: KD6 > KD5 > KD4, indicating that the distributive mixing performance increases with increasing stagger angle of the KD. Figure 4.29b shows the critical value of the arithmetic mean of log η along Z for 10%, 50%, and 90% marker particles for KD4, KD5, and KD6, respectively. Again it follows the order: KD6 > KD5 > KD4, regardless of the percentile of marker particles.

From Figure 4.29a, the arithmetic mean of log η versus Z curves fluctuates, especially at a longer axial distance. This means that not all particle surfaces are stretched along Z in a steady manner. At the initial stage of mixing, it is easy to stretch the surface area and the arithmetic mean of log η increases linearly with Z. As the surface area is further stretched, area stretching becomes more difficult. Meanwhile, Ishikawa et al. pointed out that the disk gap plays an important role in mixing due to the large amount of forward and backward flow [46]. The latter

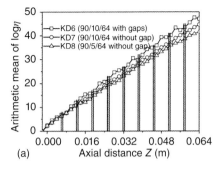
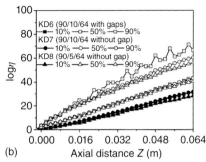

Figure 4.30 Effects of the disk gap and disk width on the axial evolution of log η: (a) arithmetic mean of log η, (b) log η for given percentiles of marker particles. Screw speed: 150 rpm, feed rate: 17.8 kg h^{-1}. The vertical lines in Figure 4.34 a correspond to the locations of the disk gaps of KD6 of Figure 4.23. (Zhang et al. 2009 [42]. Reproduced with permission of Wiley.)

may bring about stretch and shrink of surfaces. Figure 4.29b shows that stretch and shrink of surfaces occur primarily at a high area stretch ratio.

Figure 4.30a shows the effects of the disk gap and disk width on the arithmetic mean of log η as a function of Z for 90° KDs. The arithmetic mean of log η of a KD with a gap (KD6) is higher than that of a corresponding KD without a disk gap (KD7). On the other hand, a KD with a smaller disk width (KD7) has a higher arithmetic mean of log η than a corresponding KD with a larger disk width (KD8), indicating that the distributive mixing performance decreases with increasing disk width. This result is in agreement with that put forth by Andersen [47]. It is also noted that, unlike KD7 and KD8 whose arithmetic mean of log η increases linearly and smoothly with Z, the mixing performance of KD6 increases linearly but in a fluctuating manner. In fact, each fluctuation of the arithmetic mean of log η corresponds to a passage through a gap. Figure 4.30b shows the critical values for 10%, 50%, 90% marker particles. There are no fluctuations at low percentiles such as 10% and 50%. Fluctuations are visible only at high percentiles 90%.

In a closed system such as a batch mixer, the time-average efficiency exhibits three characteristic features: $\langle e_\eta \rangle$ decays as t^{-1} for flow with no reorientation; $\langle e_\eta \rangle$ decays as t^{-1} but oscillates for flow with moderate reorientation; $\langle e_\eta \rangle$ oscillates without decay for flow with strong reorientation [45, 48, 49]. For an open system such as an extruder or a Kenics mixer, Z^{-1} is equivalent to t^{-1}. Figure 4.31a shows the evolution of the arithmetic mean of the time-average efficiency as a function of Z for KD4, KD5, and KD6. The three curves have periodic oscillations without decay as Z^{-1}, indicating that all the three KDs bring about strong reorientation. Among the three KDs, the arithmetic mean of the time-average efficiency is the highest for KD6, indicating that KD6 provides strongest reorientation. Figure 4.31b shows the time-average efficiency as a function of Z for different percentiles of marker particles. At low percentiles such as 10% and 50%, the critical values of the time-average efficiency are similar for the three types of KDs. At high percentiles, such as 90%, KD6 yields significantly higher time-average efficiency. This indicates that at low percentiles the three types of KDs are similar in time-average efficiency. However, at high percentiles,

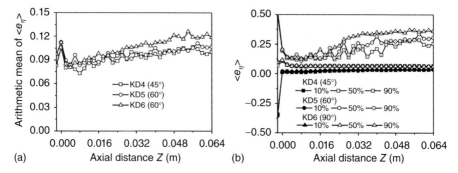

Figure 4.31 Axial evolution of time-average efficiency $\langle e_\eta \rangle$: (a) arithmetic mean of $\langle e_\eta \rangle$, (b) $\langle e_\eta \rangle$ for given percentiles of marker particles. Screw speed: 150 rpm, feed rate: 17.8 kg h^{-1}. (Zhang et al. 2009 [42]. Reproduced with permission of Wiley.)

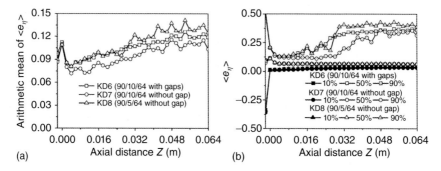

Figure 4.32 Axial evolution of time-average efficiency $\langle e_\eta \rangle$: (a) arithmetic mean of $\langle e_\eta \rangle$, (b) $\langle e_\eta \rangle$ for different percentiles of marker particles. Screw speed: 150 rpm, feed rate: 17.8 kg h^{-1}. (Zhang et al. 2009 [42]. Reproduced with permission of Wiley.)

KD6 provides the highest time-average efficiency. It is also noted that when marker particles start moving, most of them are stretched. Nevertheless, some of them shrink. Therefore, they exhibit negative time-average efficiency at the initial stage. They are all stretched eventually.

As for the three 90° KDs (KD6, KD7, and KD8), they have similar RTD curves (see Figure 4.28) but differ in $\log \eta$ (see Figure 4.31). KD6 provides the highest $\log \eta$, and therefore the highest distributive mixing performance. Figure 4.32 compares these three types of KDs in terms of $\langle e_\eta \rangle$. Clearly, KD8 provides the highest mixing efficiency and not KD6. The reason is that $\langle e_\eta \rangle$ is not only a function of $d(\log \eta)/dt$ but also of $D:D$ and RTD. Figure 4.33 shows the evolution of $d(\log \eta)/dt$ and $(D:D)^{1/2}$ as a function of Z for KD6, KD7, and KD8, respectively. The arithmetic mean of $d(\log \eta)/dt$ is very close for the three types of KDs. On the other hand, the three types of KDs differ more or less in $(D:D)^{1/2}$, the energy dissipated. The value of $(D:D)^{1/2}$ is the lowest for KD8, indicating that the fraction of the energy dissipated for the area stretch is the highest for KD8. To summarize, KD8 provides the lowest distributive mixing performance (characterized by $\log \eta$) but highest mixing efficiency (characterized by $\langle e_\eta \rangle$).

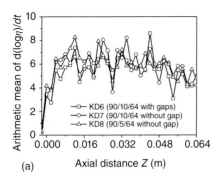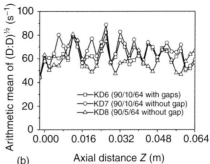

Figure 4.33 Axial evolution of (a) the arithmetic mean of d(log η)/dt and (b) that of $(D{:}D)^{1/2}$. Screw speed: 150 rpm, feed rate: 17.8 kg h^{-1}. (Zhang et al. 2009 [42]. Reproduced with permission of Wiley.)

4.5 Summary

This chapter has presented an in-line fluorescent instrument to measure in real time the global and partial RTD in screw extruders. Their residence time decreases with increasing screw speed and feed rate, irrespective of the test location. However, the influence of screw speed on the width of the RTD seems to be only slightly affected. The RTD curves significantly narrow with increasing feed rate, which is different from increasing screw speed. The in-line fluorescent instrument allows assessing the local RTD of a kneading zone and the effect of its geometry on the RTD based on a statistical theory. The local RTD of a kneading zone depends very much on the geometry (staggering angle) of the KDs. Both mean residence time and the axial mixing quality characterized by the width of the RTD, follow the order 30° < 45° ≤ 60° < 90° < gear disks.

It is shown theoretically and experimentally that specific throughput $\dot{Q}/N$ is a key process parameter for controlling the dimensionless time distribution, RRD and RVD. For a given value of $\dot{Q}/N$, the overall, partial, and local RTD are different when varying $\dot{Q}$ and N. However, the corresponding dimensionless RTD as well as the RRD and RVD all fall on single master curves, respectively. This is because when the value of $\dot{Q}/N$ is fixed, the mean degree of fill and complete FL are fixed, irrespective of the values of $\dot{Q}$ and N.

A 3-D model is developed and validated using local RTD. It is used to predict distributive mixing parameters such as the area stretch ratio of material surface η, instantaneous mixing efficiency e_η, and time-averaged mixing efficiency $\langle e_\eta \rangle$ using the interface tracking techniques. These parameters are then used to compare the distributive mixing performance and efficiency of different types of KDs. The disk gap and disk width play an important role in distributive mixing performance and efficiency. A KD with a disk gap provides a higher area stretch ratio and a higher time-average efficiency than a corresponding one without disk gap. A KD with a larger disk width has a lower distributive mixing performance than a corresponding one with a smaller disk width. However, its distributive mixing efficiency could be higher due to lower energy dissipation.

References

1 Hu, G.H., Hoppe, S., Feng, L.F., and Fonteix, C. (2009) Reactive compounding, in *Mixing and Compounding of Polymer: Theory and Practice*, 2nd edn (ed. I. Manas-Zloczower), Hanser Publishers, Munich, pp. 1019–1080.

2 Hu, G.H., Kadri, I., and Picot, C. (1999) On-line measurement of the residence time distribution in screw extruders. *Polym. Eng. Sci.*, **39** (5), 930–939.

3 Nauman, E.B. and Buffham, B.A. (1983) *Mixing in Continuous Flow System*, John Wiley & Sons, Inc., New York.

4 Kiani, A., Heidemeyer, P., and Pallas, R. (1997) *Study of Flow and RTD in a ZSK Twin Screw Extruder*. SPE ANTEC, Toronto, America, April 27 – May 2, pp 94–99.

5 Sun, Z., Jen, C.K., Shih, C.K., and Denelsbeck, D.A. (2003) Application of ultrasound in the determination of fundamental extrusion performance: residence time distribution measurement. *Polym. Eng. Sci.*, **43** (1), 102–111.

6 Wetzel, M.D., Shih, C.K., and Sundararaj, U. (1997) *Determination of Residence Time Distribution During Twin Screw Extrusion of Model Fluids*. SPE ANTEC, Toronto, America, April 27 – May 2, pp 3707–3712.

7 Unlu, E. and Faller, J.F. (2001) Geometric mean vs. arithmetic mean in extrusion residence time studies. *Polym. Eng. Sci.*, **41** (5), 743–751.

8 Chen, T., Patterson, W.I., and Dealy, J.M. (1995) On-line measurement of residence time distribution in a twin-screw extruder. *Int. Polym. Process.*, **10** (1), 3–9.

9 Laske, S., Witschnigg, A., Selvasankar, R.K., and Holzer, C. (2014) Measuring the residence time distribution in a twin screw extruder with the use of NIR-spectroscopy. *J. Appl. Polym. Sci.*, **131** (6), 39919.

10 Gerstorfer, G., Lepschi, A., Miethlinger, J., and Zagar, B.G. (2013) An optical system for measuring the residence time distribution in co-rotating twin-screw extruders. *J. Polym. Eng.*, **33** (8), 683–690.

11 Carneiro, O.S., Covas, J.A., Ferreira, J.A., and Cerqueira, M.F. (2004) On-line monitoring of the residence time distribution along a kneading block of a twin-screw extruder. *Polym. Test.*, **23** (8), 925–937.

12 Fang, H., Mighri, F., Ajji, A., Cassagnau, P., and Elkoun, S. (2011) Flow behavior in a corotating twin-screw extruder of pure polymers and blends: characterization by fluorescence monitoring technique. *J. Appl. Polym. Sci.*, **120** (4), 2304–2312.

13 Hu, G.H. and Kadri, I. (1999) Preparation of macromolecular tracers and their use for studying the residence time distribution of polymeric systems. *Polym. Eng. Sci.*, **39** (2), 299–311.

14 Zhang, X.M., Xu, Z.B., Feng, L.F., Song, X.B., and Hu, G.H. (2006) Assessing local residence time distributions in screw extruders through a new in-line measurement instrument. *Polym. Eng. Sci.*, **46** (4), 510–519.

15 Sun, Y.J., Hu, G.H., Lambla, M., and Kotlar, H.K. (1996) In situ compatibilization of polypropylene and poly(butylene terephthalate) polymer blends by one-step reactive extrusion. *Polymer*, **37** (18), 4119–4127.

16 Gao, J., Walsh, G.C., Bigio, D., Briber, R.M., and Wetzel, M.D. (1999) Residence-time distribution model for twin-screw extruders. *AIChE J.*, **45** (12), 2541–2549.

17 Razaviaghjeh, M.K., Nazockdast, H., and Assempour, H. (2004) Determination of the residence time distribution in twin screw extruders via free radical modification of PE. *Int. Polym. Process.*, **19** (4), 335–341.
18 Chen, L. and Hu, G.H. (1993) Applications of a statistical theory in residence time distributions. *AIChE J.*, **39** (9), 1558–1562.
19 Chen, L., Pan, Z., and Hu, G.H. (1993) Residence time distribution in screw extruders. *AIChE J.*, **39** (3), 1455–1464.
20 Chen, L., Hu, G.H., and Lindt, J.T. (1995) Residence time distribution in non-intermeshing counter-rotating twin-screw extruders. *Polym. Eng. Sci.*, **35** (7), 598–603.
21 Poulesquen, A., Vergnes, B., Cassagnau, P., Michel, A., Carneiro, O.S., and Covas, J.A. (2003) A study of residence time distribution in co-rotating twin-screw extruders. Part II: experimental validation. *Polym. Eng. Sci.*, **43** (12), 1849–1862.
22 Zhang, X.M., Feng, L.F., Hoppe, S., and Hu, G.H. (2008) Local residence time, residence revolution, and residence volume distributions in twin-screw extruders. *Polym. Eng. Sci.*, **48** (1), 19–28.
23 Elkouss, P., Bigio, D., and Wetzel, M.D. (2003) *Deconvolution of Residence Time Distribution Signals to Individually Describe Zones for Better Modeling*. SPE ANTEC, May 4–8, 344–348.
24 Gasner, G.E., Bigio, D., Marks, C., Magnus, F., and Kiehl, C. (1999) A new approach to analyzing residence time and mixing in a co-rotating twin screw extruder. *Polym. Eng. Sci.*, **39** (2), 286–298.
25 Sun, Y.J., Hu, G.H., and Lambla, M. (1995) Free radical grafting of glycidyl methacrylate onto polypropylene in a co-rotating twin screw extruder. *J. Appl. Polym. Sci.*, **57** (9), 1043–1054.
26 Hu, G.H., Sun, Y.J., and Lambla, M. (1996) Effects of processing parameters on the in situ compatibilization of polypropylene and poly(butylene terephthalate) blends by one-step reactive extrusion. *J. Appl. Polym. Sci.*, **61** (6), 1039–1047.
27 Mudalamane, R. and Bigio, D. (2004) Experimental characterization of fill length behavior in extruders. *Polym. Eng. Sci.*, **44** (3), 557–563.
28 Zhang, X.M., Zhu, S.Y., Zhang, C.L., Feng, L.F., and Chen, W.X. (2014) Mixing characteristics of different tracers in extrusion of polystyrene and polypropylene. *Polym. Eng. Sci.*, **54** (2), 310–316.
29 Xu, J.J., Zhang, X.M., Chen, W.X., and Feng, L.F. (2015) Numerical simulation of mixing performance in the miniature conical twin screw extruder. *Polym. Mater.: Sci. Eng.*, **31** (3), 128–132.
30 Vergnes, B., Della Valle, G., and Delamare, L. (1998) A global computer software for polymer flows in corotating twin screw extruders. *Polym. Eng. Sci.*, **38** (11), 1781–1792.
31 Zhu, L.J., Narh, K.A., and Hyun, K.S. (2005) Evaluation of numerical simulation methods in reactive extrusion. *Adv. Polym. Technol.*, **24** (3), 183–193.
32 Cassagnau, P., Bounor-Legare, V., and Fenouillot, F. (2007) Reactive processing of thermoplastic polymers: a review of the fundamental aspects. *Int. Polym. Process.*, **22** (3), 218–258.

33 Ilinca, F. and Hétu, J.-F. (2012) Three-dimensional numerical study of the mixing behaviour of twin-screw elements. *Int. Polym. Process.*, **27** (1), 111–120.

34 Hétu, J.-F. and Ilinca, F. (2013) Immersed boundary finite elements for 3D flow simulations in twin-screw extruders. *Comput. Fluids*, **87**, 2–11.

35 Sobhani, H., Anderson, P.D., Meijer, H.H.E., Ghoreishy, M.H.R., and Razavi-Nouri, M. (2013) Non-isothermal modeling of a non-Newtonian fluid flow in a twin screw extruder using the fictitious domain method. *Macromol. Theory Simul.*, **22** (9), 462–474.

36 Yang, K., Xin, C., Yu, D., Yan, B., Pang, J., and He, Y. (2015) Numerical simulation and experimental study of pressure and residence time distribution of triple-screw extruder. *Polym. Eng. Sci.*, **55** (1), 156–162.

37 Chella, R. (1994) Laminar mixing in miscible fluids, in *Mixing and Compounding of Polymers: Theory and Practice* (eds I. Manas-Zloczower, Z. Tadmor, and J.F. Agassant), Hanser, Munich, pp. 1–40.

38 Ottino, J.M., Ranz, W.E., and Macosko, C.W. (1981) A framework for description of mechanical mixing of fluids. *AIChE J.*, **27** (4), 565–577.

39 Ottino, J.M. and Chella, R. (1983) Laminar mixing of polymeric liquids; a brief review and recent theoretical development. *Polym. Eng. Sci.*, **23** (7), 357–379.

40 Avalosse, T. and Crochet, M.J. (1997) Finite-element simulation of mixing: 1. Two-dimensional flow in periodic geometry. *AIChE J.*, **43** (3), 577–587.

41 Avalosse, T. and Crochet, M.J. (1997) Finite-element simulation of mixing: 2. Three-dimensional flow through a kenics mixer. *AIChE J.*, **43** (3), 588–597.

42 Zhang, X.M., Feng, L.F., CHEN, W.X., and Hu, G.H. (2009) Numerical simulation and experimental validation of mixing performance of kneading discs in a twin screw extruder. *Polym. Eng. Sci.*, **49** (9), 1772–1783.

43 Erdmenger, R. (1962) Zur Entwicklung von Schneckenverdampfern. *Chem. Ing. Tech.*, **34** (11), 751–754.

44 Erdmenger, R. (1964) Mehrwellen-Schnecken in der Verfahrenstechnik. *Chem. Ing. Tech.*, **36** (3), 175–185.

45 Connelly, R.K. and Kokini, J.L. (2004) The effect of shear thinning and differential viscoelasticity on mixing in a model 2D mixer as determined using FEM with particle tracking. *J. NonNewtonian Fluid Mech.*, **123** (1), 1–17.

46 Ishikawa, T., Kihara, S.I., and Funatsu, K. (2000) 3-D numerical simulations of nonisothermal flow in co-rotating twin screw extruders. *Polym. Eng. Sci.*, **40** (2), 357–364.

47 Andersen, P. (1998) Co-rotating intermeshing twin screw extruder: Berstorff's system, in *Plastics Compounding: Equipment and Processing* (ed. D.B. Todd), Hanser, Munich, pp. 77–136.

48 Chella, R. and Ottino, J.M. (1985) Fluid mechanics of mixing in a single-screw extruder. *Ind. Eng. Chem. Fundam.*, **24** (2), 170–180.

49 Ottino, J.M. (1989) *The Kinematics of Mixing: Stretching, Chaos and Transport*, Cambridge University Press, Cambridge.

5

In-process Measurements for Reactive Extrusion Monitoring and Control

José A. Covas

University of Minho, Department of Polymer Engineering, Institute for Polymers and Composites/I3N, Guimarães 4800-058, Portugal

5.1 Introduction

Reactive extrusion combines melt extrusion and chemical reactions in a single operation to polymerize monomers, modify polymers, compatibilize immiscible polymer blends, or synthesize *in situ* nanoparticles in a polymer matrix. Current industrial reactive extrusion operations may encompass several stages, including feeding the ingredients (either together or sequentially, as pellets, powder, flakes, or liquids), melting (at least some of them), mixing (which involves distributive and dispersive mixing with specific degrees of intensity), devolatilizing and die forming for subsequent pelletization. Chemical reactions typically start to develop from the onset of melting and, depending on the type of reaction, full chemical conversion is generally obtained somewhere along the screw axis [1, 2].

The use of single-screw extruders for polymerization reactions began in the 1950s, in order to avoid the use of solvents, accomplish continuous manufacture, and minimize energy consumption (e.g., by elimination of solvent heating and cooling). However, despite the gradual improvements in screw design, these machines offer limited sequential addition capabilities, which confine their applicability to relatively simple operations, such as modification of polyolefins with peroxide or silane grafting [3]. Twin-screw extruders, or combinations of various kinds of equipment, were progressively adopted to better fulfill the specificities of some processes and chemical reactions [4, 5]. Counter-rotating intermeshing self-wiping twin-screw extruders were successfully utilized for copolymerization of styrene–*n*-butyl methacrylate, polymerization of ε-caprolactam, grafting of maleic anhydride (MA) on to polyethylene, polymerization of urethanes, or radical polymerization of methacrylates [6]. However, as they operate at low screw speeds (because flow in the calender gap of the intermeshing screws pulls these apart against the barrel wall causing abrasion), shear rate levels are moderate and so is dispersion efficiency. Likewise, non-intermeshing counter-rotating twin-screw extruders were applied to perform polycondensation reactions due to their excellent distributive mixing, but they have limited pressure generation

Reactive Extrusion: Principles and Applications, First Edition. Edited by Günter Beyer and Christian Hopmann.
© 2018 Wiley-VCH Verlag GmbH & Co. KGaA. Published 2018 by Wiley-VCH Verlag GmbH & Co. KGaA.

and dispersive mixing capabilities. Co-rotating intermeshing twin-screw extruders have been widely used for reactive extrusion on account of their modular construction (enabling the geometries of both screws and barrel to create the thermomechanical environment suitable for each specific application), high yield, and independent control over output and screw speed [7].

Extruder settings include screw speed, barrel and die temperatures, feed rate (in the case of twin-screw extruders and starve-fed single-screw extruders) and vacuum pressure (if devolatilization is taking place). Regardless of the eventual complexity of the chemo–physical–rheological phenomena that develops inside the extruder, the process parameters that are conventionally monitored comprise motor torque, melt pressure, and melt temperature (the last two being often measured at the die entrance). Although variations in time of these parameters are caused by changes in the overall flow conditions (e.g., partial clogging of the filters, progressive screw wear, perturbations in feeding, surging, viscous dissipation, equipment malfunction – e.g., a heater band not working), and/or in the characteristics of the polymer system (e.g., degree of chemical conversion, molecular weight (MW) or molecular weight distribution (MWD), viscosity, composition, morphology), isolation of the cause(s) is often difficult. Material characterization for quality control is habitually done by collecting extrudate samples at regular intervals, followed by standard off-line testing (a typical test to check a controlled degradation reaction would be the melt flow rate (MFR)). In this context, Machado *et al.* [8] developed a sampling technique that enabled the rapid collection of material from within the extruder at several barrel locations, which was subsequently used to investigate the evolution of chemistry, morphology, and rheology along the extruder axis during reactive processing of various polymer systems [9–13]. Figure 5.1 shows a representative result: the effect of screw speed on the evolution of melt temperature along the screw axis of a co-rotating twin-screw extruder and die during compounding of polylactic acid (PLA) with a chain extender, as measured by sticking a fast response thermocouple into the fresh collected melt. Melting starts at screw turn number 9, as the material flows through a kneading block extending up to screw turn 11.

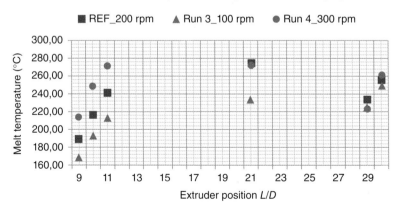

Figure 5.1 Effect of screw speed on the evolution of melt temperature along the screw axis and die of a co-rotating twin-screw extruder, upon compounding of PLA with a chain extender (unpublished data).

Varying the screw speed between 100 and 300 rpm generates melt temperature differences of up to 60°C in this region, which are attenuated downstream, as the material flows in partially filled channels separating shorter restrictive zones. Although useful – this data is not available for practical reactive extrusion operations – this approach is complicated and time consuming and may therefore bring about a delay between sample collection and test result, thus affecting its effectiveness for quality control. Moreover, in certain cases, sampling for subsequent testing may actually generate misleading data because the morphology/degree of conversion of the material may be altered during the preparation of specimens for testing (for instance, making disks for rotational rheometry). On the other hand, current process control consists in keeping the extruder settings within narrow tolerances around set values, that is, control is independent of the characteristics of the material being manufactured. If sufficiently fast and frequent data pertaining to a crucial characteristic of the material could be obtained (ideally, in real time), a different process control paradigm could be developed, whereby the extruder settings would be changed in order to keep the value of that characteristic essentially constant [14, 15]. Thus, both efficient quality and process control in reactive extrusion require swift monitoring of material attributes relating to the plasticating sequence and to the chemical reaction.

In-process (also denoted as *at-line*) measurements [16], that is, measurements of material characteristics performed by devices coupled to the extruder as reactive extrusion progresses, made without disrupting the flow and by providing data in real time, are thus highly desirable. Although these goals have not been fully attained yet, it is currently possible the viscosity, the average MW and MWD, and the level of degradation of a polymer directly on the extruder, as well as to monitor the extent of a reaction, or the particle size of a suspension. This chapter discusses the challenges faced by in-process measurements for reactive extrusion and presents the approaches and techniques that seem to show more potential for widespread application. A few practical examples of their utilization are also provided.

5.2 Requirements of In-process Monitoring of Reactive Extrusion

In-process monitoring of reactive extrusion is advantageous for:

- assembling material data that would otherwise be lengthy or problematic to acquire off-line; additionally, for collecting large amounts of data, or obtaining data at short time intervals, as the burden of manual sample collection and taking samples to the laboratory is circumvented;
- implementing a quality control scheme that is capable of assessing process stability, capable of detecting material batch-to-batch variations and uncovering process/material attribute trends that may ultimately lead to product non-conformities, and capable of achieving a short time lag between sample collection and test result;
- assessing the compatibility between the kinetics of the chemical processes and the operating conditions and residence times in the extruder;

- obtaining a better scientific understanding of specific reactive extrusion schemes, namely in terms of establishing correlations between chemical conversion, operating conditions, and material properties;
- gathering experimental data to be used for the evaluation of process modeling routines;
- supporting quicker material and/or process development, optimization, or scale-up (e.g., determining the best operating window, fine-tuning a formulation);
- assisting (automatic) process control based on a specific material attribute; for instance, since the rheological response of a polymer changes with the degree of polymerization, MW, and MWD, in-process rheometry was initially proposed for the closed loop control of the peroxide-initiated degradation of polypropylene (PP) [15].

In-process measurements to monitor polymer compounding and reactive extrusion are mostly performed using optical spectroscopy (ultraviolet–visible (UV–Vis), infrared (IR), Fourier transform infrared (FTIR), near-infrared (NIR), and Raman) as it provides molecular information or data on chemical composition and rheometry, which is sensitive to the structure and morphological features at various scales. The application of these techniques is discussed further (see a few case studies in [17]). Nonetheless, many other methods, such as other spectroscopy techniques (e.g., fluorescence), light scattering, polarimetry, turbidity, ultrasonic attenuation, and electric conductivity could also be of practical interest for quality control purposes. It is worth mentioning that, in the case of polymer blending and nanoparticle synthesis, reactive extrusion aims at creating stable and fine entities with sizes in the nanometer scale, which may defy the resolution provided by some techniques. Fluorescence spectroscopy is primarily employed to monitor residence times via the addition of fluorescent tracers in the melt [17–19], but could also be used to monitor additive concentrations. Time is a key parameter in chemical reactions and the residence time distribution (RTD) offers an estimation of the mixing history of the material. Figure 5.2 illustrates the set-up developed by Carneiro et al. [19] to measure the average residence time and the RTD at various locations along the barrel of a co-rotating twin-screw extruder. In these machines, changes in screw speed should not disturb the average residence time. However, as shown in Figure 5.3 for the same reactive extrusion operation considered in Figure 5.1, the boundaries of RTD vary with screw speed, that is, the minimum local residence time decreases with increasing screw speed (Figure 5.3a), the effect being smaller for the maximum residence time. In other words, screw speed influences the breadth of RTD and its progress along the extruder. Light scattering can estimate the size and shape of suspended particles or droplets (see [20] and references there in). It was used together with rheology to study start-up flows in a rheometer and demonstrate that the characteristic overshoot of the first normal stress difference (N_1) of uncompatibilized and compatibilized blends is similar, but shifted to shorter times for the latter, thus indicating a faster breakup process [21]. However, the technique has been rarely used in extrusion [22–25]. Teixeira et al. [20] showed that while polarized optical microscopy

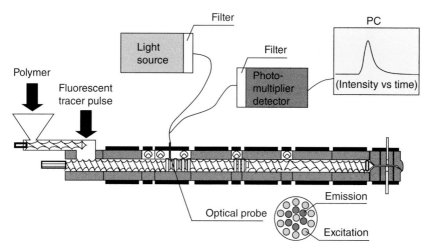

Figure 5.2 Set-up for in-process measurement of the average residence time and residence time distribution at a specific location of the extruder. The measuring probe sits in a modified barrel segment. It contacts the melt stream on one end and contains a bifurcated optical fiber bundle that transfers the light emitted by a mercury lamp source and modulated by an optical chopper, while simultaneously carrying the light emitted by the fluorescent tracer to a photodetector that converts it into a voltage signal (for details and results see [19]). (Carneiro et al. 2004 19]. Reproduced with permission of Elsevier.)

(POM) becomes ineffective below 1 μm, small-angle light scattering (SALS) allows access to the sub-micron scale, but only for very low-volume fractions before multiple scattering sets in (also, the refractive index of the blend components should not be very different). The limit concentration was found to decrease with the particle size of the dispersed phase. These results were in agreement with earlier observations for comparable non-reactive and reactive systems [24]. Turbidity measurements were used to estimate the dispersion level of fillers during the preparation of nanocomposites [26], or the concentration, particle size, and shape of the dispersed phase in immiscible polymer blends [27]. Again, limitations in the estimation of particle size and concentration may limit its application in reactive extrusion. Polarimetry can measure, in real time, polymer flow birefringence during extrusion. In turn, since at constant operating conditions birefringence increases with MW [28], the technique has the potential to monitor reactive extrusion. Reports are abundant on the use of ultrasound methods (vibrations with a frequency typically above 20 kHz) to monitor changes in polymerization, MW, density, viscosity, melt temperature, orientation and relaxation, blend composition, blend morphology, filler concentration, dispersion of solid additives, RTD, melting evolution, and so on. Application in reactive extrusion is much less frequent: Kiehl et al. [29] monitored the bulk polymerization of methyl methacrylate. The use of currently available commercial sensors for long-run operations is generally limited to 150°C and 30 bar. Thus, through-the-wall (i.e., via a buffer) measurements are generally the available option for melts, which makes the set-up more complicated and requires substantial construction accuracy in addition to the need to perform correction calculations to the results.

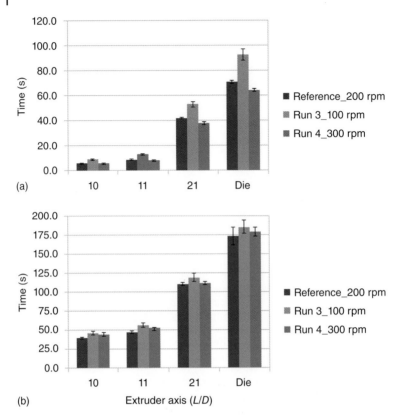

Figure 5.3 Effect of screw speed on the evolution of the minimum (a) and maximum (b) residence time along the screw axis and die of a co-rotating twin-screw extruder, upon compounding of PLA with a chain extender (unpublished data).

Therefore, despite the potential of the technique, its widespread use still requires higher robustness, accuracy, and user friendliness [30]. Finally, Perusich and McBrearty [31] used dielectric spectroscopy to monitor polymer/co-monomer composition variations during extrusion, while van Boersma and Turnhout [32] followed the morphology of a thermotropic liquid crystalline polymer (LCP)/PP blend. The technique is of recognizable relevance to monitor the manufacture of electrically conductive nanocomposites. Excellent reviews on monitoring extrusion and compounding are available, specifically focusing on rheometry [33], spectroscopic and scattering techniques [23], and pharmaceutical manufacturing [34]. In their review of the fundamental aspects of reactive processing of thermoplastics, Cassagnau et al. [35] also covered in-process monitoring.

Regardless of the technique employed, while conventional testing of a sample collected from the process performed in the laboratory is denoted as *off-line*, it is important to distinguish between in-process measurements made *in-line* and *on-line*. The former acquire data directly from the main melt stream (typically between extruder and die, or at die), for example, by using flush mounted sensors or a special measuring section. A representative example of the latter is the "partial Couette" flow rheometer developed by Dealy and coworkers [36], where a rotating drum positioned in the flow channel forms a shearing zone with the channel wall. To perform *on-line* measurements, a polymer melt stream or sample is detoured from the main flow; after testing, the material is either lost or rejoins the main process channel. Each of these two operational modes has both advantages and limitations [16, 23, 37]. Since in-line measurements are made directly in the flow channel, very short time delays exist between sampling and testing and no sampled material is wasted. In addition, in principle the results represent the bulk, because a significant portion of the material is tested. A notable exception is the optical technique using reflectance mode, where the depth of the signal penetration may be rather low. Conversely, trials are necessarily performed at the local processing temperature, which is most probably different from the set value due to viscous dissipation, and is hence unknown and not easy to determine. In the case of in-line rheological measurements, this fact hinders the comparison of sets of data obtained from experiments with either different material recipes or operating conditions. Furthermore, being dictated by the process flow rate, the shear rate cannot be set independently. Thus, either a single shear rate is probed or, for those solutions where the geometry of the measuring channel or the velocity of a moving wall can be adjusted, a narrow range of shear rates are scanned. Slit dies with flush mounted sensors (sometimes combining different types of techniques [23]) are often coupled to extruders, whose operating conditions are then tuned to access wider shear rate and temperature ranges. This type of in-line assembly is generally not applicable to industrial scale production, but is useful for the purpose of research and development.

Diverting a sample or a stream of material from the main flow channel in order to perform an on-line measurement inherently increases the time lag between sampling and measurement. Monitoring very fast reactions may even be at risk, as chemical conversion may proceed during sampling. On the positive side, testing away from the main flow channel may provide the opportunity to set autonomously the testing geometry, the test temperature, and/or the rate of deformation. Maintenance and cleaning operations and adjustments in the sensor or in the testing geometry while extrusion proceeds become easier. Still, if the test temperature is required to be considerably different from that of the material at the sampling point, thermal equilibration may extend significantly due to the low thermal conductivity of polymers, which in turn may induce alterations in the characteristics of the material (owing to further chemical conversion, polymer degradation, or coalescence of the dispersed phase) [38, 39]. Figure 5.4 illustrates this point by showing the required equilibration times prior to on-line capillary rheometry testing at 200°C, for material collected at various extruder barrel locations. Upon sampling, the quiescent melt must wait between 10 and 60 s in the reservoir of the rheometer, prior to testing. Producing a melt stream bypass or returning the diverted material after the test, may not only affect the characteristics of the material but also perturb the main flow. Sampling is often made at, or close to, the main flow channel wall, and thus might not truly represent the bulk. As will be seen further, in the case of on-line rheometry, although gear pumps can control the sampling rate, the deformation rate, and the return flow accurately, they add strain and temperature to the sample. Dealy and Rey [40] showed the importance of Taylor diffusion, that is, of the axial

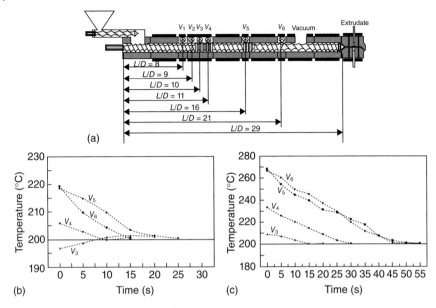

Figure 5.4 Temperature change of melt samples collected from various locations along the extruder axis (v_3 to v_6) and deposited in the reservoir of an on-line rheometer set to 200°C prior to testing. (a) extruder set-up and locations for melt sampling/measurements; (b) results for PP with extruder barrel set to 210°C; (c) results for HDPE with extruder barrel set to 230°C [39].

spreading out of the flow front due to the non-uniform velocity profile, on the dynamic response of these devices.

In-process measurements of several material characteristics (e.g., density, color, particle size, conductivity, chemical composition) or process variables (melt temperature, pressure, residence time) involve a passive interaction between sensor and sample [23, 34]. Others, such as the rheological measurement, require that the material is subjected to a controlled deformation or deformation rate, and the resulting force or change in a specific characteristic (e.g., molecular orientation, deformation of suspended droplets) is monitored [37, 41]. Given the aforementioned, designing equipment for in-process monitoring of reactive extrusion poses a number of challenges:

- The device should be able to accurately measure the relevant material response(s); data should be acquired at high rate and the test should be as rapid as possible to minimize the interval between sampling and obtaining results; output data should include both the values of the relevant material functions as well as their variation with processing time.
- The sample to be tested should be representative of the material being processed: (i) sampling and eventual thermal equilibration time for on-line measurements should not be destructive, that is should not affect the characteristics of the material, particularly its morphology; (ii) in the case of optical and vibrational measurements, sufficient path length must be ensured (this is particularly important for strong absorbing media); (iii) each measurement should be made on a fresh sample; and (iv) samples should be renewed rapidly.

- Sampling and testing should not interfere with extrusion (e.g., causing secondary flows or stagnation zones) and, concomitantly, the main flow should not interfere with the measurement [36, 42].
- Probes should be able to withstand high pressures (up to 200 MPa) and temperatures (typically up to 350°C, although more than 400°C is necessary for a few high-melting temperature polymers) and be corrosion- and abrasion-resistant.
- Probes should be easily and quickly fixed to the processing equipment. Ideally, the measuring tip of testing probes should have a geometry identical to that of conventional strain-gage melt pressure transducers of the diaphragm-type, so that they can be mounted on existing extruder barrel ports. In the case of on-line testing, melt detouring could be made through these ports.
- In-process measurements made at the die, or between screw tip and die, aim at monitoring the outcome of the reactive process, that is, to determine the degree of conversion and/or the final characteristics of the product as well as variations during production. Most in-process testing is performed in this way. Particularly during process development stages, it is advantageous to uncover the effects of type of formulation components, composition, screw design, feeding sequence, and operating conditions on the flow and reaction kinetics – and how these affect the characteristics and performance of the resulting material. Thus, measurements along the extruder axis are also important. Either ports for measurement melt pressure transducers are available, or modified barrel segments are used (examples of the latter include SALS [24], NIR spectroscopy [43], and rheometry [38, 39]).
- The data collected should be reliable and not be influenced by the mechanical vibrations of the extruder or by other interferences that arise in a typical production environment. A simple transfer of laboratory instrumentation to the process may be inadequate, as changes in construction, data acquisition time, and ensuring sufficient signal-to-noise ratio and resolution are usually necessary [23].
- The apparatus should be sufficiently compact so that it does not perturb normal operation of the compounding line, mechanically robust (this is especially relevant when using optical fibers), and guarantee proper melt sealing when applicable; user friendliness and a high level of automation are also desired.
- In-process rheological or rheo-optical testing poses additional challenges. Measurements should be performed in viscometric flows and under well-controlled conditions, in the widest possible range of deformation rates (most practical polymer melt extrusion involves shear rates between 10 and $1000\,s^{-1}$) and temperatures (the availability of isothermal rheological functions within a span of 30°C is highly desirable).

The choice of the in-process techniques and of the measurement methods depends on the purpose of the application as well as on the geometry of the probes and the fixing possibilities provided by the construction of the processing equipment. The spectra collected by optical spectroscopy may be correlated with composition as well as with the structural and morphological features. In the case of rheology, on-line capillary rheometers yield a viscosity flow curve (or the much simpler MFR), which is appropriate to characterize polymerization, chain extension, and controlled degradation reactions. Small amplitude oscillatory shear (SAOS) tests are generally considered adequate to analyze *in situ* compatibilized polymer blends, as eventual changes in the material during testing are minimized. Extensional viscosity is quite sensitive to changes in interfacial tension and thus should be a good probe for polymer blends [44]. For quality control purposes, while the absolute or quantitative value of a given material property may not be necessary, good sensitivity to changes in that property is sufficient. In addition, a moderate time lag between sampling and measurement is acceptable. In the case of process control, the same requirements for measurements apply, but only minute time delays between sampling and obtaining the result are acceptable. Gimenez *et al.* [45] attempted to control bulk ε-caprolactone polymerization in a twin-screw extruder by modeling the process dynamics via rheological modeling coupled with RTD modeling, and using pressure measurements and in-line ultraviolet fluorescence for the experimental data. They found out that the large response time relative to the process dynamics was a limiting factor for practical process control. McAfee *et al.* [46] compared in-line with on-line rheometry measurements during extrusion. They concluded that the second approach is unsuitable for the dynamic control of melt viscosity, as it could not detect changes in the melt state due to temperature or shear rate effects in the extruder and did not respond to changes in feed material within the time scale required for process control. The in-line "partial Couette" flow rheometer [36] was used to control the reactive extrusion neutralization of an ethylene-methacrylic acid copolymer to produce an ionomer [14]. Nevertheless, closed loop control with on-line capillary rheometry is applied in practice, for example, to the peroxide degradation of PP, to produce different grades of material. On the other hand, process development or optimization would benefit from precise quantitative measurements of relevant material attributes, while the time lag is of little importance (as long as it does not affect the validity of the measurements).

The high investment cost of some equipment, the limited robustness of some set-ups, the need to involve specialized personnel to operate it, as well as the existing accumulated know-how on reactive extrusion control based on monitoring the conventional process parameters, seem to be delaying the widespread adoption of in-process measurement techniques. A few attempts have been made to correlate those parameters with a major material attribute, thereby generating what is denoted as a virtual softsensor. For instance, neural networking modeling techniques were adopted to define a (non-linear) relationship between process parameters and a given material property, which was measured off-line (MFR) [47]. Then, the model was applied during extrusion in order to predict the value of that property, taking in the current real values of the process parameters. The success of this methodology depends strongly on the extent of available historical data to train and test the model. With a view to automating process control, other approaches combining partial theoretical structures with data (i.e., gray-box models) were associated with fuzzy logic, neural networks, and/or genetic algorithms [48, 49]. Recently, Nguyen *et al.* [50] developed a closed-loop control system using a conventional proportional–integral–derivative controller that keeps the viscosity constant (determined from the throughput set by a gravimetric feeder and the pressure drop measured in an in-line slit die), by automatically altering the screw speed and the temperature in the first three zones of the barrel. A distinct monitoring strategy consists of mapping the output/screw speed ratio (Q/N) against process parameters and selected material attributes, in order to capture the effects of processing on the latter [51]. Obviously, such mapping requires extensive extrusion trials under different processing conditions (various Q/N settings), which might not be practicable in a production setting.

5.3 In-process Optical Spectroscopy

Radiation at different wavelengths through the spectral range (UV–Vis, IR, and Raman) induces different molecular motions. Thus, spectroscopy can provide complementary data relating to the chemical composition and structure of the skeletal backbone of a polymer. UV–Vis radiation (wavelengths between 10 and 1000 nm) interacts with molecules and causes electronic transitions (promotion of electrons from the ground state to a higher energy state). These transitions at a given wavelength can be used to assess the concentration or amount of a particular species. Within the IR region of the electromagnetic spectrum (12 000–400 cm^{-1}), the mid-infrared (MIR) range (from 4000 to 400 cm^{-1}) is associated with molecular vibrations, whereas near-infrared (NIR) (12 000–4000 cm^{-1}) encompasses bands that result from harmonic overtones of fundamental and combination bands associated with hydrogen atoms (this being relevant for compounds containing CH, OH, and NH bonds). The Beer–Lambert law establishes a logarithmic relationship between the concentration of chemical species and absorbance, A:

$$\log A = \varepsilon c l \tag{5.1}$$

where, ε is the molar absorption coefficient of the species, c is its concentration in the sample, and l is the path length in the measurement cell. Thus, at constant l, a change in the absorbance band intensity can be associated with a change in concentration. A calibration curve is initially constructed by plotting the measured absorbance of a series of standard samples as a function of their concentration. Then, upon measuring an unknown sample by UV–Vis or MIR, the concentration of the absorbing component can be determined. Wang et al. [52] applied UV–Vis to monitor the process-induced degradation of poly-L-lactic acid (PLLA) during extrusion and found a relationship between absorption and molar mass, which was then employed to measure polymer degradation. Jakisch et al. [53] used MIR to monitor the reaction of styrene–MA copolymer with a long-chain alkyl amine in a twin-screw extruder. Although frequently utilized in many industries, application of IR to polymer melts was delayed due to the need to face high working pressures and temperatures. In-line monitoring with FTIR requires the use of the attenuated-total-reflection (ATR-IR) technique. The absorbance bands of FTIR spectra can be linked to chemical composition, but the technique requires a very short path length, as small as 50 μm [23]. Thus, as the main flow channel is not this shallow, and the measurement must

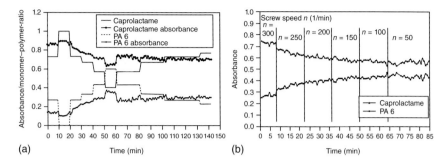

Figure 5.5 In-line monitoring of polymerization of ε-caprolactam in a twin-screw extruder by means of FTIR. Sensitivity of the measured absorbance to changes in time of the monomer/polymer ratio (a) and screw speed (b). (Haberstroh et al. 2002 [56]. Reproduced with permission of Wiley.)

be representative of the bulk, on-line configurations are preferable, as used by Fritz and Ultsch [54] to follow polyethylene grafting with organosilanes. Nevertheless, in-line FTIR was applied to monitor the copolymerization of a styrene–maleic-anhydride model system with an alky-amine during melt extrusion [53]. Interestingly, the technique was also used to track reactive processes upstream of the die, a type of study that is rarely reported. The thermal degradation of polyoxymethylene (POM) during melt extrusion was analyzed on-line, via measurement of the emission of formaldehyde (FA) gas. A dome was built in the degassing section of the extruder, and pure nitrogen gas was used to carry the FA by a heated tube from the degassing section to the FTIR [55]. Haberstroh et al. [56] used ATR-FTIR to evaluate the monomer/polymer ratio during polymerization of ε-caprolactam in a twin-screw extruder. Measurements were performed using a high-temperature and a high-pressure resistant dipping probe with a conical two-reflection ZnSe ATR crystal mounted in the intermeshing section of the extruder, in a zone where the channels operated fully filled. The penetration depth of the IR beam in the polymer melt was about 2 µm. The exercise was only partially successful, since once a high polyamide concentration was reached the signal intensities became constant, probably because the layer of polyamide melt covering the ATR crystal could not be wiped out by the lower-viscous melt that had higher caprolactam content. Figure 5.5 shows the sensitivity of the measured signal to changes in time of the monomer/polymer ratio and screw speed.

The peaks in NIR spectra show considerable overlap, making quantitative analysis more difficult. Thus, the growing interest in this technique arises from the advent of highly precise instruments with very high signal to noise ratio combined with the development of multivariate analysis (also known as *chemometrics*), which yields a calibration model. Moreover, while optical fiber technology for the mid-IR spectral range is still expensive and susceptible to damage [23], NIR can operate with inexpensive silica fiber optics, and commercial probes are available to work either in transmission or in diffusive reflectance mode (thus enabling on-line and in-line measurements, respectively). Barbas et al. [43] used the two alternatives during compounding in twin-screw extruders and obtained comparable results. In-process NIR successfully monitored the composition of PP/ethylene vinyl acetate (EVA) copolymer blends [57], the concentration of CO_2 in extrusion foaming [58], the concentration of titanium dioxide in molten PET [59], the evolution of the dispersion of organoclays in polymer melts [43, 60], and the dispersion and MFR of molten EVA copolymers [61, 62]. In reactive extrusion, Bergman *et al.* [63], aiming at subsequent implementation of in-line monitoring, obtained a good correlation between MW values of partly glycolized polyethylene terephthalate (PET) samples measured in diffuse reflection by NIR and by size exclusion chromatography (SEC). NIR was also used to follow the graft copolymerization of MA on to PP [64]. The technique was sensitive to the presence of vinyl CH, but not to simple chain scission, unlike in in-line slit rheometry, which measured a decrease of the apparent viscosity. Jakisch *et al.* [53] analyzed the conversion of a styrene/maleic-anhydride copolymer in styrene/maleimide in a twin-screw extruder. Sasic *et al.* [65] monitored the chemical modification of EVA to a copolymer of ethylene-vinyl-alcohol (EVOH), while Barres *et al.* [66] investigated the esterification of EVOH copolymer by octanoic acid, also in a twin-screw extruder. Regrettably, these studies concentrated on validating the in-process measurements rather than demonstrating their sensitivity to practical deviations in the operating conditions or to material recipe during reactive extrusion. Figure 5.6 shows an NIR set-up developed to monitor compounding operations performed in a co-rotating twin-screw extruder, using commercial

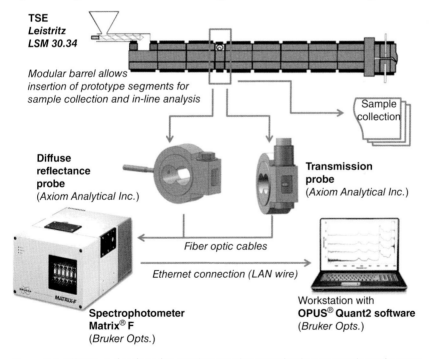

Figure 5.6 NIR set-up developed to monitor reactive extrusion in a co-rotating twin-screw extruder using commercial reflectance and transmission probes (Axiom Analytical, Inc., USA) and NIR spectrometer (Bruker Optics, Bruker Corporation). (Barbas *et al.* 2012 [43]. Reproduced with permission of Wiley.)

reflectance and transmission probes (from Axiom Analytical, Inc., USA) [43]. The former is flush mounted on the barrel of the extruder (similarly to commercial melt pressure transducers), has a sapphire window with a diameter of 5.7 mm, for an illuminated area with a diameter of 3.2 mm and a field depth of 3 mm. It uses an 80-fiber bundle cable that splits into two 40-fiber bundles to connect to the analyzer inlet and outlet channels. The transmission probe is mounted on the bypass channel of a modified barrel segment, whose flow is controlled by a rotating valve. This barrel segment can be inserted between any two original barrel segments of the extruder. The probe has a rectangular gap of 5 mm, corresponding to the optical path length. The probe's inner duct is pressurized at 4 bar with N_2 to prevent humidity condensation inside the light guide channels and uses two fiber optic cables with a core fiber diameter of 600 μm for transmitting and receiving signal. The set-up also comprises an NIR spectrometer (Bruker Optics, Bruker Corporation) with an InGaAs detector and a HeNe laser emitting at 633 nm at 1 mW (Bruker Optics) and a PC with software (OPUSVR Quant2) for data acquisition and analysis. This particular arrangement also allows the quick collection (in circa 1 s) of material samples from within the extruder at the same location where NIR spectra are measured, which are quenched and utilized for subsequent off-line characterization.

In Raman, even though the relationship between A and c in Eq. (5.1) is linear, it also depends on several sample and equipment characteristics. Consequently, it is difficult to use the absolute Raman band intensity for quantitative analysis, as the intensity of the spectral band of interest is utilized instead [16]. The technique was employed for monitoring on-line the grafting of glycidyl methacrylate (GMA) on to low-density polyethylene (LDPE) in a non-intermeshing twin-screw extruder [67].

Coates et al. [16] used the aforementioned analytical tools simultaneously to monitor the single-screw extrusion of high-density polyethylene (HDPE)/PP blends and compare their performance and suitability for in-process monitoring. On-line MIR comprised 16 scans from 4400 to 400 cm^{-1} at a resolution of 2 cm^{-1}, whereas on-line NIR involved 16 scans from 12 000 to 4000 cm^{-1} at a resolution of 4 cm^{-1}, for a collection time of approximately 30 min. The cell path length was set to 0.01 mm for the MIR measurements and to 1 mm for the NIR measurements. In-line transmission NIR with a probe protruding into the main melt stream was performed in the range 8000–4650 cm^{-1}; collated spectra comprised 32 scans at a resolution of 4 cm^{-1} and a path length of 1 mm (larger paths would cause excessive signal attenuation). Raman spectroscopy used a flush mounted single probe providing both excitation and signal collection. Scans of the melt were collected every minute, each consisting of a 25 s exposure in the 785–1080 nm wavelength range and the spectra were plotted in Raman shift (cm^{-1}) from the incident radiation (at 785 nm). The required small melt path lengths of MIR (up to ~0.01 mm) limited its suitability for in-line testing. Although NIR signals were typically an order of magnitude smaller than those of MIR for the same melt path lengths, these could be made larger (1–10 mm range) and were therefore eventually suitable for in-line testing. In fact, in this study NIR exhibited the best resolution (of ~2 wt%, i.e., changes of blend ratio of 1% were not detected). The problem remains, however, that overtones and combination bands are measured, making it challenging to quantify changes in specific bond vibrations in real time. In-line Raman scattering also showed good sensitivity, but the measurements involved a large fluorescent background related to additives, impurities, turbidity, temperature, and pressure effects.

NIR relies on chemometrics to extract as much significant and precise information as possible from the spectra. For polymer blending in an extruder using transmission probes, Rohe *et al.* [68, 69] obtained deviations between predicted and actual polymer composition below 2%. Chemometrics is a step-by-step procedure that aims at developing a calibration model that relates the spectral data to the reference characterization parameters [70, 71]. As schematized in Figure 5.7, it encompasses model development and validation. The multivariate calibration technique uses the entire spectral range (instead of a single spectral data point) to provide broader information and thus detect all the details of the measured spectra. To develop the calibration model (for a detailed explanation see, e.g., [72]) it is important: (i) to select carefully the parameter(s) that most adequately describe the reactive extrusion operation and that will be part of the model, (ii) to determine the minimum size of the population that is required to obtain a robust correlation, and (iii) to define the best validation/calibration population size ratio in order to minimize the monitoring time. A group of samples (commonly known as *training set*) is used to compute the calibration curve, which directly yields the analyte property from the respective spectra. To ensure precision, the degree of correlation between spectral and reference data should be high. For this purpose, a cross-validation step attests the quality of the adjustment of the data points to the calibration curve. A partial least squares (PLS) regression to the calibration data set is often done and the quality of the model is assessed by various measures, such as those indicated in Figure 5.7. Finally, the calibration model is ready to be used to predict the characteristics of unknown samples. Chemometrics represents both the major advantage and the constraint

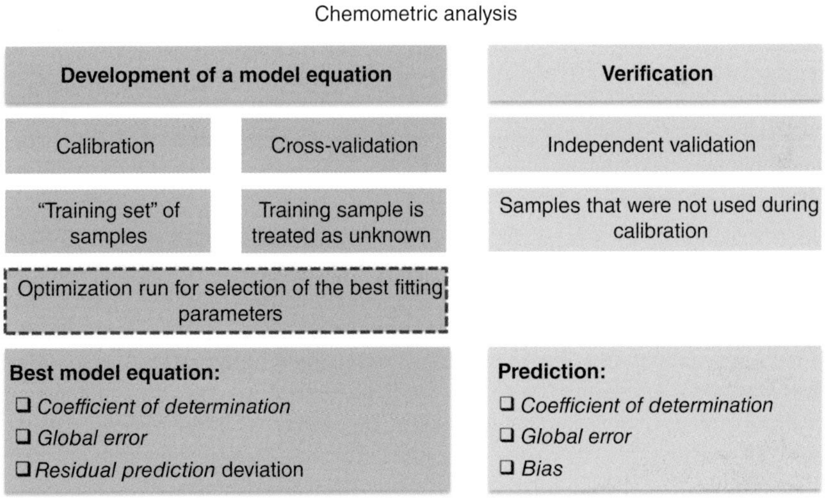

Chemometric analysis

Development of a model equation		Verification
Calibration	Cross-validation	Independent validation
"Training set" of samples	Training sample is treated as unknown	Samples that were not used during calibration
Optimization run for selection of the best fitting parameters		
Best model equation: ❏ *Coefficient of determination* ❏ *Global error* ❏ *Residual prediction deviation*		**Prediction:** ❏ *Coefficient of determination* ❏ *Global error* ❏ *Bias*

Figure 5.7 Steps for developing a chemometrical model for NIR.

of NIR. Indeed, a single calibration model could eventually provide information on multiple chemical, physical, and morphological characteristics, enabling NIR to replace several other characterization techniques. However, in practice, an adequate chemometric analysis often requires extensive experimental characterization and data treatment as well as the development of multiparameter calibration models [43, 60, 70–72]. In case of the latter, it is convenient to normalize each parameter:

$$\text{Normalized value}\,(\%) = \frac{\text{value} - \text{minimum}}{\text{maximum} - \text{minimum}} \times 100 \qquad (5.2)$$

For example, a nil score matches the initial material, while 100% refers to total chemical conversion. In turn, each model considers the average of the relevant normalized values. It should be emphasized that since NIR spectra are sensitive both to chemical and to morphological information, a change in the material recipe or in the characteristics of the processing equipment requires a new calibration of the chemometric model.

5.4 In-process Rheometry

Reactive extrusion is finding increasing application in polymer blending (to manufacture polymer blends, thermoplastic elastomers (TPEs), thermoplastic vulcanizates (TPVs)), chemical modification of polymers, (bulk and graft) polymerization, depolymerization (e.g., rubber devulcanization for recycling), or new copolymer synthesis. These operations are generally aimed at manufacturing materials that have certain ranges of MW and MWD, or at generating stable and fine multiphase morphologies. Invariably, reviews of reactive extrusion or of compatibilization of immiscible polymer blends [21, 35, 73–75] contain extensive sections dealing with rheological measurements, as they are suitable for the following:

1) Acquiring an imprint of the structural and/or morphological characteristics of the initial and final materials. Extensive experimental and theoretical work covering a wide range of real-world and model polymer systems were able to establish relationships between specific rheological responses and certain material features, and rheometrical techniques and procedures were able to obtain those responses.
2) Understanding the evolution of chemical conversion or of morphology in reactive systems, including interfacial tension and dispersion levels (due to droplet deformation and breakup, coalescence, or droplet retraction).
3) Predicting the processability of the materials obtained.
4) Enabling quality control by detecting minute changes in the material characteristics.
5) Supporting automatic process control strategies.

Rheology is very sensitive to small deviations of MW (when applied to polymerization processes, rheology is often designated as *rheo-kinetics* or *rheo-chemistry*). While a variation of MW causes a change in viscosity, fluctuations in MWD and branching affect mainly melt elasticity. The zero shear viscosity increases with MW, while in SAOS the curves of the storage (G') and loss (G'') moduli are shifted to lower frequency with increasing MW. The maximum in G', or the G', G'' cross-over point are measures of the broadness of the distribution (for a review on rheo-kinetics of linear polymerization, see [73]). The viscosity of long-branched polymers is more shear rate dependent than that of linear polymers, and the differences in elasticity can be easily detected by measuring G' and the first normal stress difference (N_1). Strain hardening effects of branched polymers is readily appraised by extensional viscosity measurements. In the case of immiscible polymer blends, Grace [76] made a momentous contribution toward establishing the connection between particle size and rheology, by correlating the value of the critical capillary number (i.e., the minimum value of the ratio between hydrodynamic and interfacial stresses for which droplet breakup occurs) with the viscosity ratio (disperse phase to matrix) in shear and extensional flows. Rheology is sensitive to changes in particle size, particle size distribution, volume fraction, and interfacial tension of suspensions. G' and N_1 can be used to estimate the size of dispersed droplets in a blend. Compatibilized blends are usually less shear-thinning than their uncompatibilized counterparts [21]. Steady state viscosity curves of compatibilized blends generally show two plateau regions, at low and moderate frequencies, the former stemming from interfacial rheology. Similarly, an additional slow relaxation process in SAOS on both physical and chemically compatibilized blends was attributed to interfacial elasticity. The interfacial tension can be calculated from the average relaxation time of the droplet relaxation [21].

While conventional rheological off-line measurements benefit from a wide range of available test geometries and rheometrical techniques, in-process experiments during reactive extrusion of polymer melts have been mostly confined to slit and capillary rheometry. A few commercial instruments are available. In-line slit rheometry is usually performed by coupling a rheological die to the extruder. The viscosity is calculated from readings of pressure differences along the slit, together with the determination of the throughput. Methods to quantify N_1 have also been proposed, namely the hole-pressure and the exit pressure methods. Although they are considered as valid (particularly the former), they entail experimental difficulties associated with the need of performing highly accurate experiments, which might be difficult to guarantee in an industrial environment [77]. The extensional viscosity can be estimated by linking two slits in series by a convergent channel [78, 79] and measuring the pressure at the entrance of the convergence, and using the well-known analysis developed by Cogswell [80] or Binding [81].

This approach has two limitations, which are typical for in-line methods. First, the viscosity is obtained at a single shear rate value, which is determined by the extrusion output. Moreover, the melt temperature might be unknown, as it is affected not only by the thermomechanical history upstream of the die, but also by viscous dissipation developing along the slit. Rauwendaal and Fernandez [82] observed that in-process measurements consistently yielded lower viscosity values than did off-line capillary and cone-and-plate rheometry. The differences were larger at higher shear rates, which was attributed to the shorter residence

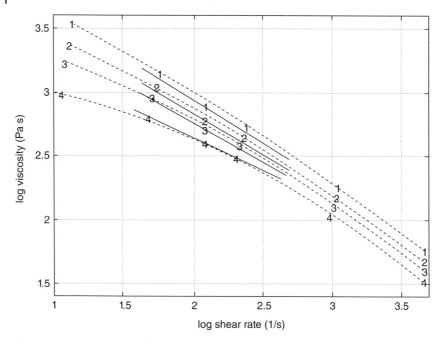

Figure 5.8 Viscosity curves of reactively degraded PP with increasing initiator concentration (curves 1–4) measured off-line (dashed lines) and in-line (solid lines) with a tapered slit. (Pabedinskas et al. 1991 [83]. Reproduced with permission of Wiley.)

times between the extruder and slit die allowing for less relaxation of the pre-shearing effects. A few constructive solutions were proposed to vary the shear rate in the measuring section, or generate different shear rates in a single measurement, under constant extrusion operating conditions. They involve the utilization of tapered slit geometries [83], of multiple-step slits with different heights but uniform width [84], and of variable gap constructions [85–87], as well as the use of flow regulators (flow valves or gear pumps) [88, 89]. In most cases, the shear rate ranges made available remain narrow. Secondly, coupling a slit die to an extruder is generally not feasible for industrial production, as it precludes the possibility of producing extrudates with non-rectangular cross-section. Thus, this type of set-up is mostly adopted for R&D purposes. Figure 5.8 compares the viscosity curves of reactively degraded PP measured off-line and in-line with a tapered slit [83].

On-line capillary or slit rheometers, either coupled to the die or between extruder and die, are preferred for industrial compounding and reactive extrusion. Comparisons between in-line, on-line, and off-line [90] and between on-line and off-line [38] rheometry have generally shown good agreement, but were performed with thermally stable polymers. The design, complexity, and compactness of these devices vary, and so do the range of attainable viscosities and shear rates (in some instruments, above $10^5 \, s^{-1}$), the time delay between

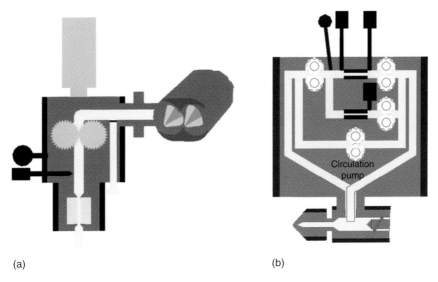

(a) (b)

Figure 5.9 On-line capillary rheometers from GÖTTFERT Werkstoff-Prüfmaschinen GmbH, Germany. (a) Compact on-line rheometer without melt return into the process (note also the purge valve); (b) on-line instrument with melt return and bypass with continuous circulating volume stream.

sample collection and measurement, and whether the sampled material is wasted or returned to the main flow (sampling rates vary typically between 0.5 to a few kilograms per hour). An on-line extensional rheometer of the filament stretching-type has also been developed [91]. Figure 5.9 displays two on-line capillary rheometers supplied by the same manufacturer (GÖTTFERT Werkstoff-Prüfmaschinen GmbH, Germany). One is a compact single-bore instrument, designed for processes with frequent product changes (so, quick die replacement is essential) and without melt return capacity. A material stream is detoured from the main channel and conditioned to the required testing temperature. A gear pump then accurately controls the flow rate in the testing chamber, where the pressure drop is measured. The other model in the figure is a twin-bore rheometer that returns the sampled melt to the main flow and has an independent material circulation system to deliver the melt to the parallel capillaries quickly (thus avoiding slower transfers for low viscosity systems). Melt flow through parallel capillaries (up to 4, depending on the model and manufacturer) of the same size but with distinct length allows performing the well-known Bagley correction for end effects, while using different diameters yields multiple points of the viscosity flow curve. Flow through these dies can be simultaneous or sequential (so that less amount of material is consumed in the measurements), in which case a melt switch or a rotating valve is used. Further, extensional viscosities can be estimated if a "zero length" die (i.e., a very short die, whose pressure drop is essentially due to entrance effects) is fixed as one of the capillaries. An optical window may also exist, so that other on-line techniques (e.g., NIR–UV–Vis) are used simultaneously. Some systems can also operate at constant pressure (corresponding to a certain shear stress) and determine the resulting flow rate (from the knowledge of the rpm of the gear pump, its volume per revolution, and the melt density at the test temperature). In this mode, it becomes possible to measure MFR: from the test load defined by the relevant standard (e.g., ASTM D-1238 or ISO 1133) the working pressure is set; knowing the flow rate, the MFR is calculated. A few apparatuses have an interchangeable measuring geometry, enabling a choice between slit and capillary viscometry. Capillaries seem to be more frequently used than slits, as they are easier to clean (no edges), can provide a wider measuring range under moderate pressures, and are geometrically more similar to the MFR test. Nonetheless, an interesting design consists of a die that divides the inlet flow into two parallel geometrically identical slits, one of them being used for the measurements [92]. The distribution of the flow is controlled by vertically moving the valves located at the entrance of each channel. The shear rate at the measuring slit can be altered by adjusting the respective valve and rotating the other valve in the opposite direction, so that the total pressure drop remains constant and so does the thermomechanical environment of the polymer in the extruder. The concept was revisited by Vergnes *et al.* [93] and was recently adopted in a modified form [87, 94].

On-line capillary rheometry measurements along the extruder are much scarcer. Covas *et al.* [38] designed a device that can be fixed at various locations along the extruder. As illustrated in Figure 5.10, it essentially comprises a modified barrel element (in this case, of a Leistritz LSM 30.34 twin-screw extruder), the measuring head of a commercial rotational rheometer, the heated body, piston, and replaceable die of the capillary instrument. The barrel element contains a lateral circular channel that links the inner barrel wall to the reservoir of the rheometer through a lateral hole in its body whereby proper alignment is

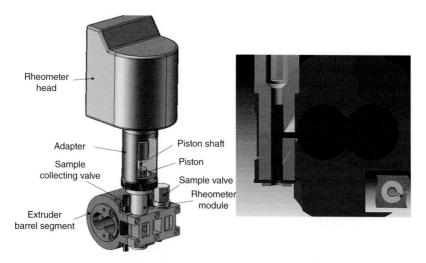

Figure 5.10 On-line capillary rheometer with modified barrel element to be inserted in between standard elements of a twin-screw extruder.

guaranteed. This is controlled by the rotation of the entire rheometer module around its vertical axis. Thus, a rheological measurement entails the following sequence: (i) fix the desired die and set test temperature; (ii) rotate the rheometer module until full alignment with the lateral channel of the barrel element is achieved; (iii) once the material flowing from within the extruder has filled the reservoir between the piston and die, rotate the module again to isolate the rheometer from the extruder; (iv) wait for thermal equilibration; (v) start the test – the rotational movement of the rheometer shaft is converted into a linear vertically descending piston movement, while the torque is recorded and used to determine the apparent wall shear rate and stress, respectively. The usual Bagley and Rabinowitsch corrections to the data (due to entrance and exit effects and due to the non-parabolic velocity profile of non-Newtonian fluids, respectively [42]) are performed in order to yield true values of the wall shear rate and stress. A modified version of the instrument allowed for the on-line measurement of MVR in accordance with the ASTM D-1238 and ISO 1133 standards [95].

This prototype was used to follow the peroxide-induced degradation of PP [96, 97]. Figure 5.11 reveals the influence of peroxide content (2,5-bis(*tert*-butylperoxy)-2,5-dimethylhexane (DHBP)) on the evolution of shear viscosity along the barrel of the extruder at constant operating conditions. As expected, the higher the peroxide content, the lower is the resulting viscosity. The degradation reaction occurs very quickly and very early in the extruder. For example, when 0.05 phr of DHBP was added to PP, as the melt reached the first kneading block at $L/D = 9$ (most pellets remained solid at $L/D = 8$) the reaction seemed to have already attained high conversion levels. Even when the peroxide content was augmented to 0.1 phr, the reaction was all but complete at the end of this first kneading block, at $L/D = 11$. The MWD of PP was measured by gel-permeation chromatography (GPC), using samples collected at the same axial locations. The results unveiled an excellent correlation between MW and viscosity changes, but showed a relative lack of sensitivity of capillary rheometry to changes in MWD.

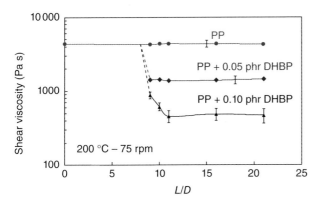

Figure 5.11 Effect of peroxide concentration on shear viscosity of PP + DHDP along a co-rotating twin-screw extruder. (Adapted from Machado *et al.* 2004 [97]. Reproduced with permission of Wiley.)

Practice demonstrated that oscillatory rheometry is the preferred method to characterize multiphase systems, because – unlike capillary rheometry – it preserves the initial material morphology (due to the small deformation amplitudes involved), is sensitive to minor variations in droplet or particle size, and can generate a rather complete rheological description. In-line vibrating probe viscometers are generally designed for installation in pipe elbows or in vessels, typically to monitor polymerization reactions (either in batch or prior to extrusion). Commercial on-line oscillatory rheometers (with parallel plate geometry) are available, but their operational window makes them suitable for studying lubricating greases, inks and paints, foodstuff, household products and cosmetic and personal care products. A Piezo Axial Vibrator was developed aiming at on-line measurements in extrusion. It generates an axial oscillation on a molten polymer sample and converts the signal into the rheological moduli G' and G''. However, preliminary validation tests demonstrated that robustness and handling should be improved [98]. Covas *et al.* [39, 99, 100] designed an on-line rotational rheometer with sensing capabilities and measuring modes similar to those of conventional laboratory instruments. With or without a modified barrel element, tests could be carried out along the axis of an extruder, or between extruder and die, respectively. Figure 5.12 depicts the set-up for measurements along the extruder, the configuration adopted for coupling to the flow channel between extruder and die, and the operating sequence. The system uses a plate–plate configuration and comprises a motor to control the position of the lower plate, a motor to open/close the valve admitting the material sampled, a linear actuator to control the movement of a ring cleaning the outer surface of the test sample, and a commercial rotational rheometer head (Figure 5.12A). The set-up is connected to the flow channel between extruder and die by a vibration-damping coupling, and the flow is controlled by a rotating valve (Figure 5.12B). As for the operating sequence, the cycle begins by assuming that the extruder is operating normally, with the rheometer ready to operate (a). Upon rotating the valve (b), melt from the extruder flows to the testing chamber. When enough material is sampled, the valve closes (c). The material is then squeezed between the two plates, in order to define the pre-set gap (d). The cleaning ring wipes out the excess material (e). After the necessary thermal equilibration time, the test starts. The preparatory steps of each experiment are controlled by a program written in LabVIEW System Design Software, whereas the rheometer software controls the test. Validation of the instrument encompassed comparisons between on-line and off-line measurements of polyolefins and a compatibilized polymer blend (polyamide 6/polyethylene-grafted-maleic anhydride (PA6/PE-*g*-MA)). Both provided very good agreement [39]. The evolution of the rheological properties of a polyamide 6/low-density polyethylene immiscible blend (PA6/LDPE) (70/30 w/w%) and of a PA6/LDPE/maleic anhydride-grafted-polyethylene (PA6/PE/PE-*g*-MA) (70/15/15 w/w%)) blend was then followed both on-line and off-line (using samples collected at the same locations and immediately quenched in liquid nitrogen) [39]. The second blend maintains the proportion of PA6, but half of the LDPE was replaced by PE-*g*-MA.

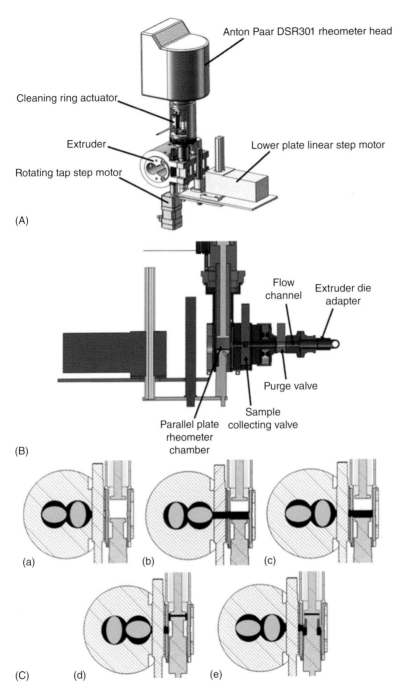

Figure 5.12 On-line rotational rheometer. From A to C: global set-up for measurements along the extruder; configuration adopted for coupling to the flow channel between extruder and die; operating sequence. (Mould et al. 2012 [100]. Reproduced with permission of Carl Hanser Verlag GmbH & Co. KG.)

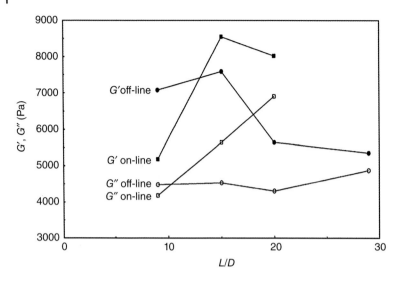

Figure 5.13 Evolution of G' and G'' (at 1 s^{-1}) of a PA6/PE/PE-g-MA (70/15/15 w/w%) blend along the axis of a co-rotating twin-screw extruder, measured off-line and on-line. (Covas et al. 2008 [39]. Reproduced with permission of Elsevier.)

Therefore, the amine groups of PA6 react with the MA groups of the modified PE to form a copolymer at the interface, which acts as a compatibilizer. In the case of the PA6/LDPE blend, the extra thermal cycles imparted on the off-line samples (due to manufacturing of a disk by compression molding and heating it for the rheological test) had a dramatic effect on their thermal stability, unlike that observed in the data obtained on-line. The evolution of the viscoelastic response was also quantitative and qualitatively different. Micrograph images of the material after sampling and after compression molding exhibited quite different morphologies. In the case of the compatibilized blend, the on-line and off-line responses were again dissimilar, although the differences are smaller. This was undoubtedly due to the morphology stabilization effect of the copolymer, which minimized the influence of compression molding. Nevertheless, as perceived in Figure 5.13, which shows the evolution along the extruder of G' and G'' at fixed frequency (1 s^{-1}), the extra thermal cycles suffered by off-line samples artificially reduce the differences along the extruder. Indeed, G'' is nearly constant in the samples tested off-line, whereas on-line tests detected a significant increase between $L/D = 9$ and $L/D = 20$. G' seems to increase and then gradually decrease, probably due to thermal degradation. Although more experiments are necessary to support these observations, on-line rheology often detects a progressive degradation of some polymers once the chemical reaction has achieved high levels of conversion. Figure 5.14 plots the complex viscosity, (η^*) and G' at three locations along the extruder for a PLA plus chain extender system. The reaction was fast and took place upon melting of the polymer and a gradual decrease in viscosity was detected.

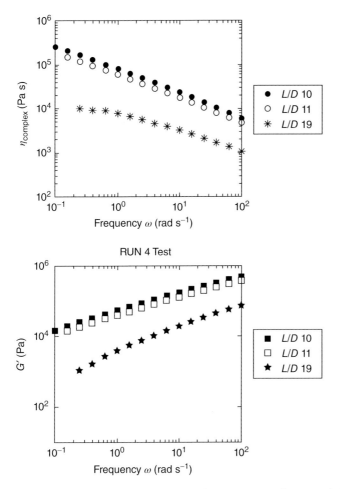

Figure 5.14 Evolution of the linear viscoelastic response of a PLA and chain extender system along the axis of a co-rotating twin-screw extruder (unpublished data).

5.5 Conclusions

The potential of in-process monitoring has not been fully explored. However, mainly in the past two decades, a significant effort has been made by the academic and industrial communities to develop in-process monitoring techniques that are capable of obtaining qualitative or quantitative data on a wide range of characteristics and properties of polymeric systems at various scales. Signal to noise ratio, precision, time lag between sampling and result, reproducibility, robustness, and operational window have largely improved. The choice of an in-process method for a given reactive extrusion process from the array of current possibilities that are available must take into consideration the purpose of the application, the geometry of the probe, and the mounting possibilities provided by the processing equipment. Thus the utilization of these techniques to support R&D activity, quality control, and, progressively, process control, is expected to rise substantially.

Acknowledgment

The author acknowledges Prof. Ana V. Machado for the interesting discussions concerning in-process optical spectroscopy.

References

1 van Duin, M., Machado, A.V., and Covas, J.A. (2001) A look inside the extruder: evolution of chemistry, morphology and rheology along the extruder axis during reactive processing and blending. *Macromol. Symp.*, **170**, 21.
2 Covas, J.A. and Machado, A.V. (2005) Monitoring reactive processes along the extruder. *Int. Polym. Proc.*, **XX**, 121.
3 Isac, S.K. and George, K.E. (2001) Reactive processing of polyethylenes on a single screw extruder. *J. Appl. Polym. Sci.*, **81**, 2545.
4 Tzoganakis, C. (1989) Reactive extrusion of polymers: a review. *Adv. Polym. Tech.*, **9**, 321.
5 Mack, W.A. and Herter, R. (1976) Extruder reactors for polymer production. *Chem. Eng. Progr.*, **72**, 64.
6 Van Der Goot, A.J., Hettema, R., and Jansen, L.P.B.M. (1997) The working domain in reactive extrusion. Part I: the effect of the polymer melt viscosity. *Polym. Eng. Sci.*, **37**, 511.
7 Kohlgruber, K. (ed.) (2008) *Co-rotating Twin-Screw Extruders*, Carl Hanser, Munich.
8 Machado, A.V., Covas, J.A., and van Duin, M. (1999) Evolution of morphology and of chemical conversion along the screw in a corotating twin screw extruder. *J. Appl. Polym. Sci.*, **71**, 135.
9 Machado, A.V., Covas, J.A., and van Duin, M. (1999) Chemical and morphological evolution of PA6/EPM/EPM-*g*-MA blends in a twin screw extruder. *J. Polym. Sci.*, **37**, 1311.
10 Machado, A.V., Covas, J.A., and van Duin, M. (2001) Monitoring polyolefin modification along the axis of a twin screw extruder. I – Effect of peroxide concentration. *J. Appl. Polym. Sci.*, **81**, 58.
11 Machado, A.V., Yquel, V., Covas, J.A., Flat, J.-J., Ghamri, N., and Wollny, A. (2006) The effect of the compatibilization route of PA/PO blends on the physico-chemical phenomena developing along a twin-screw extruder. *Macromol. Symp.*, **233**, 86.
12 Canto, L.B., Mantovani, G.L., Covas, J.A., Hage, E. Jr., and Pessan, L.A. (2007) Phase morphology development during processing of compatibilized and uncompatibilized PBT/ABS blends. *J. Appl. Polym. Sci.*, **104**, 102.

13 Yquel, V., Machado, A.V., Covas, J.A., and Flat, J.J. (2009) The contribution of the melting stage to the evolution of morphology and chemical conversion of immiscible PA/PE blends in twin screw extruder. *J. Appl. Polym. Sci.*, **114**, 1768.
14 Broadhead, T.O., Patterson, W.I., and Dealy, J.M. (1996) Closed loop viscosity control of reactive extrusion with an in-line rheometer. *Polym. Eng.Sci.*, **36**, 2840.
15 Fritz, H.-G. and Stöhrer, B. (1986) Polymer compounding process for controlled peroxide-degradation of PP. *Int. Polym. Proc.*, **1**, 31.
16 Coates, P.D., Barnes, S.E., Sibley, M.G., Brown, E.C., Edwards, H.G.M., and Scowen, I.J. (2003) In-process vibrational spectroscopy and ultrasound measurements in polymer melt extrusion. *Polymer*, **44**, 5937.
17 Barnes, S.E., Sibley, M.G., Edwards, H.G.M., and Coates, P.D. (2007) Process monitoring of polymer melts using in-line spectroscopy. *Trans. Inst. Measur. Control*, **29**, 453.
18 Migler, K.B. and Bur, A.J. (1998) Fluorescence based measurement of temperature profiles during polymer processing. *Polym. Eng. Sci.*, **38**, 213.
19 Carneiro, O.S., Covas, J.A., Ferreira, J.A., and Cerqueira, M.F. (2004) On-line monitoring of the residence time distribution along a kneading block of a twin-screw extruder. *Polym. Testing*, **23**, 925.
20 Teixeira, P.F., Covas, J.A., Maia, J.M., and Hilliou, L. (2014) In-line particle size assessment of polymer suspensions during processing. *Polym. Testing*, **37**, 68.
21 Van Puyvelde, P.V. and Moldenaers, P. (2005) Rheology and morphology development in immiscible polymer blends. *Rheol. Rev.*, **2005**, The British Soc. Rheology, 101.
22 Migler, K.B., Hobbie, E.K., and Qiao, F. (1999) In line study of droplet deformation in polymer blends in channel flow. *Polym. Eng. Sci.*, **39**, 2282.
23 Alig, I., Steinhoff, B., and Lellinger, D. (2010) Monitoring of polymer melt processing. *Meas. Sci. Technol.*, **21**, 062001.
24 Schlatter, G., Serra, C., Bouquey, M., Muller, R., and Terrisse, J. (2002) Online light scattering measurements: a method to assess morphology development of polymer blends in a twin-screw extruder. *Polym. Eng. Sci.*, **42**, 1965.
25 Xian, G.-M., Qu, J.-P., and Zeng, B.-Q. (2008) An effective on-line polymer characterization technique by using SALS image processing software and wavelet analysis. *J. Aut. Meth. Manag. Chem.*, **2008**, ID 838412.
26 Bur, A.J., Lee, Y.-H., Roth, S.C., and Start, P.R. (2005) Measuring the extent of exfoliation in polymer/clay nanocomposites using real-time process monitoring methods. *Polymer*, **46**, 10908.
27 Cunha Santos, A.M., Cáceres, C.A., Calixto, L.S., Zborowski, L., and Canevarolo, S.V. (2014) In-line optical techniques to characterize the polymer extrusion. *Polym. Eng. Sci.*, **54**, 386.
28 Soares, K., Cunha Santos, A.M., and Canevarolo, S.V. (2011) In-line rheo-polarimetry: a method to measure in real time the flow birefringence during polymer extrusion. *Polym. Testing*, **30**, 848.

29 Kiehl, C., Chu, L.-L., Leitz, K., and Min, K. (2001) On-line ultrasonic measurements of methyl methacrylate polymerization for application to reactive extrusion. *Polym. Eng. Sci.*, **41**, 1078.

30 Kotzé, R., Wiklund, J., and Haldenwang, R. (2013) Optimisation of pulsed ultrasonic velocimetry system and transducer technology for industrial applications. *Ultrasonics*, **53**, 459.

31 Perusich, S. and McBrearty, M. (2000) Dielectric spectroscopy for polymer melt composition measurements. *Polym. Eng. Sci.*, **40**, 214.

32 Boersma, A. and van Turnhout, J. (1999) Dielectric on-line spectroscopy during extrusion of polymer blends. *Polymer*, **40**, 5023.

33 Dealy, J.M. and Broadhead, T.O. (1993) Process rheometers for molten plastics: a survey of existing technology. *Polym. Eng. Sci.*, **33**, 1513.

34 Saerens, L., Vervaet, C., Remon J.-P., De Beer, T. (2014) Process monitoring and visualization solutions for hot-melt extrusion: a review, *J. Pharm. Pharmacol.*, **66**, 180.

35 Cassagnau, P., Bounor-Legaré, V., and Fenouillot, F. (2007) Reactive processing of thermoplastic polymers: a review of the fundamental aspects. *Int. Polym. Proc.*, **XXII**, 219.

36 Broadhead, T.O., Nelson, B.I., and Dealy, J.M. (1993) An in-line rheometer for molten plastics: design and steady state performance characteristics. *Int. Polym. Proc.*, **8**, 104.

37 Dealy, J.M. (1990) Challenges in process rheometry. *Rheol. Acta*, **29**, 519.

38 Covas, J.A., Nóbrega, J.M., and Maia, J.M. (2000) Rheological measurements along an extruder with an on-line capillary rheometer. *Polym. Testing*, **19**, 165.

39 Covas, J.A., Maia, J.M., Machado, A.V., and Costa, P. (2008) On-line rotational rheometry for extrusion and compounding operations. *J. Non-Newt. Fluid Mech.*, **148**, 88.

40 Dealy, J.M. and Rey, A. (1996) Effect of Taylor diffusion on the dynamic response of a on-line capillary viscometer. *J. Non-Newt. Fluid Mech.*, **62**, 225.

41 Dealy, J.M. (1984) Viscometers for online measurement and control. *Chem. Eng.*, **91**, 62.

42 Macosko, C.W. (1994) *Rheology – Principles, Measurements, and Applications*, Wiley-VCH.

43 Barbas, J.M., Machado, A.V., and Covas, J.A. (2012) In-line near-infrared spectroscopy: a tool to monitor the preparation of polymer-clay nanocomposites in extruders. *J. Appl. Polym. Sci.*, **127**, 4899.

44 Levitt, L., Macosko, C.W., Schweizer, T., and Meissner, J. (1997) Extensional rheometry of polymer multilayers: a sensitive probe of interfaces. *J. Rheol.*, **41**, 671.

45 Gimenez, J., Boudris, M., Cassagnau, P., and Michel, A. (2000) Control of bulk ε-caprolactone polymerization in a twin screw extruder. *Polym. React. Eng.*, **8**, 135.

46 McAfee, M., Thompson, S., and McNally, G.M. (2004) *In-Process Viscosity Monitoring for Extrusion Control.* SPE-ANTEC 2014, 1274.

47 Yiagopoulos, T., Schmidt, C.-U., and Marelli, E. (2001) Development of a softsensor for on-line MFI monitoring in reactive polypropylene extrusion. *Dech. Monogr.*, Wiley-VCH Verlag, **138**, 305.
48 Liu, X., Li, K., McAfee, M., and Deng, J. (2010) *'Soft-sensor' for Real-time Monitoring of Melt Viscosity in Polymer Extrusion Process.* 49th IEEE Conference Decision and Control, December 15–17, Atlanta, GA, USA, 3469.
49 Liu, X., Li, K., McAfee, M., Nguyen, B.K., and McNally, G.M. (2012) Dynamic grey-box modeling for on-line monitoring of polymer extrusion viscosity. *Polym. Eng. Sci.*, **52**, 1332.
50 Nguyen, B.K., McNally, G.M., and Clarke, A. (2014) Real time measurement and control of viscosity for extrusion processes using recycled materials. *Polym. Degrad. Stab.*, **102**, 212.
51 Wetzel, M.D., Avgousti, M., Denelsbeck, D.A., and Latimer, S.L. (2007), *Reactive Extrusion Process Characterization: Q/N Mapping Method.* SPE-ANTEC **2007**, 464.
52 Wang, Y., Steinhoff, B., Brinkmann, C., and Alig, I. (2008) In-line monitoring of the thermal degradation of poly(L-lactic acid) during melt extrusion by UV–vis spectroscopy. *Polymer*, **49**, 1257.
53 Jakisch, L., Fischer, D., Stephan, M., and Eichhorn, K.J. (1995) In-line-analytik von polymerreaktionen. *Kunststoffe*, **85**, 1338.
54 Fritz, H.-G. and Ultsch, S. (1989) Sensorentwicklung und Automatisierungstendenzen bei der Kunststoffaufbereitung. *Kunststoffe*, **79**, 785.
55 Rajan, V.V., Waber, R., and Wieser, J. (2010) Online monitoring of the thermal degradation of POM during melt extrusion. *J. Appl. Polym. Sci.*, **115**, 2394.
56 Haberstroh, E., Jakisch, L., Henuge, E., and Schwarz, P. (2002) Real-time monitoring of reactive extrusion processes by means of in-line infrared spectroscopy and infrared temperature measurement. *Macromol. Mat. Eng.*, **287**, 203.
57 Fischer, D., Bayer, T., Eichhorn, K.-J., and Otto, M. (1997) In-line process monitoring on polymer melts by NIR-spectroscopy. *J. Analyt. Chem.*, **359**, 74.
58 Nagata, T., Tanigaki, M., and Ohshima, M. (2000) On-line NIR sensing of CO_2 concentration for polymer extrusion foaming processes. *Polym. Eng. Sci.*, **40**, 1843.
59 Batra, J., Khettry, A., and Hansen, M.G. (1994) In-line monitoring of titanium dioxide content in poly(ethylene terephthalate) extrusion. *Polym. Eng. Sci.*, **34**, 1767.
60 Barbas, J.M., Machado, A.V., and Covas, J.A. (2013) Evolution of dispersion along the extruder during the manufacture of polymer-organoclay nanocomposites. *Chem. Eng. Sci.*, **98**, 77.
61 Vedula, S. and Hansen, M.G. (1998) In-line fiber-optic near-infrared spectroscopy: monitoring of rheological properties in an extrusion process. Part I. *J. Appl. Polym. Sci.*, **68**, 859.
62 Vedula, S. and Hansen, M.G. (1998) In-line fiber-optic near-infrared spectroscopy: monitoring of rheological properties in an extrusion process. Part II. *J. Appl. Polym. Sci.*, **68**, 873.

63 Bergmann, B., Becker, W., Diemert, J., and Elsner, P. (2013) On-line monitoring of molecular weight using NIR spectroscopy in reactive extrusion process. *Macromol. Symp.*, **333**, 138.

64 Moghaddam, L., Rintoul, L., Halley, P.J., George, G.A., and Fredericks, P.M. (2012) In-situ monitoring by fibre-optic NIR spectroscopy and rheometry of maleic anhydride grafting to polypropylene in a laboratory scale reactive extruder. *Polym. Testing*, **31**, 155.

65 Sasic, S., Kita, Y., Furukawa, T., Watari, M., Siesler, H.W., and Ozaki, Y. (2000) Monitoring the melt-extrusion transesterification of ethylene-vinylacetate copolymer by self-modeling curve resolution analysis of on-line near-infrared spectra. *Analyst*, **125**, 2315.

66 Barres, C., Bounor-Legaré, V., Mélis, F., and Michel, A. (2006) In-line near infrared monitoring of esterification of a molten ethylene-vinyl alcohol copolymer in a twin screw extruder. *Polym. Eng. Sci.*, **46**, 1613.

67 Jarukumjorn, K. and Min, K. (2000) *On-Line Monitoring of Free Radical Grafting in a Model Twin Screw Extruder*. SPE ANTEC 2000, 2064.

68 Rohe, T., Becker, W., Krey, A., Nägele, H., Kölle, S., and Eisenreich, N. (1998) In-line monitoring of polymer extrusion processes by NIR spectroscopy. *J. Near Infrared Spectrosc.*, **6**, 325.

69 Rohe, T., Becker, W., Kolle, S., Eisenreich, N., and Eyerer, P. (1999) Near infrared (NIR) spectroscopy for in-line monitoring of polymer extrusion processes. *Talanta*, **50**, 283.

70 Blanco, M. and Villarroya, I. (2002) NIR spectroscopy: a rapid-response analytical tool. *Trends Anal. Chem.*, **21**, 240.

71 Siesler, H.W., Ozaki, Y., Kawata, S., and Heise, H.M. (2002) *Near-Infrared Spectroscopy: Principles, Instruments, Applications*, Willey-VCH.

72 Barbas, J.M., Machado, A.V., and Covas, J.A. (2012) In-line near-infrared spectroscopy for the characterization of dispersion in polymer-clay nanocomposites. *Polym. Testing*, **31**, 527.

73 Cioffi, M., Ganzeveld, K.J., Hoffmann, A.C., and Janssen, L.P.B.M. (2002) Rheokinetics of linear polymerization. A literature review. *Polym. Eng. Sci.*, **42**, 2383.

74 Van Puyvelde, P., Velankar, S., and Moldenaers, P. (2001) Rheology and morphology of compatibilized polymer blends. *Curr. Opin. Colloid Interface Sci.*, **6**, 457.

75 Cassagnau, P., Gimenez, J., Bounor-Legaré, V., and Michel, A. (2006) New rheological developments for reactive processing of poly(ε-caprolactone). *C. R. Chimie*, **9**, 1351.

76 Grace, G.H. (1982) Dispersion phenomena in high viscosity immiscible fluid systems and application of static mixers as dispersion devices in such systems. *Chem. Eng. Commun.*, **14**, 225.

77 Teixeira, P.F., Hilliou, L., Covas, J.A., and Maia, J.M. (2013) Assessing the practical utility of the hole-pressure method for the in-line rheological characterization of polymer melts. *Rheol. Acta*, **52**, 661.

78 Martyn, M.T., Nakason, C., and Coates, P.D. (2000) Measurement of apparent extensional viscosities of polyolefin melts from process contraction flows. *J. Non-Newt. Fluid Mech.*, **92**, 203.

79 Kopplmayr, T., Luger, H.-J., Burzic, I., Battisti, M.G., Perko, L., Friesenbichler, W., and Miethlinger, J. (2016) A novel online rheometer for elongational viscosity measurement of polymer melts. *Polym. Testing*, **50**, 208.

80 Cogswell, F.N. (1972) Converging flow of polymer melts in extrusion dies. *Polym. Eng. Sci.*, **12**, 64.

81 Binding, D.M. (1988) An approximate analysis for contraction and converging flows. *J. Newt. Fluid Mech.*, **27**, 173.

82 Rauwendaal, C. and Fernandez, F. (1985) Experimental study and analysis of a slit die viscometer. *Polym. Eng. Sci.*, **25**, 765.

83 Pabedinskas, A., Cluett, W.R., and Balke, S.T. (1991) Development of an in-line rheometer suitable for reactive extrusion processes. *Polym. Eng. Sci.*, **31**, 365.

84 Horvat, M., Azad Emin, M., Hochstein, B., Willenbacher, N., and Schuchmann, H.P. (2013) A multiple-step slit die rheometer for rheological characterization of extruded starch melts. *J. Food Eng.*, **116**, 398.

85 Kalyon, D.M., Gokturk, H., and Bozan, I. (1997) *Adjustable Gap In-line Rheometer*. SPE-ANTEC 1997, 2286.

86 Kalyon, D.M., Gevgilili, H., Kowalczyk, J.E., Prickett, S.E., and Murphy, C.M. (2006) Use of adjustable-gap on-line and off-line slit rheometers for the characterization of the wall slip and shear viscosity behavior of energetic formulations. *J. Energetic Mat.*, **24**, 175.

87 Hochstein, B., Kizilbay, Z., Horvat, M., Schuchmann, H.P., and Willenbacher, N. (2015) Innovatives inline-rheometer zur bestimmung des starkeabbaus im extruder. *Chem. Ing. Tech.*, **87**, 1.

88 Chiu, S.-H., Yiu, H.-C., and Pong, S.-H. (1997) Development of an in-line viscometer in an extrusion molding process. *J. Appl. Polym. Sci.*, **63**, 919.

89 Coates, P.D., Rose, R.M., and Barghash, M.A. (2001) *A "Bleed" on Line Rheometer for Polymer Melts*. SPE-ANTEC 2001.

90 Kelly, A.L., Woodhead, M., Coates, P.D., Barnwell, D., and Marin, K. (2000) In-process rheometry studies of LDPE compounds. *Int Polym. Proc.*, **XV**, 355.

91 Donato, N. (2012) A contribution to the on-line extensional characterization of polymer melts. MSc thesis. University of Minho, Portugal.

92 Springer, P., Broadkey, R.S., and Lynn, R.E. (1975) Development of an extrusion rheometer suitable for on-line rheological measurements. *Polym. Eng. Sci.*, **15**, 583.

93 Vergnes, B., Della Valle, G., and Tayeb, J. (1993) A specific slit die rheometer for extruded starchy products – design, validation and application to maize starch, *Rheol. Acta*, **32**, 465.

94 Robin, F., Engmann, J., Tomasi, D., Breton, O., Parker, R., Schuchmann, H.P., and Palzer, S. (2010) Adjustable twin-slit rheometer for shearviscosity measurement of extruded complex starchy melts. *Chem. Eng. Tech.*, **33**, 1672.

95 Covas, J.A., Carneiro, O.S., Costa, P., Machado, A.V., and Maia, J.M. (2004) *On-line monitoring techniques for studying the evolution of physical,*

rheological and chemical effects along the extruder. *Plast. Rubber Compos. Macromol. Eng.*, **33**(1), 55.

96 Maia, J.M., Carneiro, O.S., Machado, A.V., and Covas, J.A. (2002) On-line rheometry for twin-screw extrusion (along the extruder) and its applications. *Appl. Rheol.*, **12**, 18.

97 Machado, A.V., Maia, J.M., Canevarolo, S.V., and Covas, J.A. (2004) Evolution of peroxide-induced thermomechanical degradation of polypropylene along the extruder. *J. Appl. Polym. Sci.*, **91**, 2711.

98 Zschuppe, V., Geilen, T., Maia, J.M., Covas, J.A., and Petri, H.-M. (2006) The piezo axial vibrator – new perspectives for on-line viscosity measurements on polymers. Thermo Electron Corporation, Technical Note LR-57, 2006.

99 Mould, S., Barbas, J.M., Machado, A.V., Nóbrega, J.M., and Covas, J.A. (2011) Measuring the rheological properties of polymer melts with on-line rotational rheometry. *Polym. Testing*, **30**, 602.

100 Mould, S., Barbas, J.M., Machado, A.V., Nóbrega, J.M., and Covas, J.A. (2012) Monitoring the production of polymer nanocomposites by melt compounding with on-line rheometry. *Int. Polym. Proc.*, **XXVII**, 527.

Part IV

Synthesis Concepts

6

Exchange Reaction Mechanisms in the Reactive Extrusion of Condensation Polymers

Concetto Puglisi and Filippo Samperi

Institute for Polymers, Composites and Biomaterials (IPCB) Sede Secondaria. di Catania, CNR, Via P. Gaifami 18, 95126 Catania, Italy

6.1 Introduction

Blends of condensation polymers such as polyesters, polyamides, polycarbonates, and, in general, all polymers bearing reactive functional groups inside the chains or at chain ends, may yield interchange reactions (often referred as *transreactions*) when they are mixed in the molten state [1–79]. This establishes the concept of *reactive blending* or reactive melt mixing. Some attempts of reactive blending in viscous solution at high temperature are also reported in literature [56, 57].

The occurrence of the interchange reactions leads to *in situ* formation of copolymers with different molecular architectures (block, segmented, random, grafted) that may drastically change the blends' properties. The rearrangement of the structural units and their molar composition depend on the degree of extent of the reactions that occur. The interchange processes may proceed by two different mechanisms (Figure 6.1): a direct interchange reaction between functional groups located inside the polymer chains, often induced by specific catalysts, or by attack of reactive chain ends to functional groups. The exchange reactions can occur between the same polymer chains (intramolecular exchange) or between different polymers chains (intermolecular exchange).

The intramolecular processes produce low molar mass (MM) cyclic and linear oligomers by involving either the reaction between the midchain functional groups of the same homopolymer chain (reaction 1a) or the reaction of the reactive chain ends with the functional groups of the same polymer backbone (reaction 1c).

Interchange reactions involving the midchain functional groups of two homopolymers of different structure (Figure 6.1, reaction b) yield two linear copolymer chains. The copolymer chains thus generated can, in turn, undergo inter- or intramolecular interchange reactions according to the reaction 1a or 1b. When a sufficient number of interchanges have occurred, linear and cyclic random copolymer chains would be present in the mixture.

Reactive Extrusion: Principles and Applications, First Edition. Edited by Günter Beyer and Christian Hopmann.
© 2018 Wiley-VCH Verlag GmbH & Co. KGaA. Published 2018 by Wiley-VCH Verlag GmbH & Co. KGaA.

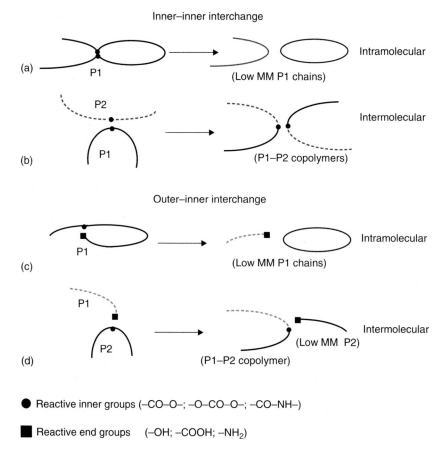

Figure 6.1 Schematic representation of the interchange reaction mechanisms that can occur in polymers containing reactive functional groups: (a) the interchange involves the inner functional groups of the same chain, (b) the interchange involves the inner functional groups of two chains, (c) the interchange involves the outer and inner reactive groups of the same chain, and (d) the interchange involves the outer and inner reactive groups of different chains.

Intermolecular outer–inner reaction of a polymer chain bearing active chain ends with the reactive inner group of the other polymer (Figure 6.1, reaction d) yields a block copolymer plus a shorter linear homopolymer chain. The progress of the intermolecular outer–inner reaction depends on the reactivity of end groups generated in the second polymer chain (P2). If these chain ends are reactive the reaction proceeds and the initial block copolymer evolves toward a random distribution of the sequences with a composition close to that of the initial homopolymers. If these chain ends are unreactive, the intermolecular reaction will stop as soon as the reactive chain ends, initially present are consumed and a block copolymer is generally produced [2, 13]. The reactive end groups can be present in the initial polymers or can be generated *in situ* by the addition of low concentrations (0.1–2 wt%) of a specific reactant [51, 56, 71] or by degradation reactions (hydrolysis, thermal degradation processes) that can occur in the melt mixing condition.

Often, end groups of polymer blends have not been taken into account during the processing of polymer blends, and some of the basic chemical reactions leading to interchange have been ignored, which indicates that no attempts have been made to control the chemical reactions from occurring.

Interchange reactions in the blends of condensation polymers lead to significant variations in their MM distribution, chemical structure, composition, and functionality. The synthesis of new copolymers with well-known and reproducible composition and architecture (i.e., block, segmented, random, grafted, or alternate), and the compatibilization of polymer blends are considered to be the main practical applications of these reactions. Although technological ability is available for the production of target materials with predetermined properties as well as for the preparation of new types of copolymers, these are restricted because the methods used control interchange reactions. The extent of exchange reactions strongly depends both on the type of polymers and on some other parameters such as the initial compatibility of blends, the nature and concentration of the reactive groups (either in the main chain or as end groups), the mixing temperature, the moisture content, the residence time in the melt, and the presence and the concentration of a catalyst or a reactant [1, 2, 5–17, 19–23, 51]. Because the melt mixing is performed at high temperature, degradative reactions can also occur.

Detailed studies about the mechanisms and the kinetics of interchange and about the degradation reactions that occur in the molten state are necessary to obtain polymeric materials with well-known and reproducible properties, to avoid detrimental reactions and to control the final properties of the products.

The interchange reactions lead first to block copolymers with high-average block lengths (L), and then, gradually, to random copolymers with the progress of the reaction. The resulting copolymers may enhance the mutual miscibility, over the original unreacted components, facilitating the compatibilization of the blends and accelerating the exchange reactions between chains [8, 12–14, 23, 31, 37, 41]. To enhance the compatibility of polymer blends, pre-synthesized copolymers (block or graft) can be also added [34, 62]. Selective coupling agents, leading to *in situ* formation of block copolymers by reaction with the specific end groups of polymer components can be also used [58–61].

A variety of experimental and theoretical approaches have been applied to determine the chemical mechanisms and reaction kinetics as well the composition and the architecture of the copolymers formed (i.e., block, segmented, random, graft, or alternating). These include NMR and FT-IR spectroscopy, mass spectrometry (MS), small-angle X-ray scattering (SAXS) methods, and so on [2, 4–6, 9, 17–19, 37, 39, 42, 49–55, 63, 64]. In most cases, the interchange reactions are also followed by DSC analysis [6, 25, 26, 32, 39, 51–54].

Generally, considerable attention is focused on the copolymer formation, and an approach relating the copolymer composition with the predominant interchange mechanism has been elaborated and applied to different reactive polymer blends [2, 4–6, 49–53, 77].

In order to distinguish between inner–inner and outer–inner processes, one may look at parameters such as (i) composition of the copolymer formed in the exchange reaction; (ii) dependence of the reaction rate on the concentration of active chain ends; (iii) dependence of the reaction rate on the molecular weight of the homopolymer reacting through its active chain ends.

Several statistical models have been proposed for better understanding the early stages of interchange reactions, to evaluate the role of reactivity and diffusion to the reaction kinetics, to reveal the specific features of reactions at the interface, and to establish the mechanism of influence of block copolymers on the structure stabilization in incompatible blends [13, 65, 66, 72]. Statistical models were also proposed to predict or define the influence of interchange reactions on the molar mass distributions (MMDs) on the melt mixed products [66–70]. In particular, Kudryavtsev and Govorun [70] derived the kinetic equation for the evolution of MMD during the course of direct interchange reactions in the melt mixing of blends. Litmanovich *et al.* [66] developed an algorithm to generate an ensemble of statistical multiblock AB copolymer chains via a polymer-analogous reaction with acceleration. They studied the influence of chain structure, chain length, and interchain interactions.

This chapter provides an overview of some of the most significant interchange reactions that take place during the processing in controlled conditions of several blends of condensation polymers such as polyester/polyester (PES/PES), polyester/polycarbonate (PES/PC), polyester/polyamide (PES/PA), polycarbonate/polyamide (PC/PA), and polyamide/polyamide (PA/PA). Powerful analytical methods such as ^{1}H NMR and ^{13}C NMR, MALDI-TOF MS and the application of specific kinetic models allowed to determine the composition and the microstructure of copolymers in the interchange reactions involving condensation polymers.

Calorimetric experiments have also been used for the characterization of the melt mixed polymer blends and the copolymers formed during the processing. Degradative reactions that occur during the melt processing of some of the aforementioned blends will be also discussed.

6.2 Interchange Reaction in Polyester/Polyester Blends

PEs/PEs blends have been studied both for industrial applications and for academic interest. In particular, poly(ethylene terephthalate) (PET)- based blends have been extensively studied [2, 3, 8–43, 56–58, 67–81]. PET is one of the most important industrial polymeric materials, and it is widely used for fibers, films, and packaging applications because of its easy processability, good creep resistance, resistance to chemical attack, and excellent optical clarity. However, its gas-barrier properties and thermal performance are not good enough to meet the requirements of some particular applications, such as packaging of oxygen-sensitive foods and hot-fill applications. To overcome these limitations, several blends with other polyesters have been commercially proposed, for example: PET/poly(butylene terephthalate) (PBT) systems are used for molded

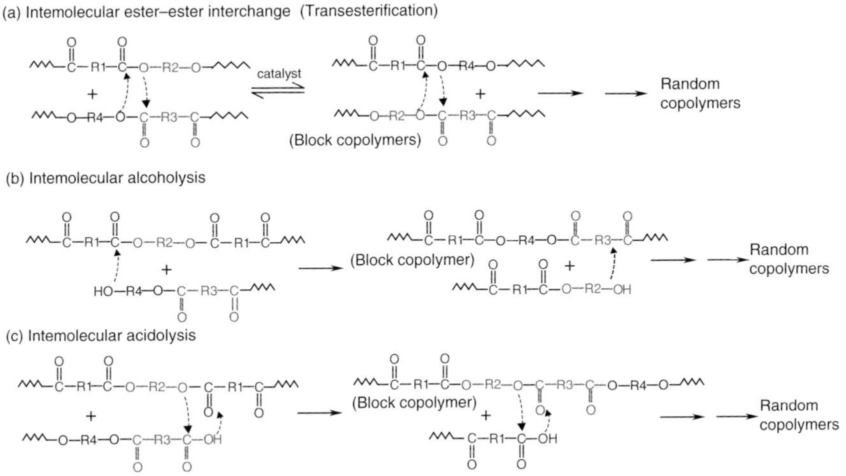

6.2 Interchange Reaction in Polyester/Polyester Blends

Scheme 6.1 Interchange reactions that occur in the melt mixing of PEs/PEs blends.

(a) Intemolecular ester–ester interchange (Transesterification)

(b) Intemolecular alcoholysis

(c) Intemolecular acidolysis

automobile parts, PET/poly(ethylene naphthalate) (PEN) blends are used as encapsulating material for solar cells, and can be also used for food-packaging [20–22, 41].

It is generally acknowledged that interchange reactions between polyesters can occur during the melt mixing in the processing conditions [2, 3, 8–12, 14–20, 27, 28, 31, 35–40]. Porter and Wang [8] have extensively studied blends of polyesters and carefully examined whether or not transreaction is a pre-requisite for forming a single-phase system for polyester blends. It is generally agreed in the literature that transesterification can lead to miscibility or enhance the compatibility between phases in many blends of polyesters.

Interchange reactions can occur by inter- or intramolecular mechanisms, and can involve ester (—CO—O—), hydroxyl (OH), and carboxyl (COOH) groups (Scheme 6.1). In particular: (i) the intermolecular interchange reaction between two ester inner groups (Scheme 6.1a) is referred as direct ester interchange and sometimes as transesterification or "esterolysis," (ii) the outer–inner interchange reaction between end groups hydroxyl and inner ester groups (Scheme 6.1b) is usually called alcoholysis, (iii) the outer–inner interchange reaction involving carboxyl (COOH) and ester groups (Scheme 6.1c) is referred as *acidolysis*. Often, these interchange reactions are referred as transesterification reactions.

Theoretically, for high MM polymers, the probability of having a direct ester–ester exchange is much higher than that of alcoholysis or acidolysis because of the lower end group concentrations. Interchange reactions in PEs/PEs blends depend strongly on their initial compatibility and on the blending conditions such as temperature (T), time (t), catalysts (including residual amount from polymerization reactions), and mixing apparatus [20–24]. They can also depend on the addition of little amount (0.1–2%) of reactive molecules such as ethylene glycol, dicarboxylyc acid, to generate *in situ* specific reactive end groups by degradation reactions, coupling agent (e.g., bis-oxazoline), or block copolymers as compatibilizer [34, 35, 56, 58, 71, 79].

Using ^{13}C NMR analysis Jacques *et al.* [9] demonstrated, that titanium alkoxide accelerates the interchange reaction during the melt mixing of PET/PBT at 275–300 °C, whereas triphenyl phosphite hinders it.

Several techniques such as NMR, FT-IR, MS, SAXS, and DSC methods have been applied to determine the mechanisms and kinetics of exchange reactions, and the composition and the molecular architecture of the copolyesters that are formed. A number of studies report on the chemical characterization of PEs/PEs reacted products by ^{1}H NMR and ^{13}C NMR analyses and by the application of statistical approaches developed by Yamadera and Murano [80] for three-component copolymers and by Devaux *et al.* [10, 11] for four-component copolymers. These models permit the determination of the microstructure of copolymers formed at different reaction time, that is, the chemical composition, the molar fractions of structural units (i.e., dyads, triads, etc.), the average sequence lengths (L), the degree of interchange (DE), and the degree of randomness (DR) which indicate the extent of interchange reactions.

The interchange reactions lead first to block copolymers with higher average block lengths (L), and then, gradually to random copolymers as the reaction time increases. Therefore, a DR value near to 0 indicates the presence of a mixture of homopolymers or block copolymers with very high average block lengths (L). As the exchange reaction proceeds, the block character of copolymers decreases and the DR value is between 0 and 1. A DR value of 1 indicates the formation of a random copolymer, whereas a value of 2 is characteristic of an alternating copolymer. A trend of values such as $1 \ll DR < 2$ indicates that the copolymer microstructure is evolving toward alternating sequences.

Within PEs/PEs blends one of the most studied are PET/PEN mixtures [15–36, 38, 41, 58]. Although a number of papers are present in literature about the transreaction that occurs in PET/PEN blends, the mechanisms and the kinetics of exchange are still debated [15–19, 23, 27, 28, 31, 33]. Mixtures of PET and PEN have been transesterified at temperatures between 280 and 300 °C [16–18]. The partially transesterified polymers were analyzed using NMR, and the rate constants for transesterification were obtained using a kinetic expression based on a second-order reversible reaction mechanism. The hydroxyl end groups have a significant effect on the kinetics of the reaction by comparing rate constants obtained for the same polyester mixture but with hydroxyl end groups quantitatively esterified by trifluoroacetic acid. Carboxyl end groups do not have such a significant role in transesterification, and from the variation of rate constant at one temperature for a series of PET polymers with different carboxyl and hydroxyl end groups, it was also affirmed that the direct ester–ester interchange reaction is very small.

The role of the hydroxyl end groups in the interchange reactions of PET/PEN (80/20 w/w) system at the processing temperature (285–300 °C) has been studied [19], by using a PET sample with unmodified end groups and another sample with the hydroxyl end groups esterified with benzoyl chloride (referred as *PET-Bz*). All samples have been characterized by means of MALDI-TOF MS and ^{1}H NMR techniques. PET and PEN samples are essentially terminated with methyl ester groups (about 68–70%mol), and with a minor amount of OH

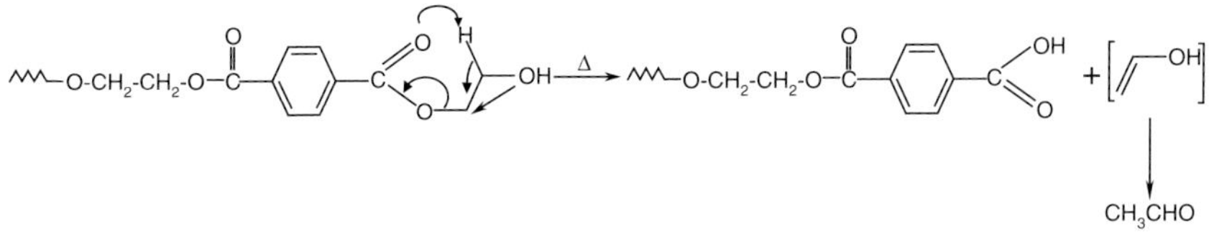

Scheme 6.2 Formation of acetaldehyde from 2-hydroxyethyl end groups of PET during processing [83].

(20–22%mol) and COOH (8–10%mol) groups, whereas hydroxyl end groups are absent in the PET-Bz sample.

Isothermal thermogravimetric (TGA) analysis carried out at 300 °C for 120 min showed that PET looses about 4–5% (wt%) of the initial weight, most likely due to the degradation reaction depicted in Scheme 6.2. In fact, at about 280–300 °C acetaldehyde (CH_3CHO) can be formed during the processing of the PET, due to the thermal degradation of hydroxyethyl end groups, which leads to the formation of carboxylic ends (Scheme 6.2) [82–84]. The esterified PET-Bz sample does not show this weight loss and therefore it is more thermally stable than the corresponding unmodified PET. Both PET/PEN and PET-Bz/PEN blends (80/20 w/w) were processed at 300 °C for different time (30, 60, and 120 min) with Ti(isopropoxide)$_4$ as catalyst, under nitrogen flow to avoid oxidative degradation. The melt mixed materials were analyzed by ^{1}H NMR for determining the composition and the microstructure of the copolymers that were formed. The sequence structure of PET/PEN copolyesters was evaluated from the resonance of the methylene protons belonging to the triad TET (ethylene unit between two terephthalate groups), NEN (ethylene unit between two naphthalate groups), and TEN (ethylene unit between terephthalate and naphthalate groups) (Figure 6.2), according to the Devaux model [10, 11]. A DR of 0.39 was determined in the melt mixing of the unmodified blend and 0.29 in the case of the blend with PET-Bz, at 30 min of mixing. For reaction times of 60 min, a DR value of 0.58 for the unmodified system and of 0.39 in the case of the capped system. After 120 min of reaction, the unmodified system is fully randomized (DR = 1), whereas the modified system has still a DR of 0.75. This is direct evidence that the presence of the hydroxyethyl and carboxyl chain ends of PET and PEN affects the mechanism and the kinetics of transesterification despite the presence of a transesterification catalyst [19].

A sample of PET with all chain ends capped (PET-capped) and therefore without any reactive end chains (OH and COOH) was melt mixed with the PEN (50/50 w/w) at the same condition discussed earlier [3]. The ester–ester interchange reaction of PEN/PET-capped blends melt mixed without a catalyst does not occur (Figure 6.3) whereas in the presence of a catalyst a copolymer with a DR 0.72 is formed.

In a recent work, Röhner et al. [71] found that the transesterification of PET and PEN at 270 °C is strongly enhanced by the addition of 1.0% zinc acetate. The catalyzed PET–PEN copolymer showed a higher DR (DR = 0.35) than the un-catalyzed sample (DR = 0.15). This attempt was made to establish the fact that the processing condition influences the transesterification.

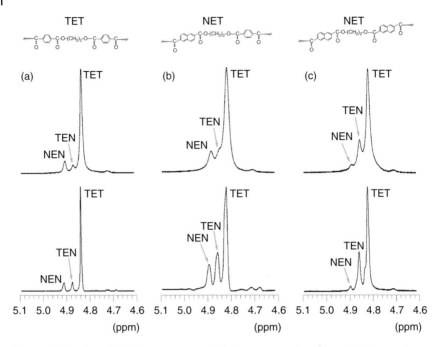

Figure 6.2 Section of the ^{1}H NMR spectra of PET/PEN (bottom) and PET-Bz/PEN (upper) blends melt mixed for: (a) 30 min, (b) 60 min, and (c) 120 min. (Blanco *et al.* 2012 [19]. Reproduced with permission of Wiley.)

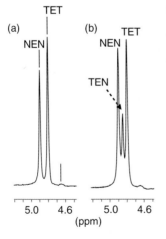

Figure 6.3 Section of ^{1}H NMR spectra of all capped PET/PEN (50/50 w/w) melt mixed at 285 °C for 60 min without (a) and with (b) Ti(isopropoxide)$_4$ as catalyst. (Blanco *et al.* 2012 [19]. Reproduced with permission of Wiley.)

Gunes *et al.* [30] observed that ultrasonic treatment at short residence time in melt (260–300 °C) led to the enhancement of transesterification reaction in the PEN/PET blend, as shown by ^{1}H NMR and MALDI-TOF analyses. The authors found that the extent of transesterification was increased by 31% in treated blend when compared to the blend without treatment. Analyses by MALDI-TOF MS have also revealed greater amounts of hydroxyl and carboxyl-terminated oligomers in ultrasonically treated PET and PEN, indicating their greater reactivity.

Khonakdar et al. [72] have developed a non-isothermal kinetic model based on mass balance and Arrhenius laws using a second-order reversible reaction capable of predicting the extent of transesterification reaction and the DR in PET/PEN blends prepared by a twin screw micro-compounder over a full composition range under different processing conditions. Good agreement was found between the theoretical and the experimental data determined by ^{1}H NMR and then affirmed that their model can be used for each polyester melt mixed blend. Some authors [67] have verified that the differences in the extent of transreactions arising from different processing histories can influence the rheological characteristics such as viscosity, elasticity, and relaxation time spectra. They observed that an increase (%) in hetero-triad TEN led to a systematic increase in viscosity of the blends, whereas a decrease of the sequence block lengths (L_{PET}, L_{PEN}) led to an increase in the elasticity and relaxation time spectra due to improvement of the blend interface with increase in extent of copolymer formation.

6.3 Interchange Reaction in Polycarbonate/Polyester Blends

Several works report that interchange reactions can occur during the melt mixing of polycarbonate/polyester (PC/PEs) blends, usually in the temperature range of 270–300 °C [1–7, 10, 11, 44–48, 85–90]. Mostly, the studies have been focused on the blend of PC with aliphatic–aromatic polyesters such as PET, PBT, and PEN [5–7, 10, 11, 44–48, 85–87], because these blends have received some industrial interest. PC and PEs give incompatible blends, which become compatible during their melt mixing owing to the interchange reactions that may occur.

The interchange reactions in PC/PEs blends can occur by inter- or intramolecular mechanisms as in PEs/PEs blends, involving: (i) ester and carbonate functional groups, (ii) hydroxyl and carboxyl end groups belonging to the polyesters, and (iii) phenol end groups of PC (in principle).

It is generally perceived that the phenol end groups do not give any interchange reaction owing to their low nucleophilic character with respect to primary aliphatic hydroxyl groups [5–7, 44, 45]. Montaudo et al. [5] carried out reactive melt mixing, at 270 °C, of PC with tailored samples of PET and PBT terminated only with hydroxyl or carboxyl groups at both end chains and in the absence of catalysts. It has been observed that the composition and the block length of the block copolymers generated did not change as the reaction time increases. At the beginning of the reaction the reactive aliphatic OH or COOH end groups were consumed leading to the formation of PC–PEs block copolymer and to low MM PC oligomers terminated with phenol end groups that are not able to react in the processing conditions used. Moreover, using PBT samples with different concentrations of hydroxyl end groups, at different M_n (3000, 6000, and 18 000 g mol^{-1}), it has been evaluated that the composition of the block copolymer depends only on the MM of the polyester and it remains constant as the reaction time increases. Furthermore, using end capped-PBT, it has been observed that the direct ester–carbonate exchange reaction of midchain functional groups does not occur at temperatures below 300 °C, in the absence of transesterification catalysts [5–7, 44, 45]. Therefore, melt mixed PC/PEs blends at temperatures below 300 °C generate copolymers sequences, with the block length gradually decreasing up to random copolymers as the reaction time increases, only in the presence of suitable catalysts that are able to promote the direct ester–carbonate exchange reaction.

The catalytic activity of various lanthanide compounds (based on europium, cerium, samarium, terbium, and erbium), titanium- and calcium/antimony-based catalysts was investigated in the reactive blending of PC/PET blends [7, 45]. Titanium showed a higher catalytic activity; however, lanthanide catalysts, especially those based on samarium, europium, and cerium, allowed the easy control of block length in the PC/PET block copolymers, and at the same time did not promote the side reactions that occurred with the other catalysts.

In the case of PC/PBT blend melt mixed in the range of 260–300 °C, block of ester–carbonate copolymers are formed, at the beginning of the exchange reaction, in the presence of $Ti(OBu)_4$ as catalyst [5, 10, 11]. The catalyzed ester–carbonate interchange reactions then evolve toward the formation of random copolymers (DR between 0 and 1) as the mixing time increases. The microstructural parameters (dyads composition, average sequence lengths, DR, DE) for the copolymers were evaluated by ^1H NMR analysis by applying the statistical model developed by Devaux et al. [10, 11] which takes into account the molar composition of the initial homopolymers and the molar fraction of the structural units formed by ester–carbonate interchange. The formation of cyclic butylene carbonate (BuC) was observed in the melt mixing of PC/PBT blends at temperatures higher than 300 °C [91], most likely generated by the thermal degradation of BuC units of the ester–carbonate copolymers (Scheme 6.3).

The same exchange reactions are expected to occur during the catalyzed melt mixing of PC/PET and PC/PEN systems (Scheme 6.4), but the occurrence of some side reactions have been observed, as the mixing time increases [5, 6, 10, 11, 92]. The aliphatic ethylene carbonate (ETC) units of copolymer chains formed during the melt mixing of PC/PET and PC/PEN blends are thermally unstable at the processing temperature (270–300 °C) and undergo elimination of ETC or CO_2, as depicted in Scheme 6.4, leading to new additional aromatic–aliphatic ether sequences (**BPA-Et-N** and **BPA-dEt-N**).

Samperi et al. [5, 6] developed a modified kinetic model, similar to that developed by of Devaux [9], that allowed the determination of the copolymers molar fractions of all structural units by combining ^1H NMR and ^{13}C NMR data of PC/PET and PC/PEN blends melt mixed at 270 and 280 °C, respectively, in the presence of $Ti(OBu)_4$ as catalyst. This model also allowed to determine the molar amount of EC and CO_2 lost from the reacting systems, in agreement with thermogravimetric data [5, 6].

In Figure 6.4 shows the molar fractions of the six structural units formed (Scheme 6.4), at different mixing time, during the reactive blending of an equimolar (with respect to repeat units) PC/PEN blend. A four component copolymer is formed after 2 min of reaction (**BPA-C, E-N, BPA-N**, and

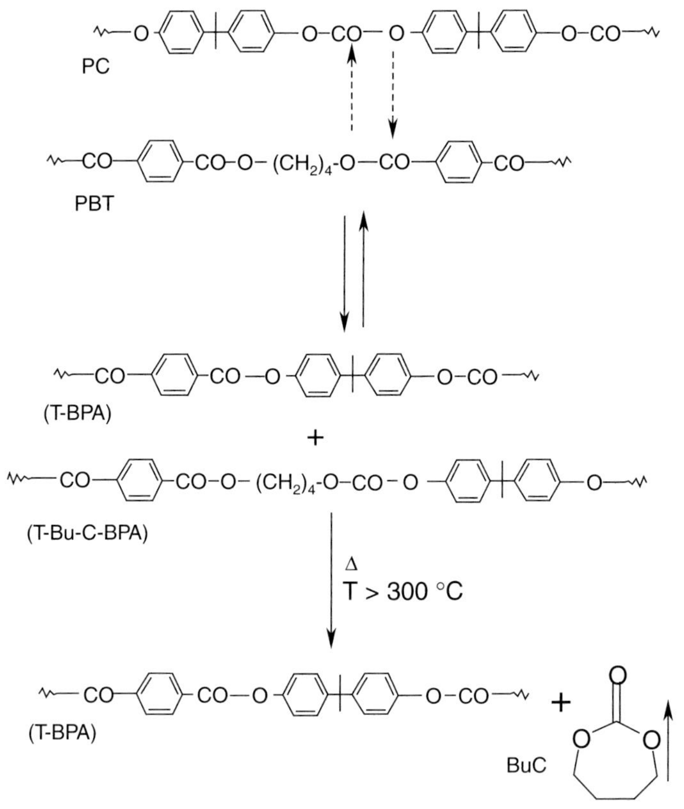

Scheme 6.3 Schematic representation of the reaction that occurs during the melt mixing of PC/PBT blends in the presence of Ti(OBu)$_4$ as catalyst at processing temperature >300 °C.

BPA-C-E) with an RD of 0.04 and an average block length (L) of 25.2 for **BPA-C** (PC) units and 50.0 for **E-N** (PEN) ones. After 5 min of mixing, the the side reactions (Scheme 6.4) produce a five component copolymer with the addition of **BPA-Et-N** units, and become a six component copolymer after 12 min of mixing with the appearance of **BPA-dEt-N** units. The DR increases with the reaction time, and a random copolymer is formed within 45 min of reaction. A random four component (**E-N**, **BPA-N**, **BPA-Et-N**, and **BPA-dEt-N**) copoly(ester-ether) has been formed at 60 min of melt mixing, because of the complete elimination of carbonate units. In Figure 6.5 the DSC analysis of melt mixed PC/PEN blends samples are reported. Table 6.1 summarizes the thermal properties (i.e., glass transition temperature T_g, cold crystallization temperature T_{cc}, normalized enthalpy of cold crystallization ΔH_{cc}, melting temperature T_m, and normalized enthalpy of melting ΔH_m) of the polymer models and of the copolymers formed by reactive blending of equimolar PC/PEN blend. The initial PC/PEN blend shows two distinct glass transition temperatures (T_g) at 125 and 149 °C, which are almost identical to those of pristine homopolymers, confirming that these polymers are incompatible in the molten state. Copolymers formed after 2 and 5 min of reactive blending show a semi-crystalline behavior exhibiting

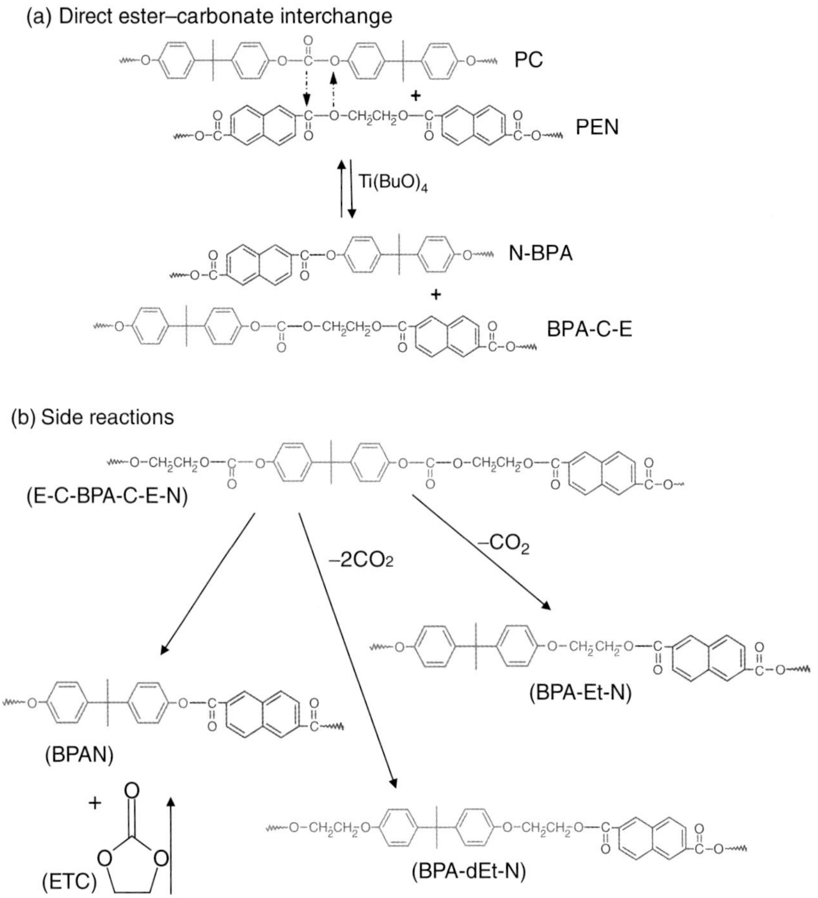

(a) Direct ester–carbonate interchange

[Scheme showing PC and PEN structures reacting via Ti(BuO)$_4$ to form N-BPA and BPA-C-E]

(b) Side reactions

[Scheme showing (E-C-BPA-C-E-N) undergoing side reactions: losing $-2CO_2$ to form (BPAN) + (ETC), or losing $-CO_2$ to form (BPA-Et-N), and (BPA-dEt-N)]

Scheme 6.4 Schematic representation of the reaction that occurs during the melt mixing of PC/PEN blends [6].

a broad melting peak with a maximum at about 258 and 255 °C, and also a T_{cc} (cold crystallization temperature) with a maximum at about 196 and 190 °C. The normalized enthalpy of both melting and cold crystallization processes decreases as the mixing time increases and amorphous copolymers were formed at mixing time higher than 8 min. Copolymers formed at 2 min of mixing show two T_g transitions at 125 and 135 °C, whereas the ones at 5 min present a single T_g at 127 °C, indicating the presence of a homogenous phase. As the reaction proceeds, the T_g values increase from 128 °C (8 min) up to 150 °C (60 min) because of the chemical evolution of the reacting PC/PEN system (Scheme 6.4). The theoretical T_g of each copolymer have been calculated by a modified Johnston equation for a random six component copolymers (Eq. (6.1)) [6, 93], taking into account the molar composition of each structural unit in the copolymers, and the T_g of the corresponding polymers relative to all six

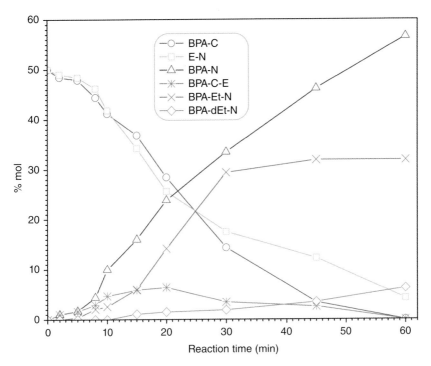

Figure 6.4 Molar amount (mol%) of structural units in the copolymers formed by reactive blending of PC/PEN (1 : 1 mol/mol with respect to the repeat units) at different reaction time. (Samperi et al. 2015 [6]. Reproduced with permission of Elsevier.)

structural units formed in the exchange process (Scheme 6.4).

$$1/T_g = \sum_{i=1}^{n} W_i/T_{gi} \quad (6.1)$$

W_i is the weight fraction of each structural unit composing the copolymer and T_{gi} is the glass transition temperature of polymer (homopolymer or copolymer) corresponding to the structural unit i.

$$W_i = \frac{X_i \cdot m_i}{\sum_{i=1}^{n} X_i \cdot m_i} \quad (6.2)$$

where X_i and m_i are the molar fraction and the MM of each structural unit, respectively.

For a six component copolymer, the equation can be explicated as in the follows:

$$\frac{1}{T_g} = \frac{W_{BPA-C}}{T_{g_{PC}}} + \frac{W_{E-N}}{T_{g_{PEN}}} + \frac{W_{BPA-N}}{T_{g_{p(BPA-N)}}} + \frac{W_{BPA-E-C}}{T_{g_{p(BPA/E-C)}}} + \frac{W_{BPA-Et-N}}{T_{g_{p(BPA-Et-N)}}} + \frac{W_{BPA-dEt-N}}{T_{g_{p(BPAdEt-N)}}}$$

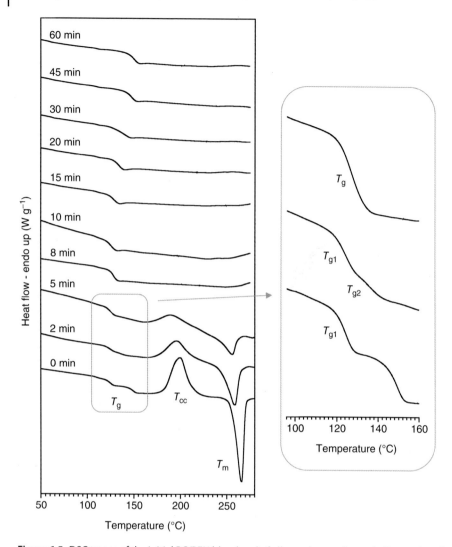

Figure 6.5 DSC traces of the initial PC/PEN blend and of all copolymers formed. (Samperi *et al.* 2015 [6]. Reproduced with permission of Elsevier.)

Calculated T_g values are reported in Table 6.1 and are in good agreement with the experimental data for copolymers formed from 30 up to and 60 min mixing, confirming the random distributions of sequences.

6.4 Interchange Reaction in Polyester/Polyamide Blends

Intensive studies have been carried on the polyester/polyamide (PESt/PA) blends [49–51, 59–61, 94–114] in order to obtain new materials that will possess certain

Table 6.1 Thermal properties (T_g, T_m, T_{cc}) data of the polymer models and of the copolymers formed by reactive blending of equimolar PC/PEN blend.

Samples	DSC data					
	T_{cc} (°C)	ΔH_{cc} (J g^{-1})	T_m (°C)	ΔH_m (J g^{-1})	$T_{gexp.}$ (°C)	$T_{gcalc.}$[a] (°C)
PC	—	—	—	—	153	—
PEN	200	36	262	38	125	
p(BPAN)	—	—	—'	—	190	—
p(BPA/ET C)	—	—	—	—	99	—
p(BPA-Et-N)	—	—	—	—	116	—
p(BPA-dEt-N)	—	—	—	—	94	—
PC/PEN[b]	200	17.4	255	17.8	125; 149	—
cop-2[c]	196	6.13	258	14.6	125; 135	134.8
cop-5[c]	190	2.20	255	6.47	127	134.7
cop-8[c]	—	—	—	—	128	133.3
cop-10[c]	—	—	—	—	128	136.4
cop-12[c]	—	—	—	—	130	137.1
cop-15[c]	—	—	—	—	131	137.5
cop-20[c]	—	—	—	—	135	140.5
cop-30[c]	—	—	—	—	139	141.3
cop-45[c]	—	—	—	—	146	144.5
cop-60[c]	—	—	—	—	150	149

a) Calculated using Eq. (6.1).
b) Melt mixed blend taken a time zero.
c) Copolymers formed at different reaction time.
Source: Samperi et al. 2015 [6]. Reproduced with permission of Elsevier.

unique properties with respect to blend components. Polyamides are commonly named nylons (Ny). Much research has been focused on PET/PA6 (PET/PA6) [50, 96, 100], PBT/PA6 [49, 59, 60], PET/PA6,6 [99, 104], and PET/poly(*m*-xylylene adipamide) (PET/MXD6) [51, 104, 105, 114] systems.

The interchange reactions that might occur in the melt mixing of PEs/PA blends might involve: (i) direct amide and ester exchange, (ii) amine and carboxyl end groups belonging to the polyamides, and (iii) hydroxyl and carboxyl end groups belonging to the polyesters.

PA6 and PBT give biphasic blends in the molten state, presenting therefore a reduced contact surface. However, when the interchange reactions take place even at low content and may partially compatibilize the blends, then enhancement of the rate of the interchange reactions occurs at the interface between phases. [49, 59, 60]. Montaudo et al. [49] have investigated the kinetics of the interchange reactions during the melt mixing of equimolar PBT/PA6 blends at a temperature range of 260–290 °C, by the combination of ^{13}C NMR and MALDI-TOF MS analyses, applying the kinetic model of Devaux [9, 10] and other statistical models [115–118]. Samples containing essentially polymer chains with only one reactive end group (e.g., PA6-COOH, PBT-OH,

PBT-COOH, PA6-NH$_2$), or homopolymers with capped end chains, were used for each set of experiments.

The direct ester amide interchange does not occur in PBT/PA6 blends melt mixed for 1 h in the temperature range of 260–290 °C in the presence of high MM homopolymers with the absence of aromatic carboxyl end groups, as well as in the presence of catalysts such as Ti(OBu)$_4$, SnBu$_2$O, and Zn(Ac)$_2$/Sb$_2$O$_3$ [49]. However, a limited extent of interchange reaction was observed in the system melt mixed at 290 °C for 1 h, most likely due to the thermal degradation of PBT [119] (onset of thermal degradation 280 °C), which generates carboxyl-terminated chains that are able to react with PA6. Similarly, interchange reactions due to amine and carboxyl end groups of PA6, hydroxyl end groups of PBT, occur with very low rate in the range of 260–290 °C, without the influence of temperature on the extent of the reactions. The formation of copolymer in the melt mixing of PBT-COOH/PA6 blends indicated that carboxyl end groups of PBT samples are the most reactive chain ends through an outer–inner interchange reaction, according to Schemes 6.5 and 6.6, with the extent of

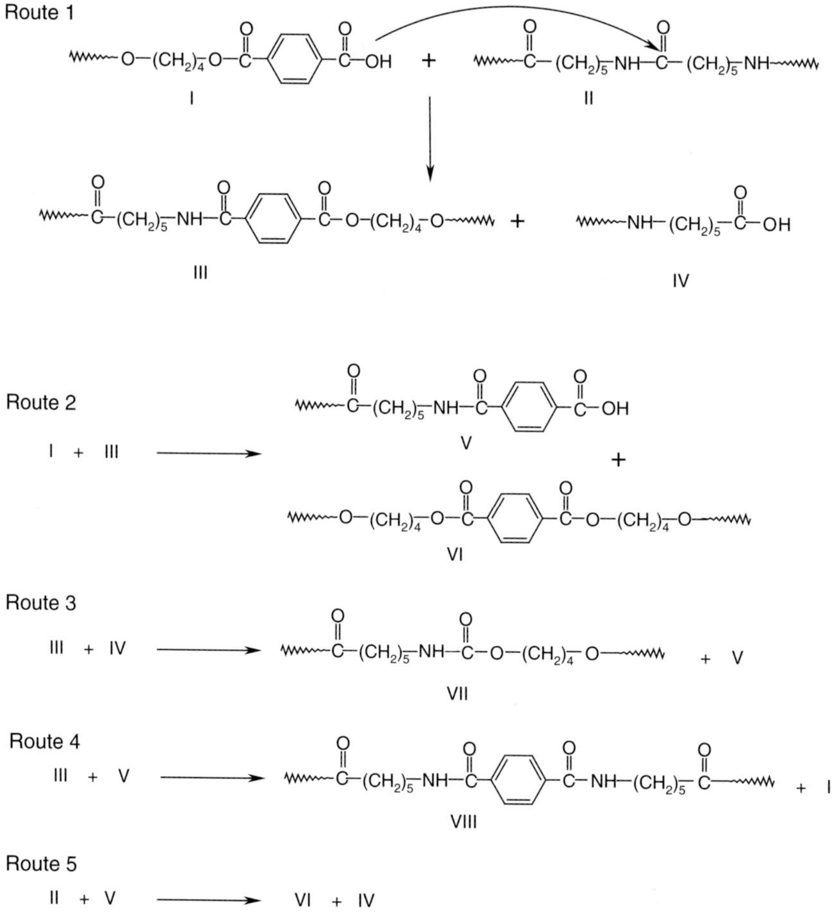

Scheme 6.5 Sequence of reactions responsible for the formation of copoly(ester-amide) by melt mixing of PBT-COOH/PA6 blends at 280 °C [49].

6.4 Interchange Reaction in Polyester/Polyamide Blends

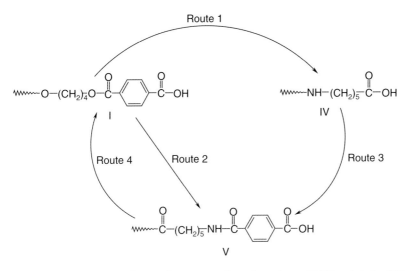

Scheme 6.6 A cycle that depicts the regeneration of reactive —COOH end chains [49].

interchange reaction increasing as the temperature of mixing increased. The composition of the copoly(ester-amide) formed at 260 °C (for 1 h) showed a higher amount of PA6 units with respect to the initial blend composition and a high value of the PA6 average sequence length, indicating the presence of block copolymers [119]. At 280 °C and a reaction time of 30 min the exchange reaction was completed and the composition of the copolymers was comparable to the initial feed (50/50), with a random distribution of average sequence lengths [49].

The reaction of the carboxyl end groups of PBT with PA6 produces a copoly(ester amide) and a PA6 with aliphatic COOH groups (Scheme 6.5, route 1) that are unreactive toward the ester groups of PBT in the melt mixing condition. However, PA6-COOH chains can react with the block copoly-(ester amide) (routes 2–4, Scheme 6.5) producing again aromatic carboxyl end groups that can continue the exchange reaction. This is most likely due to the formation of partially compatible micro domains that promote the exchange reaction responsible for the regeneration of reactive COOH end groups, leading to the completion of the exchange process, as summarized in Scheme 6.6. Figure 6.6 shows the relative abundance of species **II**, **III**, **IV**, and **V**, obtained by MALDI-TOF MS (Schemes 6.5 and 6.6), as a function of the mixing temperature. It can be observed that the intensity of the pristine PA6 oligomers (species **II**) decreases steadily, whereas PA6 oligomers bearing carboxyl chain ends (species **IV**), nearly absent in the initial blend, become intense at 260 °C and then fade off at 270 °C. Species **V** (Scheme 6.5), which are not present in the physical blend, show a maximum at 270 °C and then disappear at 280 °C. Finally, the intensity of the copoly(esteramide) oligomers (species **III** in Figure 6.6 and Scheme 6.5) increases steadily from 260 up to 280 °C.

Some authors [95, 100, 101] stated that *p*-toluene sulfonic acid (TsOH) is the most efficient catalyst for the direct amide–ester exchange reactions in PET/PA6 and PET/PA66 blends; however, Brown [111] has suggested that TsOH catalyzes the outer–inner exchange reaction of an amino end group with the inner ester group of PET. It was shown by ^{13}C NMR and MALDI-TOF MS analyses [50]

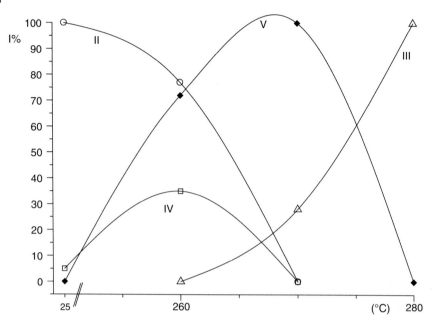

Figure 6.6 Normalized intensity/temperature profiles of the species **II**, **III**, **IV**, and **V** in Scheme 6.5, characterized by MALDI-TOF MS spectra of the PBT-COOH/PA6 blend melt mixed for 60 min at different temperatures. (Samperi et al. 2003 [49]. Reproduced with permission of American Chemical Society.)

that TsOH reacts with the polyester chains to form PET chains with carboxylic groups and *p*-toluene sulfonic esters at 285 °C. The sulfonic esters moiety suddenly undergoes hydrolysis and/or thermal degradation producing TsOH and PET with carboxyl chain ends (Scheme 6.7). Therefore, the role of TsOH can be considered to be that of a reactant instead of a catalyst. The high amount of PET with COOH chain ends promotes the interchange reactions of the PET/PA6 blends with the same mechanisms discussed for PBT/PA6 systems (Schemes 6.5 and 6.6) [50].

PET/MXD6 blends have attracted scientific and commercial interest in the area of food packaging because of the lower O_2 and CO_2 permeability of the MXD6, with respect to PET associated with the good toughness and clarity of PET. However, PET and MXD6 give incompatible blends with low optical clarity and an undesirable yellow color when processed in the molten state [112]. Compatibilized PET/MXD6 (90/10 w/w) blend was obtained through the incorporation of small amount of sodium 5-sulfoisophthalate (SIPE) in the PET matrix [104]. At 43% relative humidity the oxygen permeability was reduced by a factor of 3, in comparison with PET, for biaxially oriented film of the compatibilized blends [104]. Compatibilized blends were also obtained using ionomers [113]. Xie *et al.* reported that the interchange reaction in PET/MXD6 blends (90/10 w/w) in the molten state (290–310 °C) occurs only in the presence of sodium *p*-toluenesulfonate (*p*-TSONa) [114].

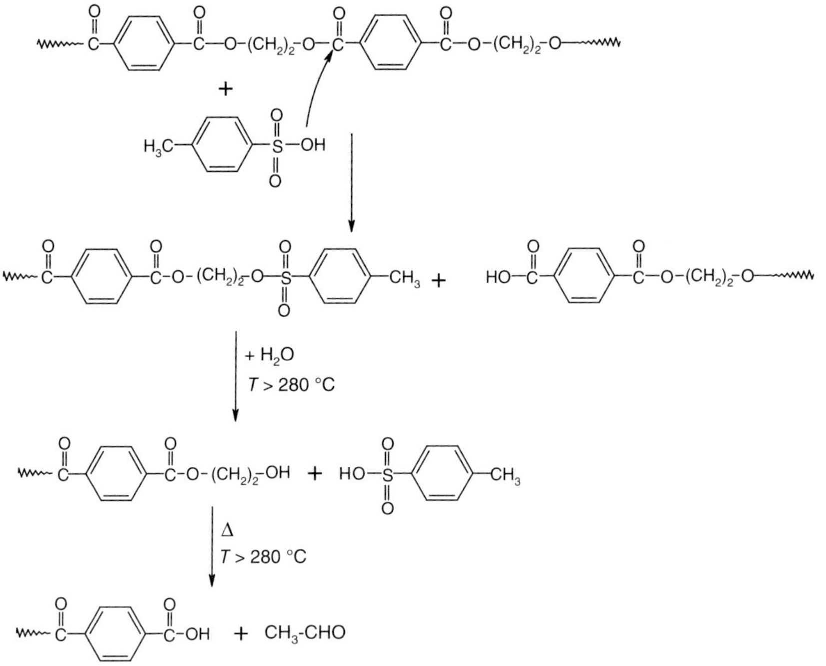

Scheme 6.7 Thermal-degradation mechanism of PET heated at 285 °C in the presence of 0.5 wt% of TsOH [50].

The melt mixing of PET/MXD6 blends (50/50 m/m) was carried out at 285 °C up to 120 min in the presence of terephthalic acid (TA) (1 wt%) [51]. The exchange reactions and the role of carboxyl end groups were studied by NMR analyses and by MALDI-TOF/MS. PET and MXD6 chains terminated with aromatic carboxyl groups were formed by the reaction of acidolysis induced by TA (Scheme 6.8). As in the case of PBT/PA6 blends, discussed earlier, these functionalized polymer chains quickly react with ester and amide inner groups to produce a PET-MXD6 copolymer (Scheme 6.8) that may compatibilize the initial biphasic blends that produce microdomains where the reaction of the aliphatic carboxyl-terminated MXD6 chains with the ester–amide copolymers may occur easily. Data in Table 6.2 indicate that block copolymers was formed at lower reaction time (15–20 min). The average sequence lengths of the PET and MXD6 blocks decrease as the mixing time rises, and a random copolymer was formed after 120 min of mixing with a composition typical for a Bernoullian distribution of copolymer sequences [10, 11, 80].

Chains terminated with cyclopentanone units are observed in the MALDI-TOF spectra, most likely formed by the thermal degradation of adipic acid end units belonging to the MXD6 sequences (Scheme 6.9). This is a typical thermal degradation reaction of polymers containing adipic acid moiety, for example, PA66 [115].

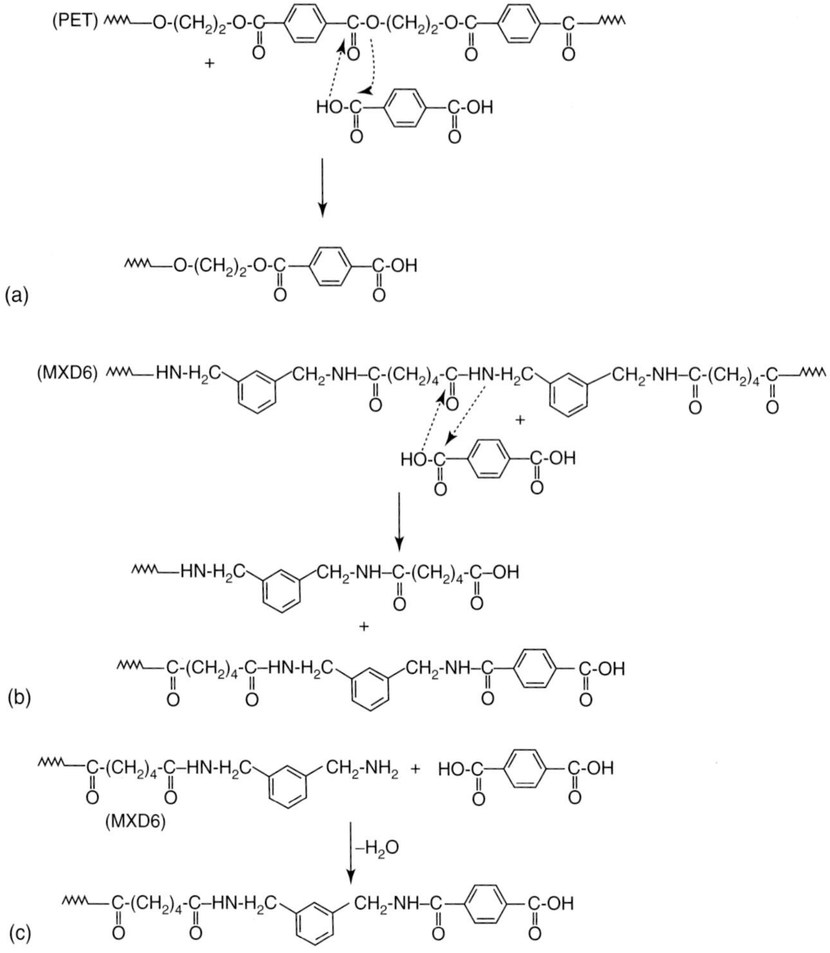

Scheme 6.8 Reactions occurring by melt mixing (285 °C) of PET/MXD6 in the presence of terephthalic acid [51].

The DSC traces of PET/MXD6 blends (Figure 6.7) showed two second-order transitions (T_g) at 80 and 89 °C, close to the initial values of PET and MXD6 homopolymers. Blends processed in the presence of TA at 285 °C between 15 and 120 min, show only one glass transition temperature (T_g) that decreases from 81 °C, at 15 min, to 63.2 °C, at 120 min. The experimental values of T_g fit with the values calculated by the modified Johnston equation (Eq. (6.1)), as in the case of the PC/PEN system.

Blends of PBT-PA6 and PBT-PA66 have been compatibilized *in situ*, by melt mixing, at temperatures lower than 275 °C, with 2 phr of a cyclo(organo) phosphazene bearing epoxy groups (namely CP-2EPOX) [59], which can react with carboxyl and amino end groups leading to the formation of block or graft copolymers. The blends were processed in a co-rotating twin-screw extruder using an appropriate temperature profile. The morphology of blends was

Table 6.2 Average sequence lengths of dyads, degree of exchange (DE) and degree of randomness (DR) calculated from ^{13}C NMR spectra of PET/MXD6- (1 wt%) terephthalic acid melt mixed blend at 285 °C for different time.

Time (min)	Average Sequence Lengths				DE	DR
	L_{E-T}	L_{M-A}	L_{E-A}	L_{M-T}		
0	—	—	—	—	—	—
15	17	24.5	1.04	1.06	5	0.1
20	18	21.5	1.06	1.05	5.1	0.102
30	9.09	10.4	1.11	1.12	10.3	0.206
45	5.06	6	1.2	1.25	17.2	0.364
60	3.4	3.5	1.4	1.42	29	0.58
75	2.83	2.88	1.53	1.55	0.35	0.7
90	2.45	2.48	1.68	1.69	40.6	0.81
105	2.18	2.18	1.84	1.84	46	0.92
120	2.07	2.07	1.93	1.94	48.3	0.97

Source: Samperi *et al.* 2010 [51]. Reproduced with permission of Wiley.

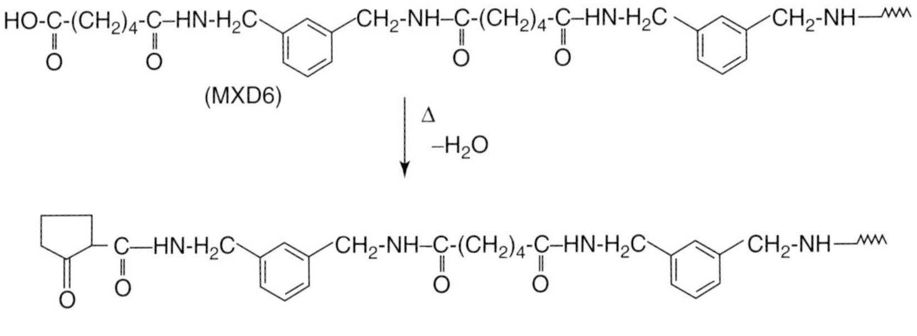

Scheme 6.9 Formation of cyclopentanone end groups during the melt mixing of MXD6 [51].

investigated by SEM and the images recorded for PBT/PA6 blends (25/75 w/w) are reported in Figure 6.8. The melt mixed blend without additive showed a gross morphology with a poor adhesion between the phases (Figure 6.8a), whereas in the presence of CP-2EPOX it showed a better adhesion between the phase and the particle dimensions decrease (Figure 6.8b). A similar behavior was also found for PBT/PA66/CP-2EPOX systems [59].

6.5 Interchange Reaction in Polycarbonate/Polyamide Blends

Polycarbonate/polyamide blends and in particular poly(bisphenol-A carbonate)/PA6 (PC/PA6) have received academic and industrial interest [62, 120–129] to obtain materials with good properties such as resistance to organic solvents,

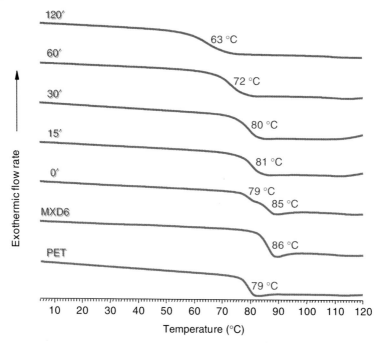

Figure 6.7 DSC trace in the T_g region for the PET/MXD6 melt mixed for different times in the presence of terephthalic acid. (Samperi et al. 2010 [51]. Reproduced with permission of Wiley.)

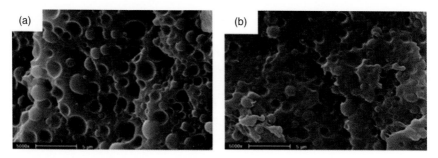

Figure 6.8 SEM images (magnification 5000×) of PBT/PA6 blends (25/75 w/w): (a) binary, (b) added with CP-2EPOX. (Scaffaro et al. 2006 [59]. Reproduced with permission of Elsevier.)

insensible to moisture, and durability under various weathering conditions [121–124]. It has been pointed out that PA6 and PC are thermodynamically immiscible over the entire composition range, but if suitable mixing conditions are applied, their blends show evidence that some degree of compatibility has been induced [120–123].

The melt mixing of blends of PC with PA6 samples having different molecular weights and different terminal groups shows that the increasing content of NH_2 PA6-terminal groups plays an important role in the change of the morphology and in the improvement of the mechanical properties. These results were attributed to the chemical reactions that take place during melt mixing of such

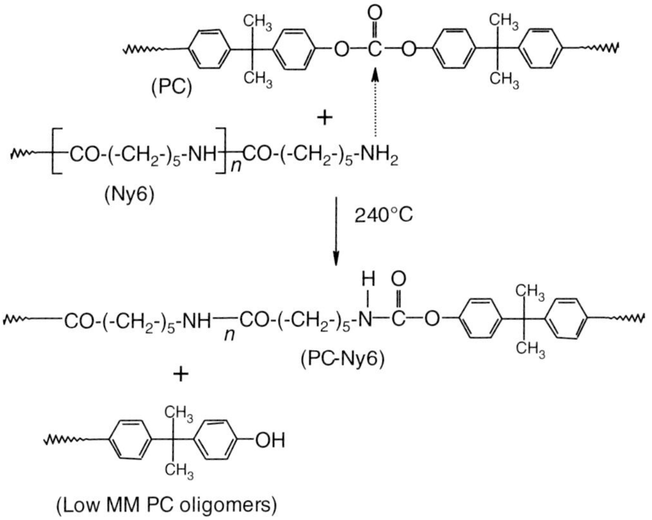

6.5 Interchange Reaction in Polycarbonate/Polyamide Blends

Scheme 6.10 Aminolysis interchange reaction that occurs in the PC/PA6 blends [120].

groups with polycarbonate by formation of PC-PA6 copolymer chains acting as interfacial agents between incompatible PA6 and PC [123].

Interchange reaction between carbonate group of PC and amide groups of PA6 and reactions between carbonate groups and the terminals of polyamide ($-NH_2$ and $-COOH$), should, in principle, be possible. However, it is generally acknowledged that the aminolysis interchange reaction (Scheme 6.10) is the main process in the molten state [120, 121, 126] and in viscous solution [62]. For the successive chemical characterization (^1H NMR and ^{13}C NMR analysis), Montaudo et al. [120] followed the kinetics of the exchange processes that occur during the melt mixing of PC/PA6 blends at 240 °C (i.e., the lowest temperature used to process PA6) performing a selective solvent extraction of the copolymers produced. Equimolar mixture of PC/PA6 blends by using high MM PA6 (M_w 50 000 g mol^{-1}) with a concentration of amino end groups ([NH_2]) of 13.5 mmol kg^{-1}, and a low-MM PA6 (M_w 7930 g mol^{-1}) having a [NH_2] of 252 mmol Kg^{-1} were melt mixed at 240 °C, for different times, under N_2 flow. The selective solvent extraction is carried out in sequences: $CHCl_3$ dissolves only unreacted PC, then trifluoroethanol (TFE) extracts PA6 and a PA6-rich copolymer leaving an insoluble fraction constituting polycarbonate-rich PC-PA6 copolymer (Figure 6.9). The formation of both TFE-soluble polymer and insoluble fractions is almost immediate in the low MM PA6 blend (Figure 6.9), whereas in the high MM PA6, formation of both TFE-soluble polymer and insoluble fractions were observed after 30 min of mixing (Figure 6.9). The composition of the TFE-soluble copolymers decreases as expected for an exchange reaction that occurs through active chain ends, whereas that of insoluble fraction (PC-rich copolymer) increases with the mixing times with a medium of 30 mol% of PA6 units, similar to the composition of PA6 in the blend.

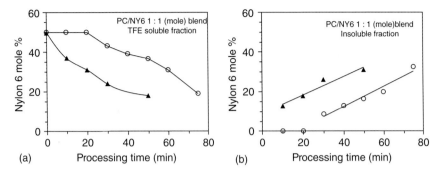

Figure 6.9 Mol% of PA6 units versus processing time in: (a) TFE-soluble fractions and (b) insoluble fractions of melt-mixed PC/PA6 (1 : 1 mol/mol) blends. PA6 M_w: (▲) 7930, (○) 50 000 g mol^{-1}. (Montaudo et al. 1994 [120]. Reproduced with permission of Wiley.)

The CHCl$_3$-soluble fraction constitutes low MM PC oligomers terminated with phenol groups, formed by the aminolysis interchange reaction (Scheme 6.10). The M_n of PC in the blend decreases with the processing time according to the random scission law (Eq. (6.3)) valid for the random scission processes:

$$1/DP = 1/DP_0 + Kt \tag{6.3}$$

where DP indicates the degree of polymerization (DP = M_n/254.2; 254.2 g mol^{-1} is the mass of the PC repeat unit).

The reaction of amino groups of PA6 with polycarbonate produces copolymers with PC and PA6 blocks linked together by urethane groups (Scheme 6.10), which increases with the processing time, indicating a reduction of the block size as a function of the extent of interchange. Ideally, the aminolysis reaction should proceed until the reactive amino chain ends are consumed (Scheme 6.10) because phenol end groups of PC are not able to react. However, the weight of copolymers increases with mixing time and the copolymer molar composition and PA6 block sizes change as the reaction time increases indicating that the exchange reaction is in progress. This fact can be explained by the hydrolysis of the PA6 amide groups during the melt mixing producing amino end groups and carboxyl chain ends (Scheme 6.11). Therefore, in the case of the blend constituting high MM PA6, the production of amino terminals by hydrolysis is the rate-determining step in the overall kinetics. The carboxyl ends of PA6 do not react, as in the case of PBT/PA6 blends.

The compatibilizing effect of the PC-PA6 block copolymers in PC-PA6 blends was also studied by Montaudo et al. [62]. Diblocks PA6-PC copolymers were prepared by aminolysis reactions of amino-terminated PA6 with PC at 130 °C in anhydrous dimethylsulfoxide (DMSO). The morphology of PA6/PC blend (75/25 w/w) melt mixed at 240 °C for 5 min in the presence of 2% by weight of above block copolymers, was studied by scanning electron microscopy (SEM) (Figure 6.10), which showed the presence of only one phase (Figure 6.10b), in contrast to the same blend without compatibilizing copolymer, which is clearly biphasic (Figure 6.10a).

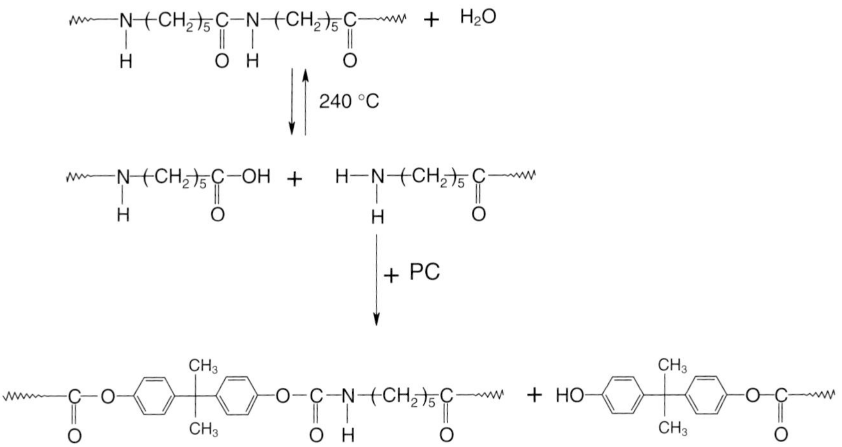

Scheme 6.11 Hydrolyis reaction of high molar mass PA-6 and consecutive reaction of amino end groups with the inner carbonate groups of PC [120].

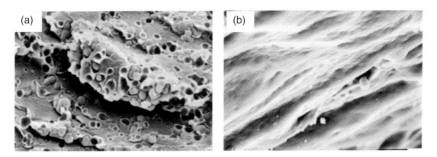

Figure 6.10 SEM images of a PC/PA6 blend (25/75 w/w) melt mixed at 240 °C for 5 min (a) without and (b) 2wt% of PC-PA6 di-block copolymer. (Montaudo et al. 1996 [62]. Reproduced with permission of Wiley.)

6.6 Interchange Reaction in Polyamide/Polyamide Blends

Some attempts have been made to define the kinetics and the extent of interchange reactions that occur during the melt mixing of PA/PA blends [3, 52–55, 130–138]. It has been acknowledged that the direct amide–amide interchange (often referred as *amidolysis*) does not occur at temperatures below 320 °C that is usually used to process these blends [52, 53, 131]. At these processing temperatures, only aminolysis or acidolysis outer–inner interchange reactions can occur. Some studies report that amide exchange reactions are acid-catalyzed and are a complex function of the water content [3, 131]. A similar occurrence has also been claimed in literature for some other catalysts, such as phosphite compounds [54] and H_3PO_4 [133]. Recently, transamidation between PA10,12 and

PA6,12 was studied by combining DSC, rheometry, NMR, and variable temperature FT-IR methods, but the data do not give clear evidence of the reactions that occur [136]. Ersel *et al.* [54, 55] have characterized the polyamide 4,6/polyamide 6I (polyhexamethylene isophthalamide) melt mixed at 315 °C at different time by ^{13}C NMR analysis and by applying the Devaux approach [9–11]. On the basis of their studies the authors have affirmed that reaction mechanism consists of amide hydrolysis followed by recombination (referred as *direct transamidation*). Several investigations by DSC, NMR, and FT-IR techniques report that transamidation reactions take place during the melt mixing of the polyamide blends. However, often, the authors omit the role of end groups and do not report a detailed study on the kinetics and mechanism of the reactions that occur. Yao *et al.* [139] have used the Monte Carlo method to simulate the time evolution of MMD and copolymer composition during the interchain process between polyamides with AA (i.e., PA6) and BC (i.e., PA$x,y$ such as PA6,6, PA4,6, PA6,10) structure. The authors concluded that the interchange reaction to produce copolymers with MMD depends on the difference of DP of the initial polyamides.

The role of end groups in the melt mixing of equimolar blends of PA6, PA6,6, PA6,10, and PA4,6 with PA samples bearing specific amino or carboxyl end groups (i.e., PA6-NH$_2$, PA6,6-NH$_2$, PA6-COOH, PA6,6-COOH) has been investigated by the combination of NMR, DSC, and MALDI-TOF/MS experiments [52, 53].

The reaction of amino end groups of polyamides involves condensation reactions with carboxyl end chains. Moreover, the reaction between two amino end groups, with consequent formation of secondary amino groups along the chains and elimination of NH$_3$, has been observed during the melt mixing of PA66 (Scheme 6.12a) [115, 130]. The secondary amino groups can further react with carboxylic acid end groups leading to the formation of branching (Scheme 6.12b).

The reaction of carboxyl end groups produces a random cleavage of the amides functional groups (acidolysis) as illustrated in Scheme 6.13. The role of carboxyl end groups has been investigated by melt mixing at 310 °C as a function of time and for four blends based on PA6,6-COOH/PA6,10, PA6-COOH/PA4,6, PA6-COOH/PA6,10, and PA6-COOH/PA6,6. The carbonyl resonance signals of the formed copolyamides are sensitive to the sequence length distributions when polyamides blends are dissolved in sulfuric acid or fluorosulfonic acid, allowing the determination of copolyamides microstructure [52–55, 130]. Figure 6.11 reports some sections of the ^{13}C NNR spectra of the equimolar PA6-COOH/PA4,6 blends melt mixed at 310 °C for 0 min (a), 30 min (b), 60 min (c), and 120 min (d) [53]. Two new carbonyl peaks (b and c), in the 178–180 ppm range, are clearly observed in the carbon spectra of melt mixed blends, with respect to that of the initial blend (peaks a and d). A similar behavior is observed in the region between 43 and 44 ppm where resonate the methylene carbons linked with the nitrogen of the amide groups. The relative intensity of signals ϵ and ω (between 33 and 34 ppm) due to the methylene linked with the carbonyl of the amide groups, indicates that molar compositions of the blends do not change as the reaction time increases. The same behavior was also observed in the ^{13}C NMR spectra of both the PA6-COOH/PA6,10 and PA6-COOH/PA6,6

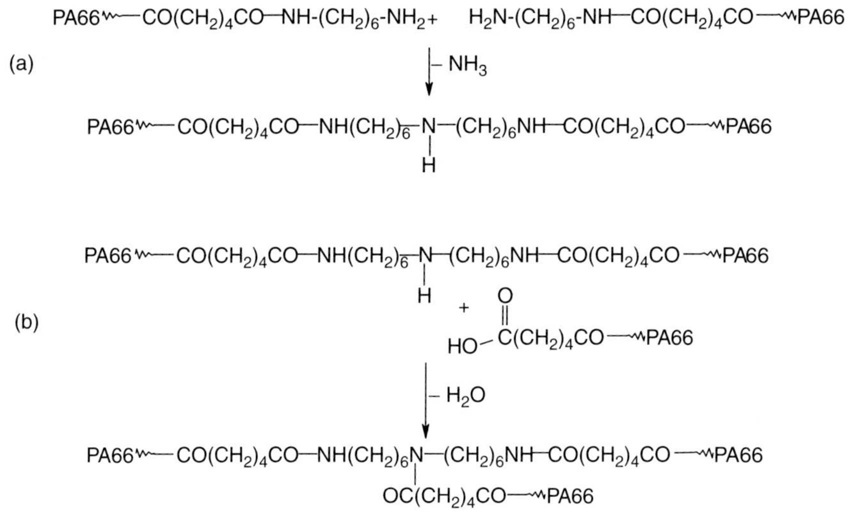

6.6 Interchange Reaction in Polyamide/Polyamide Blends

(a)

$$PA66\text{ww}-CO(CH_2)_4CO-NH-(CH_2)_6-NH_2 + H_2N-(CH_2)_6-NH-CO(CH_2)_4CO-\text{ww}PA66$$

$$\downarrow -NH_3$$

$$PA66\text{ww}-CO(CH_2)_4CO-NH(CH_2)_{\overline{6}}\underset{H}{N}-(CH_2)_6NH-CO(CH_2)_4CO-\text{ww}PA66$$

(b)

$$PA66\text{ww}-CO(CH_2)_4CO-NH(CH_2)_{\overline{6}}\underset{H}{N}-(CH_2)_6NH-CO(CH_2)_4CO-\text{ww}PA66$$

$$+$$

$$HO-\overset{O}{\underset{\|}{C}}(CH_2)_4CO-\text{ww}PA66$$

$$\downarrow -H_2O$$

$$PA66\text{ww}-CO(CH_2)_4CO-NH(CH_2)_6\underset{|}{N}-(CH_2)_6NH-CO(CH_2)_4CO-\text{ww}PA66$$
$$OC(CH_2)_4CO-\text{ww}PA66$$

Scheme 6.12 Formation of (a) secondary amino groups and (b) branching along the PA66 chains during the processing [115].

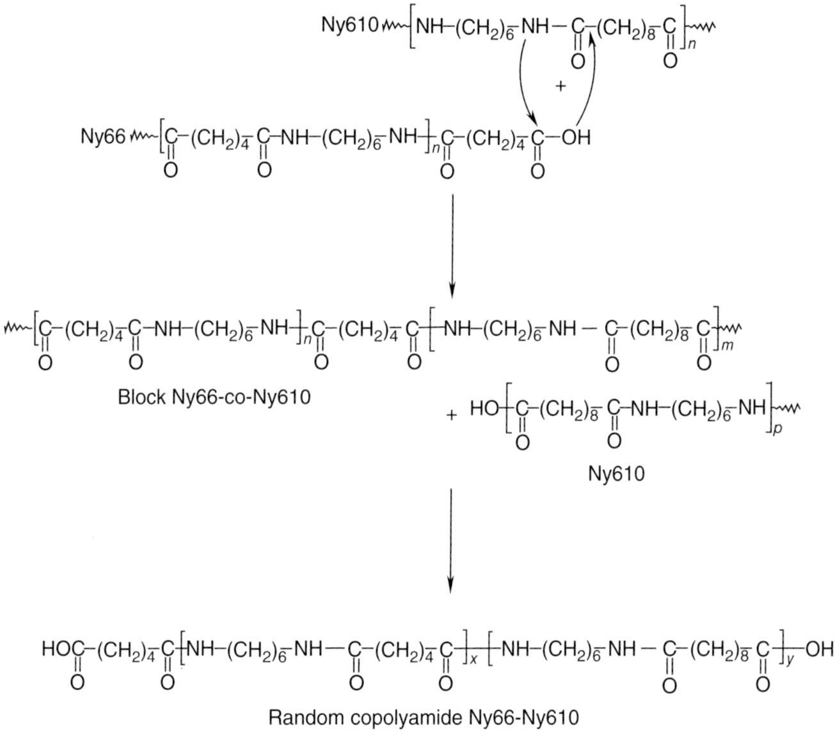

$$Ny610\text{ww}-[NH-(CH_2)_{\overline{6}}-NH-\underset{O}{\overset{\|}{C}}-(CH_2)_{\overline{8}}-\underset{O}{\overset{\|}{C}}]_n-\text{ww}$$

$$+$$

$$Ny66\text{ww}-[\underset{O}{\overset{\|}{C}}-(CH_2)_{\overline{4}}-\underset{O}{\overset{\|}{C}}-NH-(CH_2)_{\overline{6}}-NH]_n-\underset{O}{\overset{\|}{C}}-(CH_2)_{\overline{4}}-\underset{O}{\overset{\|}{C}}-OH$$

$$\downarrow$$

$$\text{ww}-[\underset{O}{\overset{\|}{C}}-(CH_2)_{\overline{4}}-\underset{O}{\overset{\|}{C}}-NH-(CH_2)_{\overline{6}}-NH]_n-\underset{O}{\overset{\|}{C}}-(CH_2)_{\overline{4}}-\underset{O}{\overset{\|}{C}}-[NH-(CH_2)_{\overline{6}}-NH-\underset{O}{\overset{\|}{C}}-(CH_2)_{\overline{8}}-\underset{O}{\overset{\|}{C}}]_m-\text{ww}$$

Block Ny66-co-Ny610

$$+ HO-[\underset{O}{\overset{\|}{C}}-(CH_2)_{\overline{8}}-\underset{O}{\overset{\|}{C}}-NH-(CH_2)_{\overline{6}}-NH]_p-\text{ww}$$

Ny610

$$\downarrow$$

$$HO\underset{O}{\overset{\|}{C}}-(CH_2)_{\overline{4}}-\underset{O}{\overset{\|}{C}}-[NH-(CH_2)_{\overline{6}}-NH-\underset{O}{\overset{\|}{C}}-(CH_2)_{\overline{4}}-\underset{O}{\overset{\|}{C}}]_x-[NH-(CH_2)_{\overline{6}}-NH-\underset{O}{\overset{\|}{C}}-(CH_2)_{\overline{8}}-\underset{O}{\overset{\|}{C}}]_y-OH$$

Random copolyamide Ny66-Ny610

Scheme 6.13 Acidolysis interchange reactions that occur during the melt mixing of the aliphatic polyamide blends such as PA6,6/PA6,10 [52].

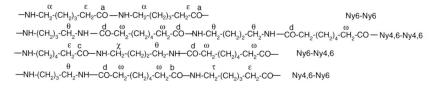

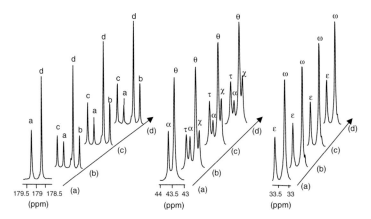

Figure 6.11 Sections of the ^{13}C NNR spectra of the PA6-COOH/PA4,6 melt mixed blends at 310 °C for (a) 0 min, (b) 30 min, (c) 60 min, and (d) 120 min. (Samperi et al. 2004 [53]. Reproduced with permission of American Chemical Society.)

melt mixed products [53, 130]. The molar composition, the dyads sequence distributions, the average sequence length, and the DR of the copolyamides formed, have been calculated from the intensities of carbonyl peaks in the ^{13}C NMR spectra, using an appropriate statistical model taking into account that diacid moiety in PA6,10, in the PA4,6, and in the PA6,6 units, have two carbonyls, whereas only one is present in the PA6 moieties (Eqs (6.4)–(6.11)).

$$2(I_{AA} + I_{AB}) = C_A/C_B \cdot (I_{BB} + I_{BA}) \tag{6.4}$$

$$f_{AA} = C_A \cdot I_{AA}/I_{AA} + I_{AB} \tag{6.5}$$

$$f_{AB} = C_A \cdot I_{AB}/I_{AA} + I_{AB} \tag{6.6}$$

$$f_{BA} = C_B \cdot 2I_{BA}/I_{BB} + I_{BA} \tag{6.7}$$

$$f_{BB} = C_B \cdot I_{BB} - I_{BA}/I_{BB} + I_{BA} \tag{6.8}$$

$$L_{A-A} = 1/f_{AB} \cdot C_A \tag{6.9}$$

$$L_{B-B} = 1/f_{BA} \cdot C_B \tag{6.10}$$

$$DR = f_{AB}/C_A + f_{BB}/C_B \tag{6.11}$$

where C_A and C_B indicate the molar concentration of component A (PA6) and of component B (PA6,10, PA4,6, or PA6,6), respectively, in the copolyamides; f_{AA}, f_{AB}, f_{BA}, f_{BB}, stand for the molar fractions of the AA, AB, BA, and BB dyads, respectively; I_{AA}, I_{AB}, I_{BA}, and I_{BB} indicate the intensity of their corresponding carbonyl peaks in the ^{13}C NMR spectra; DR indicates the DR, L_{A-A} and L_{B-B} stand for the average sequence length.

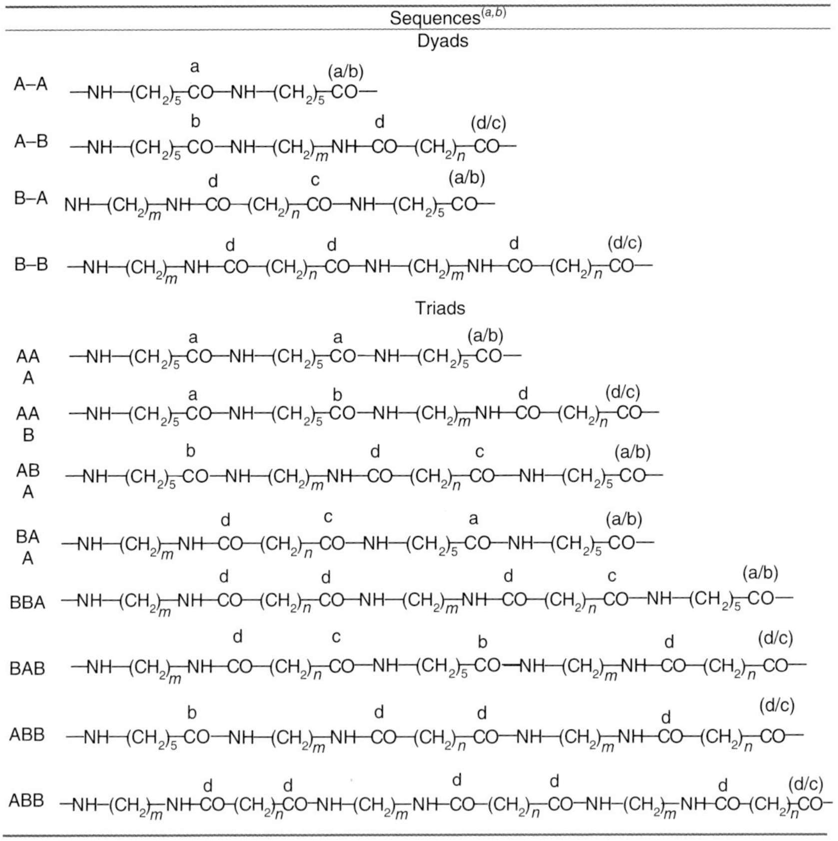

6.6 Interchange Reaction in Polyamide/Polyamide Blends | 163

	Sequences[a,b]
	Dyads

A–A	$\quad$a$\qquad\qquad$(a/b) —NH—(CH$_2$)$_5$—CO—NH—(CH$_2$)$_5$—CO—	
A–B	$\quad$b$\qquad\qquad\quad$d$\qquad\qquad$(d/c) —NH—(CH$_2$)$_5$—CO—NH—(CH$_2$)$_m$—NH—CO—(CH$_2$)$_n$—CO—	
B–A	$\quad$d$\qquad\qquad\quad$c$\qquad\qquad$(a/b) NH—(CH$_2$)$_m$—NH—CO—(CH$_2$)$_n$—CO—NH—(CH$_2$)$_5$—CO—	
B–B	$\quad$d$\qquad\qquad\quad$d$\qquad\qquad\quad$d$\qquad\qquad$(d/c) —NH—(CH$_2$)$_m$—NH—CO—(CH$_2$)$_n$—CO—NH—(CH$_2$)$_m$—NH—CO—(CH$_2$)$_n$—CO—	

	Triads

AA A	$\quad$a$\qquad\qquad\quad$a$\qquad\qquad$(a/b) —NH—(CH$_2$)$_5$—CO—NH—(CH$_2$)$_5$—CO—NH—(CH$_2$)$_5$—CO—
AA B	$\quad$a$\qquad\qquad\quad$b$\qquad\qquad\quad$d$\qquad\qquad$(d/c) —NH—(CH$_2$)$_5$—CO—NH—(CH$_2$)$_5$—CO—NH—(CH$_2$)$_m$—NH—CO—(CH$_2$)$_n$—CO—
AB A	$\quad$b$\qquad\qquad\quad$d$\qquad\qquad\quad$c$\qquad\qquad$(a/b) —NH—(CH$_2$)$_5$—CO—NH—(CH$_2$)$_m$—NH—CO—(CH$_2$)$_n$—CO—NH—(CH$_2$)$_5$—CO—
BA A	$\quad$d$\qquad\qquad\quad$c$\qquad\qquad\quad$a$\qquad\qquad$(a/b) —NH—(CH$_2$)$_m$—NH—CO—(CH$_2$)$_n$—CO—NH—(CH$_2$)$_5$—CO—NH—(CH$_2$)$_5$—CO—
BBA	$\quad$d$\qquad\qquad$d$\qquad\qquad\quad$d$\qquad\qquad\quad$c$\qquad\qquad$(a/b) —NH—(CH$_2$)$_m$—NH—CO—(CH$_2$)$_n$—CO—NH—(CH$_2$)$_m$—NH—CO—(CH$_2$)$_n$—CO—NH—(CH$_2$)$_5$—CO—
BAB	$\quad$d$\qquad\qquad$c$\qquad\qquad\quad$b$\qquad\qquad\quad$d$\qquad\qquad$(d/c) —NH—(CH$_2$)$_m$—NH—CO—(CH$_2$)$_n$—CO—NH—(CH$_2$)$_5$—CO—NH—(CH$_2$)$_m$—NH—CO—(CH$_2$)$_n$—CO—
ABB	$\quad$b$\qquad\qquad\quad$d$\qquad\qquad\quad$d$\qquad\qquad\quad$d$\qquad\qquad$(d/c) —NH—(CH$_2$)$_5$—CO—NH—(CH$_2$)$_m$—NH—CO—(CH$_2$)$_n$—CO—NH—(CH$_2$)$_m$—NH—CO—(CH$_2$)$_n$—CO—
ABB	$\quad$d$\qquad\qquad$d$\qquad\qquad\quad$d$\qquad\qquad\quad$d$\qquad\qquad\quad$d$\qquad\qquad$(d/c) —NH—(CH$_2$)$_m$—NH—CO—(CH$_2$)$_n$—CO—NH—(CH$_2$)$_m$—NH—CO—(CH$_2$)$_n$—CO—NH—(CH$_2$)$_m$—NH—CO—(CH$_2$)$_n$—CO—

(a) The symbol A correspond to the Ny6 units, where as the symbol B correspond to the Ny6,10, Ny4,6, or Ny6,6 units, for Ny6-Ny6,10, Ny6-Ny4,6, and Ny-Ny6,6 copolymers, respectively.
(b) Where: $m = 6$ and $n = 8$ for the Ny6,10 units; $m = 4$ and $n = 4$ for Ny4,6 units; $M = 6$ and $n = 4$ for Ny6,6 moiety.

Figure 6.12 Types of carbonyl resonance signals due to the dyads and triads sequences in the PA6/PA6,10, PA6/PA6,6, and PA6/PA4,6 copolyamides. (Samperi et al. 2004 [53]. Reproduced with permission of American Chemical Society.)

The results showed that all copolyamides formed have a molar composition very close to that of the initial physical blends, and that the segmented copolyamides formed at relatively low reaction time at 310 °C evolve quickly toward a random distribution of the dyad sequences, as the reaction time increases. The NMR data agree with those obtained by MALDI-TOF MS analysis of PA6/PA6,10 and PA6/PA4,6 reacted systems [53].

To validate the microstructural model reported earlier, a theoretical model has been generated taking into account the carbonyl types belonging to all possible dyads and triads sequences in the copolyamides (as summarized in Figure 6.12) according to chain statistics [2, 4–7, 52, 53, 64, 77, 78] for an equimolar AB random copolymer. Figure 6.13a–d shows the comparison between the experimental and calculated ^{13}C NMR spectral intensities of the carbonyl peaks

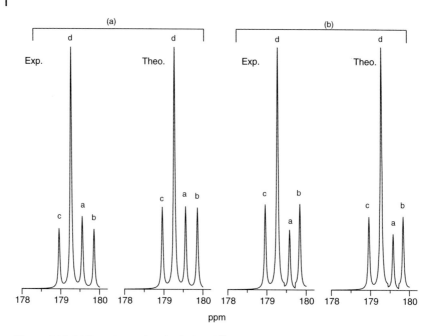

Figure 6.13 (a) Experimental and calculated ^{13}C NMR spectra of PA6-COOH/PA6,10 blends reacted at 310 °C (a) for 30 min and (b) for 60 min. (Samperi *et al.* 2004 [53]. Reproduced with permission of American Chemical Society.)

corresponding to the PA6/PA6,10 copolyamides obtained after 30 and 60 min heating at 310 °C. The experimental intensities of the carbonyl peaks of the blend melt mixed at 30 min (Figure 6.13a) agree with those calculated for the random copolyamide (Figure 6.13b), whereas the random distribution of sequences appears a bit altered for the copolyamide formed at 60 min (Figure 6.13c).

The MALDI-TOF MS analysis of these blends showed also the formation of copolymer chains terminated with cyclopentanone moiety generated by thermal degradation reaction of the adipic acid terminal units in accordance to Scheme 6.9.

The DSC analysis of PA6/PA4,6 and PA6/PA6,10 melt mixed products indicated that semi-crystalline copolyamides with a degree of crystallinity ranging from 22% up to 30% were formed with a single melting temperature (T_m) [53].

The copolyamides formed by reactive blending of PA6,6/PA6,10 blends cannot be characterized by ^{13}C NMR because the dyad sequences generated by interchain reactions do not affect the chemical surrounding of carbonyls and methylenes, with respect to the homopolymers. In this case, the microstructural characterization was obtained by MALDI-TOF MS analysis applying the statistical model MACO4 to mass spectra [4, 52, 64]. Figure 6.14 displays the mass spectra in the mass range of 850–1850 Da of an equimolar PA6,6-COOH/PA6,10 blend melt mixed at 290 °C for (a) 10 min, (b) 15 min, and (c) 30 min. A copolyamide containing 64 mol% of PA6,6 units is formed after 10 min heating, which decreases at 59 mol% after 15 min of reactions. After 30 and 60 min processing time, copolyamides show a composition close

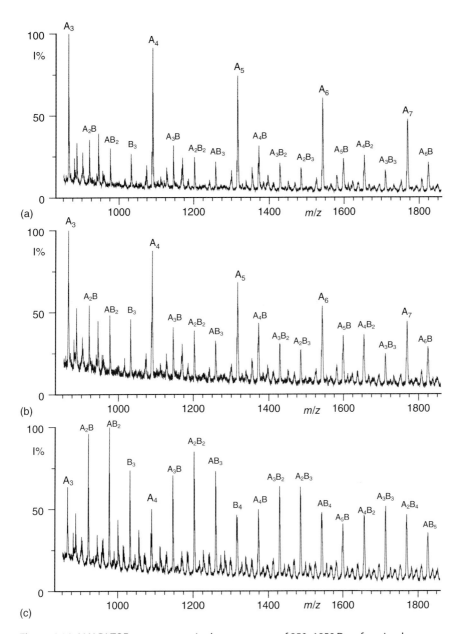

Figure 6.14 MALDI-TOF mass spectra, in the mass range of 850–1850 Da, of equimolar PA66-COOH/PA6,10 blend melt mixed at 290 °C for (a) 10 min, (b) 15 min, and (c) 30 min. Labels A and B indicate the PA6,6 and PA6,10 repeat units, respectively. (Puglisi et al. 2003 [52]. Reproduced with permission of American Chemical Society.)

to that of the initial blend composition, that is: 50/50 (mol/mol), as expected for an interchange process carried out to completion [2]. The DE and the DR values are very high even at lower heating time (10 min) indicating a high rate of interchange and the formation of a random copolyamide, due to the presence of a high concentration of reactive carboxyl chain ends. Also, in this case, copolyamides obtained at 30 min mixing (Figure 6.14c) show chains terminated with cyclopentanone generated by thermal degradation of adipic acid terminal units (Scheme 6.9).

The interchange reaction of a mixture of high MM PA6,6 and PA6,10, both containing a very low amount of carboxylic chain ends, produced segmented copolyamides even at high heating time (150 min) [52].

The PA66/PA6,10 melt mixed products were also characterized by DSC and WAXS (wide-angle X-ray spectroscopy) techniques showing the formation of semi-crystalline copolyamides. The initial physical blend showed two T_g transitions that get closer with heating time and finally gave a single T_g transition at 59 °C after 30 min at 290 °C, indicating that a random PA6,6/PA6,10 copolyamide was formed. This value is very close to that calculated (58 °C) by the empirical Fox equation: $1/T_g = W_1/T_{g1} + W_2/T_{g2}$, where W_1 (45%), W_2 (55%), T_{g1} (48 °C), and T_{g2} (70 °C) are the weight fractions and the glass transitions of PA6,6-COOH and PA6,10 initial components, respectively.

6.7 Conclusions

The application of an analytical approach, which is based on the determination of the microstructure (molar composition, sequence distribution, average sequence length, DR) of the copolymers formed in the melt mixing of polymer blends has allowed to obtain detailed information on the mechanisms of reactions that occur during the melt mixing of condensation polymer blends. To obtain reliable information on the mechanisms and the kinetics of the interchange reactions that occur in the reactive blending in the molten state and in solution, it is important to know: (i) the nature of the end groups that are either initially present or are formed *in situ* during the processing, (ii) the presence of eventual residue of catalyst used in the polymerization, (iii) control the atmosphere (inert or air), and (iv) the processing parameters (T, t, mixing rate).

On the basis of the data reported here it can be deduced that the interchange reactions can be controlled by monitoring the composition of melt mixed products as a function of the processing parameters (time, temperature, catalyst, kinds of end groups, reactant). Therefore, polymeric materials (i.e., copolymers, mixture of copolymers and homopolymers) with reproducible architecture and composition, and consequently with tailored and well-known properties can be produced by melt mixing blends of polymer containing specific active chain ends and/or catalysts.

Remarkably, at temperatures below 300 °C the interchange reactions between functional groups located inside the polymer chains only occurs in the presence of specific catalysts. At higher temperature the direct exchange is in principle possible; however, thermal degradation reactions or hydrolysis processes may produce reactive terminal groups that are able to produce extensive exchange reactions.

It was observed that carboxyl end groups are the most reactive moieties in the polyester/polyamide and PA/PA blends. Carboxyl and hydroxyl end groups of polyester give interchange reactions in PE/PC blends in the molten state. However, in this system the ester–carbonate interchange reactions in the presence of catalysts such as Ti(OBu)$_4$ dominate. The microstructural analysis has also permitted the characterization of the consecutive degradative reactions that occur during the melt mixing of PC/PET and PC/PEN blends, which lead to the elimination of ETC and CO_2 with consequent changing of the architecture and the composition of the copolymers formed. Moreover, in the case of polyester/polyamide and PA/PA melt blending, the MALDI-TOF MS analysis has revealed the formation the cyclopentanone end chains by thermal degradation of the copolyamides and copolyesteramides terminated with adipic acid.

The role of some reactants that promote the formation of *in situ* of reactive polymer end chains was also established in some blends like PC/PET in the presence of TsOH.

References

1 Fakirov, S. (ed.) (1999) *Transreactions in Condensation Polymers*, Wiley-VCH, Weinheim.
2 Montaudo, G., Puglisi, C., and Samperi, F. (1999) Copolymer composition: a Key to the mechanisms of exchange in reactive polymer blending, in *Transreactions in Condensation Polymers* (ed. S. Fakirov), Wiley-VCH, Weinheim, pp. 159–193.
3 Kotliar, A.M. (1981) Interchange reactions involving condensation polymers. *J. Polym. Sci. Macromol. Rev.*, **16**, 367–395.
4 Montaudo, G., Montaudo, M.S., Scamporrino, E., and Vitalini, D. (1992) Mechanism of exchange in polyesters. Composition and microstructure of copolymers formed in the melt mixing process of poly(ethylene terephthalate) and poly(ethylene adipate). *Macromolecules*, **25**, 5099–5107.
5 Montaudo, G., Puglisi, C., and Samperi, F. (1998) Mechanism of exchange in PBT/PC and PET/PC blends. Composition of the copolymer formed in the melt mixing process. *Macromolecules*, **31**, 650–661.
6 Samperi, F., Battiato, S., Recca, G., Puglisi, C., and Mendichi, R. (2015) Reactive melt mixing of PC/PEN blend. Structural characterization of reaction products. *Polymer*, **74**, 108–123.
7 Pilati, F., Fiorini, M., and Berti, C. (1999) Effect of catalysts in the reactive blending of bisphenol A polycarbonate with pol(alkylene terephthalate)s, in *Transreactions in Condensation Polymers* (ed. S. Fakirov), Wiley-VCH, Weinheim, pp. 79–123.
8 Porter, R.S. and Wang, L.H. (1992) Compatibility and transesterification in binary polymer blends. *Polymer*, **33**, 2019–2029.

9 Jacques, B., Devaux, J., Legras, R., and Nield, R. (1996) NMR study of ester-interchange reactions during the melt mixing of poly(ethylene terephthalate) and poly(butylene terephthalate) blends. *J. Polym. Sci. A Polym. Chem.*, **34**, 1189–1194.

10 Devaux, J., Godard, P., and Mercier, J.P. (1982) Bisphenol-A polycarbonate–poly(butylene terephthalate) transesterification. I. Theoretical study of the structure and of the degree of randomness in four-component copolycondensates. *J. Polym. Sci. Polym. Phys. Ed.*, **20**, 1875–1880.

11 Devaux, J. (1999) Model studies of transreactions in condensation polymers, in *Transreactions in Condensation Polymers* (ed. S. Fakirov), Wiley-VCH, Weinheim, pp. 125–158.

12 Litmanovich, A.D., Platé, N.A., and Kudryavtsev, Y.V. (2002) Reactions in polymer blends: interchain effects and theoretical problems. *Prog. Polym. Sci.*, **27**, 915–970.

13 Plate, N.A., Litmanovich, A.D., and Kudryavtsev, Y.V. (2004) Reactions in polymer blends: experiment and theory – a review. *Polym. Sci. Ser. A*, **46** (11), 1108–1140.

14 Kudryavtsev, Y.V., Chertovich, A.V., Guseva, D.V., and Litmanovich, A.D. (2007) Early stages of interchange reactions in polymer blends. *Macromol. Symp.*, **254**, 188–195.

15 Stewart, M.E., Cox, A.J., and Naylor, D.M. (1993) Reactive processing of poly(ethylene 2,6-naphthalene dicarboxylate) poly(ethylene terephthalate) blends. *Polymer*, **34**, 4060–4067.

16 Kenwright, A.M., Peace, S.K., Richards, R.W., Bunn, A., and MacDonald, W.A. (1999) Transesterification in poly(ethylene terephthalate) and poly(ethylene naphthalene 2,6-dicarboxylate) blends; the influence of hydroxyl end groups. *Polymer*, **40**, 5851–5856.

17 Collins, S., Kenwright, A.M., Pawson, C., Peace, S.K., Richards, R.W., MacDonald, W.A., and Mills, P. (2000) Transesterification in mixtures of poly(ethylene terephthalate) and poly(ethylene naphthalene-2,6-dicarboxylate): an NMR study of kinetics and end group effects. *Macromolecules*, **33**, 2974–2980.

18 Collins, S., Peace, S.K., Richards, R.W., MacDonald, W.A., Mills, P., and King, S.M. (2001) Transesterification in poly(ethylene terephthalate)–poly(ethylene naphthalene-2,6-dicarboxylate mixtures: a comparison of small-angle neutron scattering with NMR. *Polymer*, **42**, 7695–7700.

19 Blanco, I., Cicala, G., Restuccia, C.L., Latteri, A., Battiato, S., Scamporrino, A., and Samperi, F. (2012) Role of 2-hydroxyethyl end group on the thermal degradation of poly(ethylene terephthalate) and reactive melt mixing of poly(ethylene terephthalate)/poly(ethylene naphthalate) blends. *Polym. Eng. Sci.*, **52**, 2498–2505.

20 Tharmapuram, S.R. and Jabarin, S.A. (2003) Processing characteristics of PET/PEN blends, part 1: extrusion and transesterification reaction kinetics. *Adv. Polym. Tech.*, **22**, 137–146.

21 Tharmapuram, S.R. and Jabarin, S.A. (2003) Processing characteristics of PET/PEN blends, part 2: rheology and degradation kinetics. *Adv. Polym. Tech.*, **22**, 147–154.

22 Tharmapuram, S.R. and Jabarin, S.A. (2003) Processing characteristics of PET/PEN blends, part 3: injection molding and free blow studies. *Adv. Polym. Tech.*, **22**, 155–167.

23 Shi, Y. and Jabarin, S.A. (2001) Transesterification reaction kinetics of poly(ethylene terephthalate)/poly(ethylene 2,6-naphthalate) blends. *J. Appl. Polym. Sci.*, **80**, 2422–2436.

24 Patcheak, T.D. and Jabarin, S.A. (2001) Structure and morphology of PET/PEN blends. *Polymer*, **42** (21), 8975–8985.

25 Shi, Y. and Jabarin, S.A. (2001) Glass-transition and melting behavior of poly(ethylene terephthalate)/poly(ethylene 2,6-naphthalate) blends. *J. App. Polym. Sci.*, **81**, 11–22.

26 Shi, Y. and Jabarin, S.A. (2001) Crystallization kinetics of poly(ethylene terephthalate)/poly(ethylene 2,6-naphthalate) blends. *J. App. Polym. Sci.*, **81**, 23–37.

27 Medina, R.M., Likhatchev, D., Alexandrova, L., Sánchez-Solís, A., and Manero, O. (2004) Mechanism and kinetics of transesterification in poly(ethylene terephthalate) and poly(ethylene 2,6-naphthalene dicarboxylate) polymer blends. *Polymer*, **45**, 8517–8522.

28 Jun, H.W., Chae, S.H., Park, S.S., Myung, H.S., and Im, S.S. (1999) Relationship between sequence distribution and transesterification of PEN/PET random/block copolyesters. *Polymer*, **40**, 1473–1480.

29 Okamoto, M. and Kotaka, T. (1997) Phase separation and homogenization in poly(ethylene naphthalene-2,6-dicarboxylate)/poly(ethylene terephthalate) blends. *Polymer*, **38**, 1357–1361.

30 Gunes, K., Isayev, A.I., Li, X., and Wesdemiotis, C. (2010) Fast *in situ* copolymerization of PET/PEN blends by ultrasonically-aided extrusion. *Polymer*, **51**, 1071–1081.

31 Krentsel, L.B., Makarova, V.V., Kudryavtsev, Y.V., Govorun, E.N., Litmanovich, A.D., Markova, G.D., Vasnev, V.A., and Kulichikhin, V.G. (2009) Interchain exchange and interdiffusion in blends of poly(ethylene terephthalate) and poly(ethylene naphthalate). *Polym. Sci. Ser. A*, **51** (11–12), 1241–1248.

32 Tao, W., Wei, W., Yu, C., Ren, W., and Qiaoling, L. (2013) Influence of the mixing time on the phase structure and glass-transition behavior of poly(ethylene terephthalate)/poly(ethylene-2,6-naphthalate) blends. *J. Appl. Polym. Sci.*, **130**, 673–679.

33 Lee, S.C., Yoon, K.H., Park, I.H., Kim, H.C., and Son, T.W. (1997) Phase behaviour and transesterification in poly(ethylene 2,6-naphthalate) and poly(ethylene terephthalate) blends. *Polymer*, **38**, 4831–4835.

34 Ida, S.-I., Yamamoto, H., and Ito, M. (2009) The Changes of Transesterification Level in PET/PEN Blends with the Addition of PET–PEN Copolymer. *J. Appl. Polym. Sci.*, **112**, 2716–2723.

35 Yang, H., Ma, J., Li, W., and Liang, B. (2002) Reactive blending of poly(ethylene terephthalate) (PET)/poly(ethylene 2,6-naphthalate) (PEN). 1: effect of mixing conditions on chain structure. *Polym. Eng. Sci.*, **42**, 1629–1641.

36 Samperi, F., Puglisi, C., and Battiato, S. (2016) Unpublished results.
37 Backsona, S.C.E., Richards, R.W., and King, S.M. (1999) Small angle neutron scattering investigation of transesterification in poly(ethylene terephthalate)–poly(butylene terephthalate) mixtures. *Polymer*, **40**, 4205–4211.
38 Chang, Y.K., Youk, J.H., Jo, W.H., and Lee, S.C. (1999) The effect of transreactions on phase behavior in poly(ethylene 2,6-naphthalate) and poly(ethylene isophthalate) blends. *J. App. Polym. Sci.*, **73**, 1851–1858.
39 Woo, E.M., Hou, S.-S., Huang, D.-H., and Lee, L.-T. (2005) Thermal and NMR characterization on trans-esterification-induced phase changes in blends of poly(ethylene-2,6-naphthalate) with poly(pentylene terephthalate). *Polymer*, **46**, 7425–7435.
40 Zhou, C., Ma, J., Pan, L., and Liang, B. (2002) Transesterification kinetics in the reactive blends of liquid crystalline copolyesters and poly(ethylene terephthalate). *E. Polym. J.*, **38**, 1049–1053.
41 Aoki, Y., Li, L., Amari, T., Nishimura, K., and Arashiro, Y. (1999) Dynamic mechanical properties of poly(ethylene terephthalate)/poly(ethylene 2,6-naphthalate) blends. *Macromolecules*, **32**, 1923–1929.
42 Castellano, M., Marsano, E., Turturro, A., and Canetti, M. (2011) Reactive blending of aromatic polyesters: thermal and X-ray analysis of melt-blended poly(ethylene terephthalate)/poly(trimethylene terephthalate). *J. App. Polym. Sci.*, **122**, 698–705.
43 Caligiuri, L., Stagnaro, P., Valenti, B., and Canalini, G. (2009) Reactive blending of poly(ethylene 2,6-naphthalate) and Vectra A. *E. Polym. J.*, **45**, 217–225.
44 Kollodge, J.S. and Porter, R.S. (1995) Phase behavior and transreaction studies of model polyester/bisphenol-A polycarbonate blends. 3. Midchain and alcoholysis reactions. *Macromolecules*, **28**, 4106–4115.
45 Fiorini, M., Pilati, F., Berti, C., Toselli, M., and Ignatov, V. (1997) Reactive blending of poly(ethylene terephthalate) and bisphenol-A polycarbonate: effect of various catalysts and mixing time on the extent of exchange reactions. *Polymer*, **38**, 413–419.
46 Fiorini, M., Berti, C., Ignatov, V., Toselli, M., and Pilati, F. (1995) Poly(ethylene terephthalate) bisphenol A polycarbonate reactive blending. *J. App. Polym. Sci.*, **55** (8), 1157–1163.
47 Pereira, P.S.C., Mendes, L.C., and Ramos, V.D. (2009) Rheological study bringing new insights into PET/PC reactive blends. *Macromol. Symp.*, **290**, 121–131.
48 Guessoum, M., Haddaoui, N., and Fenouillot-Rimlinger, F. (2008) Effects of reactive extrusion and interchange catalyst on the thermal properties of polycarbonate/poly (ethylene terephthalate) system. *Int. J. Polym. Mat.*, **57** (7), 657–674.
49 Samperi, F., Montaudo, M.S., Puglisi, C., Alicata, R., and Montaudo, G. (2003) Essential role of chain ends in the Ny6/PBT exchange. A combined NMR and MALDI approach. *Macromolecules*, **36**, 7143–7154.
50 Samperi, F., Puglisi, C., Alicata, R., and Montaudo, G. (2003) Essential role of chain ends in the nylon-6/poly(ethylene terephthalate) exchange. *J. Polym. Sci. A Polym. Chem.*, **41**, 2778–2793.

51 Samperi, F., Montaudo, M.S., Battiato, S., Carbone, D., and Puglisi, C. (2010) Characterization of copolyesteramides from reactive blending of PET and MXD6 in the molten state. *J. Polym. Sci. A Polym. Chem.*, **48** (22), 5135–5155.

52 Puglisi, C., Samperi, F., Di Giorgi, S., and Montaudo, G. (2003) Exchange reactions occurring through active chain ends. MALDI-TOF characterization of copolymers from nylon 6,6 and nylon 6,10. *Macromolecules*, **36**, 1098–1107.

53 Samperi, F., Montaudo, M.S., Puglisi, C., Di Giorgi, S., and Montaudo, G. (2004) Structural characterization of copolyamides synthesized via the facile blending of polyamides. *Macromolecules*, **37**, 6449–6459.

54 Eersel, K.L.L., Aerdts, A.M., and Groeninckx, G. (1999) Reactive melt processing of aliphatic/aromatic polyamide blends: effect on molecular structure, semicrystalline morphology, and thermal properties, in *Transreactions in Condensation Polymers* (ed. S. Fakirov), Wiley-VCH, Weinheim, pp. 267–317.

55 Aerdts, A.M., Eersels, K.L.L., and Groeninckx, G. (1996) Transamidation in melt mixed aliphatic and aromatic, polyamides. 1. Determination of the degree of randomness and number-average block length by means of ^{13}C NMR. *Macromolecules*, **29**, 1041–1045.

56 Röhner, G.B., Geyer, B., Gad'on, S., Kandelbauer, A., Chassè, T., and Lorenz, G. (2015) Tailoring block-copolyesters by reactive blending of polyethylene terephthalate and polyethylene naphthalate using statistical design of experiments. *J. Appl. Polym. Sci.*, **132** (41997), 1–12.

57 Krentsel, L.B., Markova, G.D., Kudryavtsev, Y.V., Filatova, M.P., Vasnev, V.A., Litmanovich, A.D., and Platé, N.A. (2005) Transesterification of poly(4,4′-isopropylidene-2,2′-dimethyldiphenylene terephthalate) and poly(ethylene adipate) blend in solution. *Macromol. Chem. Phys.*, **206**, 2206–2211.

58 Yang, H., He, J., and Liang, B. (2001) Transesterification kinetics of poly(ethylene terephthalate) and poly(ethylene 2,6-naphthalate) blends with the addition of 2,2′-bis(1,3-oxazoline). *J. Polym. Sci. B Polym. Phys.*, **39**, 2607–2614.

59 Scaffaro, R., Botta, L., La Mantia, F.P., Gleria, M., Bertani, R., Samperi, F., and Scaltro, G. (2006) Effect of adding new phosphazene compounds to poly(butylene terephthalate)/polyamide blends. II: effect of different polyamides on the properties of extruded samples. *Polym. Degrad. Stab.*, **91**, 2265–2274.

60 Chiou, K.C. and Chang, F.C. (2000) Reactive compatibilization of polyamide-6(PA 6)/polybutylene terephthalate (PBT) blends by a multifunctional epoxy resin. *J. Polym. Sci. Polym. Phys. Ed.*, **38**, 23–33.

61 Castellano, M., Nebbia, D., Turturro, A., Valenti, B., Costa, G., and Falqui, L. (2002) Reactive blending of polyamide 6,6 and Vectra A, 2 – role of a bifunctional epoxy coupler. *Macromol. Chem. Phys.*, **203** (10–11), 1614–1624.

62 Montaudo, G., Puglisi, C., Samperi, F., and La Mantia, F.P. (1996) Synthesis of AB and ABA block copolymers as compatibilizers in nylon 6/polycarbonate blends. *J. Polym. Sci. A Polym. Chem.*, **34**, 1283–1290.

63 Matsuda, H. and Asakura, T. (2004) Longer range sequence analysis of four-component copolyester using NMR. *Macromolecules*, **37**, 2163–2170.

64 Montaudo, M.S. and Montaudo, G. (1992) Further-studies on the composition and microstructure of copolymers by statistical modeling of their mass-spectra. *Macromolecules*, **25**, 4264–4280.

65 Chertovich, A.V., Guseva, D.V., Kudryavtsev, Y.V., and Litmanovich, A.D. (2008) Monte Carlo simulation of the interchain exchange reaction in a blend of incompatible polymers. *Polym. Sci. Ser. A*, **50** (4), 451–461.

66 Litmanovich, A.D., Podbelskiy, V.V., and Kudryavtsev, Y.V. (2010) On the ordering of statistical multiblock copolymers. *Macromol. Theory Simul.*, **19**, 269–277.

67 Khonakdar, H.A., Golriz, M., Jafari, S.H., and Wagenknecht, U. (2009) Correlation of sequence block lengths and degree of randomness with melt rheological properties in PET/PEN blends. *Macromol. Mater. Eng.*, **294**, 272–280.

68 Sudduth, R.D. (2003) Theoretical influence of polyester molecular weight distribution variation on melt viscosity during injection molding and extrusion as influenced by ester–ester interchange. *Polym. Eng. Sci.*, **43** (3), 519–530.

69 Gallardo, A., San Román, J., Dijkstra, P.J., and Feijen, J. (1998) Random polyester transesterification: prediction of molecular weight and MW distribution. *Macromolecules*, **31**, 7187–7194.

70 Kudryavtsev, Y.V. and Govorun, E.N. (2003) End-group interchain exchange reaction in polymer blends: evolution of the block weight distribution. *e-Polymers*, **3**, 1–19.

71 Geyer, B., Röhner, S., Lorenz, G., and Kandelbauer, A. (2014) Synthesis of ethylene terephthalate and ethylene naphthalate (PET-PEN) block-*co*-polyesters with defined surface qualities by tailoring segment composition. *J. Appl. Polym. Sci.*, **131** (40731), 1–15.

72 Golriz, M., Khonakdar, H.A., Jafari, S.H., Oromiehie, A., and Abedini, H. (2008) An improved non-isothermal kinetic model for prediction of extent of transesterification reaction and degree of randomness in PET/PEN blends. *Macromol. Theory Simul.*, **17**, 241–251.

73 Radmard, B. and Dadmun, M.D. (2001) Effect of transesterification on the morphology and mechanical properties of a blend containing a liquid crystalline polymer. *J. App. Polym. Sci.*, **80**, 2583–2592.

74 Ou, C.F. (1998) Interchange reactions between poly(ethylene terephthalate) and its copolyesters. *J. App. Polym. Sci.*, **68**, 1591–1595.

75 Kamoun, W., Salhi, S., Rousseau, B., El Gharbi, R., and Fradet, A. (2006) Furanic-aromatic copolyesters by interchange reactions between poly(ethylene terephthalate) and poly[ethylene 5,5′-isopropylidene-bis(2-furoate)]. *Macromol. Chem. Phys.*, **207**, 2042–2049.

76 Lorenzetti, C., Finelli, L., Lotti, N., Vannini, M., Gazzano, M., Berti, C., and Munari, A. (2005) Synthesis and characterization of poly(propylene terephthalate/2,6-naphthalate) random copolyesters. *Polymer*, **46**, 4041–4051.

77 Montaudo, M.S. (2011) Predicting and measuring the sequence distribution of condensation polymers, in *Mathematical Approaches to Polymer Sequence Analysis and Related Problems* (ed. R. Bruni), Springer, New York, pp. 227–246.

78 Tessier, M. and Fradet, A. (2003) Determination of degree of randomness in condensation copolymers containing both symmetrical and unsymmetrical monomer units: a theoretical study. *e-Polymers*, **30**, 1–16.

79 Saint-Loup, R., Jeanmarie, T., Robin, J.-J., and Boutevin, B. (2003) Synthesis of poly(ethylene terephthalate/polyε-caprolactone) copolyesters. *Polymer*, **44**, 3437–3449.

80 Yamadera, R. and Murano, M. (1967) The determination of randomness in copolyesters by high resolution nuclear magnetic resonance. *J. Polym. Sci.*, **5A-1**, 2259–2268.

81 Jang, S.S., Ha, W.S., Jo, W.H., Youk, J.H., Kim, J.H., and Park, C.R. (1998) Monte Carlo simulation of copolymerization by ester interchange reaction in miscible polyester blends. *J. Polym. Sci. B Polym. Phys.*, **36**, 1637–1645.

82 Montaudo, G., Puglisi, C., and Samperi, F. (1993) Primary thermal degradation mechanisms of PET and PBT. *Polym. Degrad. Stab.*, **42**, 13–28.

83 Samperi, F., Puglisi, C., Alicata, R., and Montaudo, G. (2004) Thermal degradation of poly(ethylene terephthalate) at the processing temperature. *Polym. Degrad. Stab.*, **83**, 3–10.

84 Khemami, K.C. (2000) A novel approach for studying the thermal degradation, and for estimating the rate of acetaldehyde generation by the chain scission mechanism in ethylene glycol based polyesters and copolyesters. *Polym. Degrad. Stab.*, **67**, 91–99.

85 Woźniak-Braszak, A.Q., Jurga, K., Nowaczyk, G., Dobies, M., Szostak, M., and Jurga, J. (2015) Characterization of poly(ethylene 2,6-naphthalate)/polycarbonate blends by DSC, NMR off-resonance and DMTA methods. *Eur. Polym. J.*, **64**, 62–69.

86 Carrot, C., Mbarek, S., Jaziri, M., Chalamet, Y., Raveyre, C., and Frédéric, P. (2007) Immiscible blends of PC and PET, current knowledge and new results: rheological properties. *Macromol. Mat. Eng.*, **292** (6), 693–706.

87 Licciardello, A., Auditore, A., Samperi, F., and Puglisi, C. (2003) Surface evolution of polycarbonate/polyethylene terephthalate blends induced by thermal treatments. *Appl. Surf. Sci.*, **203–204**, 556–560.

88 Lee, S.-S., Jeong, H.M., Jho, J.Y., and Ahn, T.O. (2000) Miscibility of poly(ethylene terephthalate)/poly(ester carbonate) blend. *Polymer*, **41**, 1773–1782.

89 Ho, J.-C. and Wei, K.-H. (1999) The kinetics of transesterification in blends of liquid crystalline copolyester and polycarbonate. *Polymer*, **40**, 717–727.

90 Jayakannan, M. and Anilkumar, P. (2004) Mechanistic aspects of ester–carbonate exchange in polycarbonate/cycloaliphatic polyester with model reaction. *J. Polym. Sci. A Polym. Chem.*, **42**, 3996–4008.

91 Montaudo, G., Puglisi, C., and Samperi, F. (1993) Chemical reactions occurring in the thermal treatment of polymer blends investigated by direct pyrolysis mass spectrometry: polycarbonate/polybuthyleneterephthalate. *J. Polym. Sci A Polym Chem.*, **31**, 13–25.

92 Montaudo, G., Puglisi, C., and Samperi, F. (1991) Chemical reactions which occur in the thermal treatment of polycarbonate/polyethyleneterephthalate blends, investigated by direct pyrolysis mass spectrometry. *Polym. Degrad. Stab.*, **31**, 291–326.

93 Johnston, N.W. (1976) Sequence distribution-glass transition effects. *J. Macromol. Sci. Rev. Macromol. Chem.*, **14B**, 215–250.

94 Fakirov, S., Sarkissova, M., and Denchev, Z. (1996) Sequential reordering in condensation copolymers .3. Miscibility-induced sequential reordering in random copolyesteramides. *Mcromol. Chem. Phys.*, **197** (9), 2889–2907.

95 Denchev, Z., Kricheldorf, H.R., and Fakirov, S. (2001) Sequential reordering in condensation copolymers, 6 – average block lengths in poly(ethylene terephthalate)-polyamide 6 copolymers as revealed by NMR spectroscopy. *Macromol. Chem. Phys.*, **202** (4), 574–586.

96 Retolaza, A., Eguiazábal, J.I., and Nazábal, J. (2005) Reactive processing compatibilization of direct injection molded polyamide-6/poly(ethylene terephthalate) blends. *J. Appl. Polym. Sci.*, **97**, 564–574.

97 Fakirov, S., Evstatiev, M., and Petrovich, S. (1993) Microfibrillar reinforces composites from binary and ternary blends of polyesters and nylon-6. *Macromolecules*, **26**, 5219–5226.

98 Evstatiev, M., Schultz, J.M., Petrovich, S., Georgiev, G., Fakirov, S., and Friedrich, K. (1998) In situ polymer/polymer composites from poly(ethylene terephthalate), polyamide-6, and polyamide-66 blends. *J. Appl. Polym. Sci.*, **67**, 723–737.

99 Fakirov, S., Evstatiev, M., and Schultz, J.M. (1993) Microfibrillar reinforced composite from drawn poly(ethylene terephthalate) nylon-6 blend. *Polymer*, **34**, 4669–4679.

100 Pillon, L.Z. and Utracki, L.A. (1984) Compatibilization of polyester/polyamide blends via catalytic ester–amide interchange reaction. *Polym. Eng. Sci.*, **24**, 1300–1305.

101 Pillon, L.Z., Utracki, L.A., and Pillon, D.W. (1987) Spectroscopic study of poly(ethylene terephthalate)/poly(amide-6,6) blends. *Polym. Eng. Sci.*, **27**, 562–567.

102 Huang, C.-C. and Chang, F.-C. (1997) Reactive compatibilization of polymer blends of poly/butylene terephthalate) (PBT) and polyamide-6,6 (PA66): 1. Rheological and thermal properties. *Polymer*, **38**, 2135–2141.

103 Retolaza, A., Eguiazabal, J.I., and Nazabal, J. (2004) Structure and mechanical properties of polyamide-6,6/poly(ethylene terephthalate) blends. *Polym. Eng. Sci.*, **44** (8), 1405–1413.

104 Prattipati, V., Hu, Y.S., Bandi, S., Schiraldi, D.A., Hiltner, A., Baer, E., and Mehta, S. (2005) Effect of compatibilization on the oxygen-barrier properties of poly(ethylene terephthalate)/poly(*m*-xylylene adipamide) blends. *J. Appl. Polym. Sci.*, **97**, 1361–1370.

105 Ozen, I., Bozoklu, G., Dalgıçdir, C., Yücel, O., Unsal, E., Çakmak, M., and Menceloğlu, Y.Z. (2010) Improvement in gas permeability of biaxially stretched PET films blended with high barrier polymers: the role of chemistry and processing conditions. *Eur. Polym. J.*, **46**, 226–237.

106 Ho, J.C. and Wei, K.H. (2000) Ester–amide exchange in blends of liquid-crystalline copolyester and polyamide 6. *J. Polym. Sci. B Polym. Phys.*, **38** (16), 2124–2135.

107 Costa, G., Meli, D., Song, Y., Turturro, A., Valenti, B., Castellano, M., and Falqui, L. (2001) Reactive blending of polyamide 6,6 and Vectra A. *Polymer*, **42** (19), 8035–8042.

108 Granado, A., Iturriza, L., and Eguiazábal, J.I. (2014) Structure and mechanical properties of blends of an amophous polyamide and an amorphous copolyester. *J. Appl. Polym. Sci.*, **131**, 40785–40792.

109 Zhen, Y., Jia-ming, S., Qiang, W., and Cao, K. (2012) Study on ester–amide exchange reaction between PBS and PA6IcoT. *Ind. Eng. Chem. Res.*, **51**, 751–757.

110 Pai, F.-C., Lai, S.-M., and Chu, H.-H. (2013) Characterization and properties of reactive poly(lactic acid)/polyamide 610 biomass blends. *J. Appl. Polym. Sci.*, **130**, 2563–2571.

111 Brown, S.B. (1992) Reactive extrusion: a survey of chemical reactions of monomers and polymers during extrusion processing, in *Reactive Extrusion* (ed. M. Xanthos), Hanser, Munich, pp. 75–199.

112 Bell, E.T., Bradley, J.R., Long, T.E., and Stafford, S.L. (2001) Polyester blend composition useful as starting material for the production of moldings comprises at least one acid terminated polyamide. US Patent 6,239,233, filed May 29, 2001.

113 Maruhashi, Y. and Iida, S. (2001) Transparency of polymer blends. *Polym. Eng. Sci.*, **41**, 1987–1995.

114 Xie, F., Kim, Y.W., and Jabarin, S.A. (2009) The interchange reaction between poly(ethylene terephthalate) and poly(*m*-xylylene adipamide). *J. Appl. Polym. Sci.*, **112**, 3449–3461.

115 Puglisi, C., Samperi, F., Di Giorgi, S., and Montaudo, G. (2002) MALDI-TOF characterisation of thermally generated gel from nylon 66. *Polym. Degrad. Stab.*, **78**, 369–378.

116 Montaudo, M.S., Puglisi, C., Samperi, F., and Montaudo, G. (1998) Structural characterization of multicomponent copolyesters mass spectrometry. *Macromolecules*, **31**, 8666–8676.

117 Montaudo, G., Montaudo, M.S., and Samperi, F. (2001) Matrix-assisted laser desorption/ionization mass spectrometry of polymers (MALDI-MS), in *Mass Spectrometry of Polymers* (eds G. Montaudo and R.P. Lattimer), CRC Press LLC, Boca Raton, FL, pp. 419–522.

118 Montaudo, G. and Montaudo, M.S. (2001) Polymer characterization methods, in *Mass Spectrometry of Polymers* (eds G. Montaudo and R.P. Lattimer), CRC Press LLC, Boca Raton, FL, pp. 41–112.

119 Samperi, F., Puglisi, C., Alicata, R., and Montaudo, G. (2004) Thermal degradation of poly(butylene terephthalate) at the processing temperature. *Polym. Degrad. Stab.*, **83**, 11–17.

120 Montaudo, G., Puglisi, C., and Samperi, F. (1994) Exchange reaction occurring through active chain ends: melt mixing of nylon 6 and polycarbonate. *J. Polym. Sci. A Polym. Chem.*, **32**, 15–31.

121 Gattiglia, E., Turturro, A., Lamantia, F.P., and Valenza, A. (1992) Blends of polyamide 6 and bisphenol-A polycarbonate. Effects of interchange reactions on morphology and mechanical properties. *J. Appl. Polym. Sci.*, **46** (11), 1887–97.

122 Gattiglia, E., Turturro, A., Pedemonte, E., and Dondero, G. (1990) Blends of polyamide 6 with bisphenol-A polycarbonate. 2. Morphology-mechanical properties relationships. *J. Appl. Polym. Sci.*, **41**, 1411–1423.

123 Valenza, A., La Mantia, F.P., Gattiglia, E., and Turturro, A. (1994) Reactive blending of polyamide 6 and polycarbonate – effects of polyamide-6 terminal groups. *Int. Polym. Process.*, **9** (3), 240–245.

124 Wang, Q., Jiang, Y., Li, L., Wang, P., Yang, Q., and Li, G. (2012) Mechanical properties, rheology, and crystallization of epoxy-resin-compatibilized polyamide 6/polycarbonate blends: effect of mixing sequences. *J. Macromol. Sci. B*, **51**, 96–108.

125 van Bennekom, A.C.M., Pluimers, D.T., Bussink, J., and Gaymans, R.J. (1997) Blends of amide modified polybutylene terephthalate and polycarbonate: transesterification and degradation. *Polymer*, **38** (12), 3017–24.

126 Wang, M., Yuan, G., and Han, C.C. (2013) Reaction process in polycarbonate/polyamide bilayer film and blend. *Polymer*, **54**, 3612–3619.

127 Walia, P.S., Gupta, R.K., and Kiang, C.T. (1999) Influence of interchange reactions on the crystallization and melting behavior of nylon 6,6 blended with other nylons. *Polym. Eng. Sci.*, **39**, 2431–2444.

128 Wang, X., Zheng, Q., Du, L., and Yang, G. (2008) Influence of preparation methods on the structures and properties for the blends between polyamide 6*co*6T and polyamide 6: melt-mixing and *in-situ* blending. *J. Polym. Sci. B Polym. Phys.*, **46**, 201–211.

129 Shibayama, M., Uenoyama, K., Oura, J.-i., Nomura, S., and Iwamoto, T. (1995) Miscibility and crystallinity control of nylon 6 and poly(*m*-xylene adipamide) blends. *Polymer*, **36**, 4811–4816.

130 Puglisi, C., Samperi, F., Montaudo, M.S., Montaudo, G. Unpublished results.

131 Miller, I.K. (1976) Amide-exchange reactions in mixtures of *N*-alkyl amides and in polyamide melt blends. *J. Polym. Sci. Polym. Chem. Ed.*, **14**, 1403–1417.

132 Lake, W.B., Kalallunnath, S., and Kalika, D.S. (2004) Crystallization, melting, and rheology of reactive polyamide blends. *J. Appl. Polym. Sci.*, **94**, 1245–1252.

133 Vannini, M., Marchese, P., Celli, A., and Lorenzetti, C. (2015) Block and random copolyamides of poly(*m*-xylylene adipamide) and poly(hexamethylene isophthalamide-*co*-terphthalamide): methods of preparation and relationships between molecular structure and phase behavior. *Polym. Eng. Sci.*, **55**, 1475–1484.

134 Endo, M., Morishima, Y., Yano, S., Tadano, K., Murata, Y., and Tsunashima, K. (2006) Miscibility in binary blends of aromatic and alicyclic polyamides. *J. Appl. Polym. Sci.*, **101**, 3971–3978.

135 Berti, C., Celli, A., Marchese, P., Sullalti, S., Vannini, M., and Lorenzetti, C. (2012) Transamidations in melt-mixed MXD6 and PA6I-6T polyamides: 1. Determination of the degree of randomness and block length by ^{1}H-NMR analysis. *E. Polym. J.*, **48**, 1923–1931.

136 Wang, L., Dong, X., Gao, Y., Huang, M., Han, C.C., Zhu, S., and Wang, D. (2015) Transamidation determination and mechanism of long chain-based aliphatic polyamide alloys with excellent interface miscibility. *Polymer*, **59**, 16–25.

137 Bennet, C., Zeng, J., Kumar, S., and Mathias, L.J. (2006) Synthesis of copolyamides containing octadecanedioic acid: an investigation of nylon 6/6, 18 in various ratios. *J. Appl. Polym. Sci.*, **99**, 2062–2067.

138 Telen, L., Van Puyvelde, P., and Goderis, B. (2016) Random copolymers from polyamide 1 and polyamide 12 by reactive extrusion: synthesis, eutectic phase behavior, and polymorphism. *Macromolecules*, **49**, 876–890.

139 Yao, Z., Sun, J.-M., Wang, Q., Zhou, C.-D., and Cao, K. (2012) Monte Carlo simulation of molecular weight distribution and copolymer composition in transamidation reaction of polyamides. *J. Appl. Polym. Sci.*, **125**, 3582–3590.

7

In situ Synthesis of Inorganic and/or Organic Phases in Thermoplastic Polymers by Reactive Extrusion

Véronique Bounor-Legaré[1,2], Françoise Fenouillot[1,3], and Philippe Cassagnau[1,2]

[1] Université de Lyon, 69003 Lyon, France
[2] Université de Lyon 1, Ingénierie des Matériaux Polymères, CNRS UMR 5223, 69622 Villeurbanne, France
[3] INSA de Lyon, Ingénierie des Matériaux Polymères, CNRS, UMR 5223, 69621 Villeurbanne, France

7.1 Introduction

One of the significant advantages of the extruder over batch reactors is its capacity to facilitate a continuous bulk reactive process, that is, high-viscous solvent-free reactive systems. Actually, reactive processing addresses both combines the difficulties of polymer processing and the problems of controlling a chemical reaction in very specific conditions: high-viscous medium ($\eta \sim 10^3$ Pa s), high temperatures ($T \sim 250\,°C$), and short residence times ($t \sim 1$ min). Many research works have therefore been devoted to reactive extrusion: twin-screw extruders have been used as chemical reactors and a number of reactive systems have been investigated by reactive extrusion, for example, chemical modification of molten polymers, bulk polymerization, reactive blending of immiscible polymer blends by reaction at the interface, and *in situ* polymerization of a minor phase [1].

This chapter focuses on these last developments and more particularly on the *in situ* synthesis of organic and inorganic phase inside a polymer matrix by reactive extrusion. First, formation of nanocomposites by sol–gel chemistry will be highlighted. In a second part, the polymerization and crosslinking of thermoplastic and thermoset minor phase under shear will be described. This chapter is organized as follows:

1) Synthesis of nanocomposites
2) Polymerization of a thermoplastic minor phase
3) Polymerization of a thermoset minor phase.

7.2 Nanocomposites

Nanocomposites have been extensively studied these past 15 years for academic purposes and for commercial applications. The development of such multifunctional advanced materials may have a major impact on future applications

Reactive Extrusion: Principles and Applications, First Edition. Edited by Günter Beyer and Christian Hopmann.
© 2018 Wiley-VCH Verlag GmbH & Co. KGaA. Published 2018 by Wiley-VCH Verlag GmbH & Co. KGaA.

in diverse fields that involve optics, electronics, ionic, mechanics, membranes, catalysis sensors, and biology. Actually, the dispersion of inorganic fillers into a molten polymer by extrusion is generally the most used process, which can be classified as a top–down process. However, the specific dispersion in terms of final applications often requires pre- or post-modification of the filler and/or of the polymer matrix more particularly with nonpolar polymers such as polyolefins [2]. An alternative approach for specific applications is the *in situ* synthesis of the inorganic filler in the molten polymer matrix by sol–gel chemistry. This bottom-up approach by *in situ* synthesis of the inorganic phase from inorganic precursors is then an elegant way to process nanocomposites or organic/inorganic hybrid structures. The main advantage of the sol–gel chemistry is to run in mild conditions, so that organic species can be mixed with the inorganic precursors to form the composites or hybrids systems. In particular, the *in situ* sol–gel synthesis of inorganic particles inside a polymeric matrix provides good dispersions of silica or metal oxides such as titanium dioxide [3, 4].

Unfortunately, one of the major drawbacks of sol–gel chemistry is that an organic solvent has to be used, which limits its application to coating of glass, metal, and polymer substrata. More recently, new strategies have been developed to extend the sol–gel chemistry to bulk materials without the use of any solvents. The objective of the present part is to present these strategies that have been developed for sol–gel chemistry in twin-screw extruder.

The sol–gel process is generally described by hydrolysis and condensation reactions (Figure 7.1) of metal alkoxides, M(OR): where M is a metal ion (Al, Ti, etc.) or silicon and R an alkyl group (R = Me, Et, etc.) [5].

The hydrolysis reaction (Figure 7.1a) consists of transforming the alkoxide groups (OR) into the hydroxyl groups (OH). The condensation reactions then involve the hydroxyl groups to produce siloxane bonds (Si—O—Si) and alcohol by-products (ROH) or water (Figure 7.1b,c). In most of the cases, the condensation reactions begin before the hydrolysis reaction is completed. Numerous studies in the literature have shown that variations in the synthesis conditions such as the molar ratio $r = H_2O/Si$, the type of catalysis, the nature of the solvent, and the temperature are parameters that are relevant to the modifications of the structure and the properties of the product obtained. These parameters of importance have been discussed in our previous review mainly to illustrate their influence on sol–gel synthesis in molten polymers [6]. The influence of other parameters must be considered in the case of *in situ* synthesis in molten polymers, such as the dispersion of the inorganic precursors, their thermal stability, and the effect of potential polymer crystallization on the final

(a) $Si(OR)_4 + nH_2O \longrightarrow (RO)_{(4-n)}Si(OH)_n + nROH$

(b) $\equiv SiOH + HOSi\equiv \longrightarrow \equiv SiOSi\equiv + H_2O$

(c) $\equiv SiOH + ROSi\equiv \longrightarrow \equiv SiOSi\equiv + ROH$

Figure 7.1 Hydrolysis–condensations reactions of a tetraalkoxysilane (M = Si, R is an alkyl group).

morphology. The appearance of a phase separation between the polymer and the growing inorganic phase can also impact the physical characteristics of the polymer, such as crystalline morphology and rate of crystallinity. Furthermore, it can be noted that two types of organic–inorganic materials produced by the sol–gel process are distinguished. The first one corresponds to materials with no covalent bonding between the organic and inorganic phases. These materials behave like thermoplastics, filled with inorganic domains and therefore the term *nanocomposite* appears to be the most appropriate word to characterize such materials. The second class of materials correspond to organic–inorganic hybrid materials with covalent bonds between the polymer matrix and the inorganic phase. This class will not be described in this chapter.

7.2.1 Synthesis of *in situ* Nanocomposites

In the literature, only a few studies have dealt with the synthesis of inorganic filler through sol–gel chemistry in a single processing step [6–10]. Kluenker *et al.* [8] created *in situ* silica in a polypropylene matrix by twin screw extrusion with a screw diameter of 26 mm and a length to diameter (L/D) ratio of 56. The inorganic precursor was a polyethoxysilane (PAOS) synthesized for this purpose. The polypropylene granules were fed to the main hopper followed by the feeding of a liquid inorganic precursor (PAOS)/catalyst premixed by using a gear pump. Subsequently, water as curing agent was injected into the polymer melt via an injection nozzle under high pressure using an HPLC pump. The by-product of the condensation and the excess water were extracted by degassing at atmospheric conditions and in vacuum at the end of the process. More recently, Miloskovska *et al.* [10] described a way to produce, in a one-step process, *in situ* nanocomposites with sufficient silica content to obtain properties that were comparable to those of conventional nanocomposites by reactive extrusion. Contrary to the recommendation of previous authors, they used commercial tetraalkoxysilane as their silica source in an EPDM rubber matrix. A specific screw with a proprietary design was used to match the requirements for efficient mixing between the rubber and the low-viscous reactive liquids (tetraethoxysilane (TEOS) and catalyst solution), and for sufficient reaction time for the silica formation. Both authors discussed the final morphology and its impact on the mechanical properties and the limitations in terms of silica concentration and appearance of aggregation phenomena. Oliveira and Machado [9] developed this approach for the production of alumina. For this purpose, two polymers, polypropylene grafted with maleic anhydride (PP-*g*-MA) and ethylene vinyl acetate (EVA), were used as organic phases and aluminum isopropoxide (Al(Pr-*i*-O)$_3$) as inorganic precursor. A finer morphology in the EVA phase with well-dispersed nanoparticles containing alumina was achieved using a batch mixer.

To illustrate the details of the relationship between the reactive extrusion conditions and the chemical creation of the filler, the synthesis by reactive extrusion of titanium dioxide TiO$_2$ into a polypropylene matrix [7] can be considered as an example. From an experimental point of view, the PP/TiO$_2$ nanocomposite was prepared using a co-rotating twin-screw extruder (Leistritz LSM 30–34, screw diameter $D = 34$ mm, length to diameter ratio $L/D = 34.5$) and a specific screw

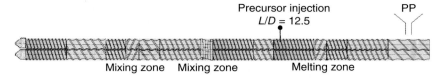

Figure 7.2 Screw profile and injection point of titanium *n*-butoxide precursor for the processing of PP/TiO$_2$ nanocomposite. Extruder: Leistritz LSM 30–34, screw diameter $D = 34$ mm, length to diameter ratio $L/D = 34.5$. (Bahloul et al. 2011 [7]. Reproduced with permission of Wiley.)

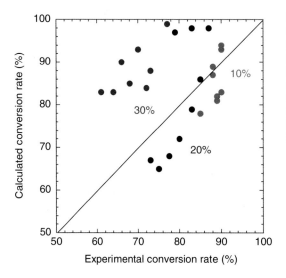

Figure 7.3 Prediction of the conversion compared with experimental data for different concentrations (12%, 20%, and 30% of inorganic precursors) under different processing conditions (screw speed and flow rate). (Bahloul et al. 2011 [7]. Reproduced with permission of Wiley.)

profile, as shown in Figure 7.2. This screw profile is composed of a conveying zone with right-handed screw elements, a left-handed element for the PP melting before titanium *n*-butoxide injection, and two mixing zones. The liquid titanium *n*-butoxide precursor was injected by a side pump and added to molten PP after the melting zone (point at $L/D = 12.5$). In these experiments, no water was added during the extrusion process.

From a simulation point of view, the Ludovic® software [11] was used. Based on a 1D approach, the evolution of the main local parameters of the extrusion process (pressure, temperature, residence time, shear rate, filling ratio, etc.) along the screws could be calculated. As a result, all the experimental results (conversion corresponding to the disappearance of Ti—O—C bonds expressed in mol% determined by FTIR) have been compared with the software predictions, for 10, 20, and 30 wt% precursors under different processing conditions (flow rate and screw speed). The results shown in Figure 7.3 clearly demonstrates that the prediction for 10 and 20 wt% precursors are satisfactory, leading to a correct value of the conversion at ±15%, irrespective of the screw speed and feed rate. However, the calculations at 30 wt% are overestimated.

Actually, at high precursor amount, the assumption made on the rheological behavior is probably not correct anymore, because the reactive medium cannot be considered homogeneous. For instance, the mixing of a low-viscous liquid (high amount of precursor) with a molten polymer leads to highly non-linear phenomena such as lubrication effects, associated with complex flows and diffusion mixing [12, 13]. The large mismatch between the viscosity of the precursor viscosity and that of the molten polymer greatly magnifies the difficulties associated with homogenizing these components in a laminar flow. Actually, this example shows one of the most complex phenomena encountered in reactive extrusion, that is, the mixing at different scales (macro- to molecular mixing) of species of different viscosities.

7.2.2 Some Specific Applications

The most ambitious goal of *in situ* synthesis of nanocomposites is to achieve a one step process to produce new functional materials by controlling the morphology, the dispersion, and the specific functionality of the created inorganic product. By judiciously selecting the functional inorganic precursors and processing conditions, specific properties such as biocide, fire retardancy, and proton conductivity can be achieved.

7.2.2.1 Antibacterial Properties of PP/TiO$_2$ Nanocomposites

In a previous work [14], the antibacterial properties of PP, PP/TiO$_2$ (anatase) and *in situ* PP/TiO$_2$ samples submitted to *Escherichia coli* and *Satphylococcus aureus* bacteria were studied. The PP/TiO$_2$ nanocomposite materials were prepared by reactive extrusion according to Figure 7.2. A 30 wt% of titanium *n*-butoxide precursor was injected in the extruder in order to achieve 9.3 wt% of titanium dioxide assuming that the hydrolysis-condensations reaction conversions are complete. Furthermore, at the die exit, the PP/TiO$_2$ nanocomposites were made to undergo a second treatment in hot water (80 °C) for 72 h in order to achieve a more condensate structure of TiO$_2$ nanoparticles. In terms of morphologies, Figure 7.4a shows that the size of the titanium dioxide in the *in situ* PP/TiO$_2$ nanocomposite prepared via sol–gel process is less than 10 nm in diameter. Furthermore, a fractal structure of these inorganic domains was highlighted with a characteristic aggregation size $d_{aggr} \approx 130$ nm [15]. On the contrary, the PP/TiO$_2$ (anatase)

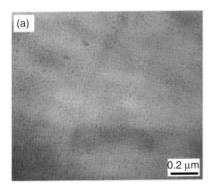
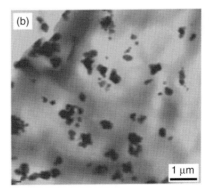

Figure 7.4 TEM image of PP/TiO$_2$ nanocomposite materials: (a) *in situ* PP/TiO$_2$ and (b) melt blended PP/TiO$_2$ (anatase). (Bahloul *et al.* 2010 [15]. Reproduced with permission of Wiley.)

7 In situ Synthesis of Inorganic and/or Organic Phases in Thermoplastic Polymers

Table 7.1 Antibacterial activity of PP, melt blended PP/TiO$_2$ (anatase) and in situ PP/TiO$_2$.

Sample	Test bacteria[a)]	Ct24 (CFU)	log(Ct24)	Activity[b)] (log)	Standard deviation
Glass reference	E. coli	33 000 000	7.52	—	0.05
	S. aureus	11 600 000	6.05	—	0.02
Samples	**Test bacteria[a)]**	**Et24 (CFU)**	**log(Et24)**	**Activity[b)] (log)**	**Standard deviation**
PP	E. coli	21 400 000	7.33	<1	0.15
	S. aureus	11 000 000	6.04	<1	0.08
PP/TiO$_2$ (anatase)	E. coli	35 400 000	7.55	<1	0.09
	S. aureus	220 000	5.34	<1	0.41
In situ PP/TiO$_2$	E. coli	30	1.48	6.04	0.05
	S. aureus	5650	3.75	2.3	1.85

a) Bacteria tested: *Escherichia coli* CIP 54 127 and *Satphylococcus aureus* CIP 4.83, 400 μl of suspension at 2.5×10^5 UFC ml^{-1}.
b) Activity = log(Ct24)−log(Et), Ct is the reference medium and Et is the medium sample.
Source: Balhoul et al. 2012 [14]. Reproduced with permission of Elsevier.

nanocomposite prepared via a usual blending method of TiO$_2$ anatase particles in molten polypropylene matrix, shows an average diameter size of 0.5 μm of TiO$_2$ particles as shown in Figure 7.4b.

As a result, it is observed (Table 7.1) that the antibacterial effect after 24 h for PP and PP/TiO$_2$ (anatase) is relatively weak, whereas this antibacterial property increases remarkably for the *in situ* PP/TiO$_2$ nanocomposite. For example, in the case of *E. coli* bacteria, only 30 CFU (colony forming units) out of 33.10^6 CFU survived after 24 h of incubation.

The source of such behavior can be found in the structure and specific surface of the *in situ* created inorganic phase. Actually, the specific surface of TiO$_2$ (anatase) is around 2 m^2 g^{-1} as compared to 100 m^2 g^{-1} for the *in situ* TiO$_2$. This leads to more interactions between bacteria and the filler as discussed by Kubacka et al. [16]. Hence the structures identified by XPS for the *in situ* synthesis evidenced a low condensation degree, around 18% with a high concentration in hydroxyl groups (Ti$_x$O$_y$(OH)$_z$-type like structure), which may participate in specific interactions with bacteria cells without photocatalysis. Such phenomenon has already been underlined by Fenoglio et al. [17] who identified radicalar species such as HO$^\bullet$, $^-$O$_2^\bullet$, and HO$_2$. These species are created by reactions between TiO$_2$ and oxygen, water, or H$_2$O$_2$ in the absence of light, which induced oxidation reactions.

7.2.2.2 Flame-Retardant Properties

With respect to flame-retardant properties, most of the works have been focused on polyamide nanocomposites based on *in situ* functionalized silica during processing. For example, Theil-Van Nieuwenhuyse et al. [18] studied

Table 7.2 Composition of the PA6-based nanocomposites prepared by extrusion process.

Materials	PA6 feed rate (kg h^{-1})	Precursor feed rate (ml min^{-1})	wt% Si measured[a]	wt% P measured[a]
PA6	3	0	0	0
PA6 + SiP	3	20	2.5	2.5
PA6 + TEOS	3	6.5	0.4	0
PA6 + SiP + TEOS (1/1 molar ratio)	3	16.5	1.9	1.2

a) Determined by elemental analysis.
Source: Theil-Van Nieuwenhuyse et al. 2013 [18]. Reproduced with permission of Elsevier.

the fire retardancy properties of polyamide 6 (PA6)/phosphorylated silica nanocomposites. Typically, PA6 (containing 0.08 wt% of residual water) was melt blended in the twin-screw extruder as described in Figure 7.2. The processing parameters were set as follows: temperature range from 220 °C at the feeder to 210 °C at the die, screw speed $N = 100$ rpm, polymer flow rate $= 3$ kg h^{-1}. Under these processing conditions, SiP (diethylphosphatotriethoxysilane) or TEOS or a mixture of both is introduced in the PA6 melt at a constant flow rate by a liquid pump. A summary of the prepared composites by reactive extrusion is shown in Table 7.2.

The extent of the condensation reactions was estimated by the ^{29}Si CP-MAS NMR on the materials obtained directly at the die of the extruder. It was clearly shown that the hydrolysis–condensation reactions occurred for the SiP precursor during the PA6 extrusion process with the formation of highly condensed species [18]. The impact of such *in situ* created phosphatosilica species on the fire-retardant behavior was mainly approached through cone-calorimeter analysis (Figure 7.5). An important modification of the heat release rate (HRR) curves versus times with the *in situ* SiP-based nanocomposites can be observed. It changed from a « non-charging » behavior for the polyamide polymer (no char formation) to a « charring » behavior (formation of a char) for the SiP-based nanocomposites. Besides this change of behavior, the total heat released is decreased to 85% of the initial value with only 2 wt% of phosphorus.

A second important point revealed by the analysis of PA6 + TEOS behavior is that the flame-retardant properties are not due to creation of the (Si—O—Si) network. Indeed, the flame-retardant properties for PA6 + TEOS are not enhanced as compared to those of the PA6 one. These results clearly evidence the role of both phosphorus and silicone atoms at the molecular scale, which has been achieved from *in situ* synthesis by reactive extrusion.

However, a weak point observed in this system is the decrease in time to ignition (TTI) between the pure PA6 and the nanocomposites. The same behavior was also observed with PA66/PA6 copolymers [19]. Further analysis performed by TGA-IR, TGA-CG-SM, and the gas identification during the cone-calorimeter test evidenced the presence of ethene (gas highly flammable) produced by SiP rearrangement reactions [19]. Based on these observations, a new inorganic precursor called SiDOPO, was synthesized as illustrated in

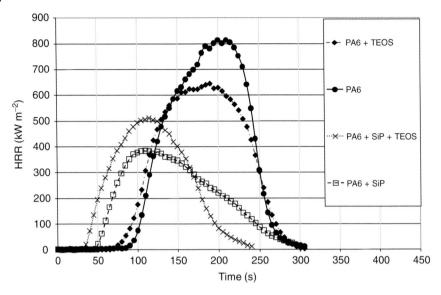

Figure 7.5 Heat release rate curves for pure PA 6 and *in situ* nanocomposites versus time at 35 kW m^{-2}. (Theil-Van Nieuwenhuyse *et al*. 2013 [18]. Reproduced with permission of Elsevier.)

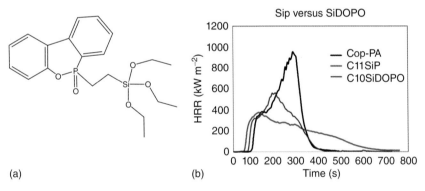

Figure 7.6 (a) Structure of the SiDOPO and (b) Heat release rate versus time for the copolymer PA66/PA6, and nanocomposites based on SiP and SiDOPO for a heat flux of 35 kW m^{-2}. (Sahyoun *et al*. 2015 [20]. Reproduced with permission of Elsevier.)

Figure 7.6a [20]. Finally, this inorganic precursor allowed to create functional silica *in situ* leading to an enhancement of the fire-retardant behavior of the polyamide copolymer (Figure 7.6b) In the last case, a large amount of char is created in addition to an increase of TTI compared to the previously tested systems.

7.2.2.3 Protonic Conductivity

The synthesis of thiol-functionalized silica/PVDF-HFP (poly(vinylidene fluoridehexafluoropropylene)) nanocomposite materials was carried out [21] by reactive extrusion through *in situ* sol–gel reactions of an alkoxysilane inorganic precursor solution composed of polydimethoxysiloxane (PDMOS)

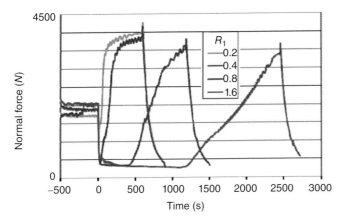

Figure 7.7 Variation of the normal force versus mixing time. PVDF-HFP/Silica-SH nanocomposite ($R_0 = 0.5$) for different R_1 ratio. Time $t = 0$ s corresponding to the injection of the pre-hydrolyzed inorganic solution. ($R_1 = 0.2$ curve on the left, $R_1 = 1.6$ curve on the right). (Seck et al. 2015 [21]. Reproduced with permission of Elsevier.)

and mercaptopropyltriethoxysilane (MPTES). Successful introduction of the functional MPTES and structural PDMOS alkoxysilanes and subsequent condensation reactions in the PVDF-HFP were achieved through pre-hydrolysis reactions of the precursors. A DSM Micro 15 laboratory vertical micro-extruder (15 cm³) with double intermeshing counter-rotating screws and a manual floodgate allowing either extrusion or recirculation was used to carry out this reactive extrusion process.

This force variation associated with the evolution of the viscosity was evaluated for different R_1 ratio (molar ratio between functional and structuring inorganic precursors). Analysis of the curves reported in Figure 7.7 clearly demonstrates that the lower the R_1 value (which decreases with increasing concentration of the structuring inorganic precursor), the greater the impact of the condensation reactions on this force. This is consistent with the rapid formation of a dense inorganic network.

Based on this work it can be deduced that the more the functional precursor in the material (higher R_1), the better the nanocomposite conduction properties when oxidation reactions of the SH lead to conductive SO_3H conductive groups. SEM studies of the nanocomposites (Figure 7.8) reveal another motivation to raise the R_1 value. Figure 7.8 clearly demonstrates that the higher the R_1 value, the smaller are the silicon phases that are generated. This dispersion behavior could be described in terms of solubility parameters. Since the δ of PDMOS and MPTES are 54.6 and 19.6 MPa$^{1/2}$, respectively, we can expect that an increasing value of R_1, that is, an increasing amount of functional groups on the surface, leads to a better compatibilization with the polymer matrix ($\delta = 15.2$ MPa$^{1/2}$), thereby resulting in smaller silica domains and a better dispersion.

A very encouraging result is obtained from the materials synthesized with $R_1 = 1.6$ corresponding to a theoretical ionic exchange capacity (IEC) of 2 meq. g^{-1} following the oxidation step. An experimental IEC value of 0.7 meq. g^{-1} and a conductivity of 9.6 mS cm^{-1} confirm that the introduced

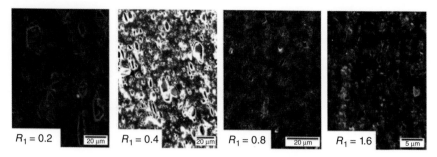

Figure 7.8 SEM images of extruded nanocomposites. Effect of R_1 ratio on the morphology: From left to right $R_1 = 0.2$; 0.4; 0.8; 1.6. The last image was magnified due to a change of scale in the particle size. (Seck et al. 2015 [21]. Reproduced with permission of Elsevier.)

SH groups were partially oxidized in SO_3H. This original approach demonstrated again the possibility of incorporating in situ functionalized silica into a molten fluorinated polymer matrix in a unique reactive extrusion procedure to synthesize the proton exchange membrane fuel cells (PEMFC) membranes.

7.3 Polymerization of a Thermoplastic Minor Phase: Toward Blend Nanostructuration

Thermoplastic polymer blends have been extensively studied during these past 20 years. Among the different strategies of blending, reactive compounding may be a viable mechanism for the in situ elaboration of the desired blend with controlled structure and morphology. Very few works have been reported on polymer blending via in situ polymerization of monomers dispersed in a thermoplastic matrix. Kye and White [22] investigated the polymerization of caprolactam/polyether sulfone solutions in a twin-screw extruder to form reactive polyamide-6/polyether sulfone blends, and Cassagnau et al. [23] theoretically and experimentally investigated the effect of microscale phenomena on the molecular and morphological properties of the new polymeric alloys produced by in situ polymerization of the dispersed phase. For this purpose, they used a alcohol/isocyanate addition reaction for the formation of a polyurethanne (PU) phase in a polyethylene (PE).

More recently, nanostructured polymer blends were processed from a careful selection of macromolecular architectures, that is, molar masses of the grafted species, grafting ratio, and position of the reactive groups along the chains [24]. Relevant results were reported by Leibler and his group [25] for self-organized materials from grafted copolymers of poly(methymethacrylate) and polyamide-6 obtained by reactive mixing in molten state [26] on polyolefin-polyamide blends. The formation of such a nanostructured material has to be related to the architecture of the copolymers based in particular on the size and the molar masses of the sequences or the grafts as well as on the grafting ratio (copolymer concentration) and the type of reactive chains (position of the functional reactive group [25, 27–29]. The concept of reactive extrusion for designing nanoblends from

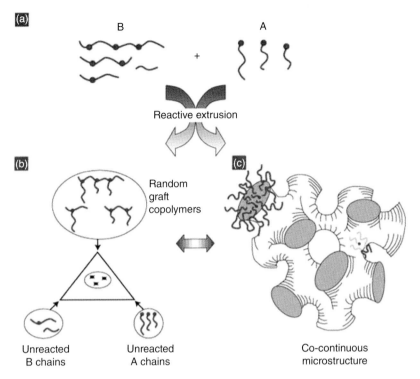

Figure 7.9 Principle of reactive extrusion for designing nanoblends from the *in situ* synthesis of grafted copolymers. (Ruzette and Leibler 2002 [25]. Reproduced with permission of Nature Publishing Group.)

the *in situ* synthesis of grafted copolymers is described in Figure 7.9, according to Ruzette and Leibler [25].

Several strategies, as described by Harrats *et al.* [30], have been developed to reach well-defined structures of polymer blends at the micro- and nanoscale. However, there remains a need for research into the synthesis of block or graft copolymers that would present the ability to give nanostructures; this work is both scientifically challenging and industrially important. Reactive extrusion [1] in particular represents an attractive way to obtain such specific nanostructures. Generally, there are three main ways of synthesis of such copolymers:

a) Living copolymerization
b) Chemical modification by post polymerization
c) Coupling between two appropriately functionalized polymer chains.

Methods (a) and (b) are associated with "the grafting from" approach and method (c) belongs to the "grafting onto" route. Structures of copolymers obtained through methods (b) and (c) are specifically relevant to reactive extrusion since they are unattainable by the classical copolymerization method. However, the main difficulty lies in the fact that the polymers/polymers or polymers/monomers systems are generally not miscible. The chemical reaction occurs at the interface and thus a large quantity of copolymer is difficult to obtain. This interfacial reaction leads to compatibilization of the blends by reactive mixing [31]. Thus, when synthesis of nanostructured copolymers or blends is required, other parameters have to be optimized in order to increase the concentration of the copolymers synthesized. In this case, the copolymers would arrange themselves at a nanometric scale but would also be organized at a larger scale, ultimately forming a fully nanostructured material.

An alternative way to prepare block or graft copolymers is to polymerize a monomer, in an extruder in the presence of either end-group-functionalized pre-polymers or polymers that would act as the starter molecules or by judicious choice of initiator [32]. Focusing specifically on the ring-opening polymerization of cyclic monomers, Stevels et al. [33] reported the polymerization of L-lactide initiated by either a hydroxyl-terminated poly-ε-caprolactone (PCL) or a PE glycol. Lee and White [34] used isocyanate-terminated telechelic polytetramethylene ether glycol (PTMEG) premixed with ε-caprolactame to feed a twin-screw extruder to form the polyetheramide triblock copolymer. The conversion of ε-caprolactame is typically around 95 mol% and could be enhanced by devolatilization. Zuniga-Martinez and Yanez-Flores [35] reported the polymerization of the ε-caprolactam monomer in the presence of a linear pre-polymer of poly(ether-esteramide). Madbouly et al. [36] studied the in situ polymerization of macrocyclic carbonates in the presence of PP-g-MA. The degree of polymerization was shown to decrease with increasing concentration of maleic anhydride (MA), which was consistent with the higher proportion of graft copolymer formed. Becquart et al. [37] reported a method for the synthesis of polylactone-grafted EVOH from the in situ polymerization of ε-caprolactone.

Besides, the specificity of some initiators for the ring-opening polymerization of cyclic ester can either enhance or control the expected copolymers structures. To illustrate that, a comparison was made between a commercial titanium tetrapropoxyde (Ti(OPr)$_4$) and the synthesized titanium tetraphenoxyde (Ti(OPh)$_4$) on in situ ε-caprolactone (CL) polymerization to create either grafted or block copolymers. According to the reactions involved, in a model reactive media (4-heptyl acetate as a model molecule of EVA polymer chains), the results obtained [38] clearly demonstrated a statistical distribution of the nature of the chain ends in the case of the polymerization initiated by titanium tetrapropoxyde while the polymer chains initiated by titanium tetraphenoxyde were mostly terminated by heptyl or acetate groups. Actually, these results predict that a more important exchange rate, and thus a higher grafting rate should be obtained with the aromatic initiator in the case of EVA chemical modification.

All of these basic studies allow to extend these concepts to the grafting of polymers able to crystallize (or with high glass transition temperature) at a temperature higher than that of the major phase in order to improve its mechanical properties at higher temperatures. It can be then considered that these grafted segments could act as physical crosslinking domains and the whole material would then have a specific behavior in terms of reversible crosslinking like in the case of a thermoplastic elastomer. As already mentioned, the formation of such nanostructured materials is connected to the architecture of the copolymers, not only to the molar masses and size of the sequences or grafted elements

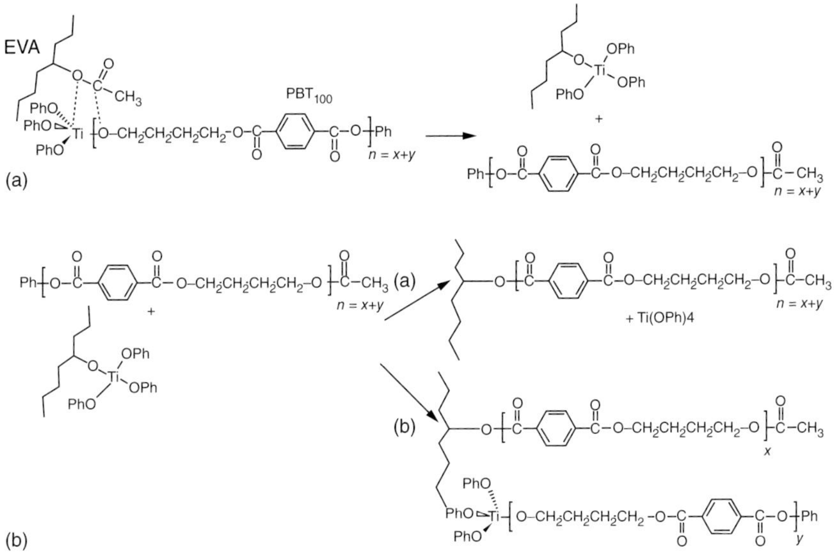

7.3 Polymerization of a Thermoplastic Minor Phase: Toward Blend Nanostructuration

Figure 7.10 Reaction of PBT_{100} grafting resulting in the EVA-g-PBT_{100} copolymer synthesis. (Bahloul et al. 2009 [32]. Reproduced with permission of Elsevier.)

but also to the rate of grafting (concentration in copolymers) and nature of the reactive groups (position of the functional group on the polymer backbone). Within this objective, we extend our studies on the $Ti(OPh)_4$ to the ring-opening polymerization CBT_{100} (cyclic butylene terephthalate) in the presence of EVA (Figure 7.10).

Following the conditions for synthesis (proportion of CBT_{100}/initiator/EVA) carried out in a batch mixer, it has been demonstrated that the average molar mass in number expected for the PBT is about $4154 \, g \, mol^{-1}$ [32]. Furthermore, a very fine morphology of this blend was confirmed by transmission electronic microscopy (TEM) as shown in Figure 7.11.

Finally, the mechanical properties of such a new blend are widely superior to those measured on the blend obtained with cBT_{160}, another commercial cyclic monomer. Actually, as shown in Figure 7.12, a clear and an important increase of the mechanical properties are observed for the reactive system $EVA28\,800/cBT_{100}/Ti(OPh)_4$, which are attributed to the formation of the copolymer EVA-g-PBT_{100} at the interface of the blend according to the scheme in Figure 7.10.

The specificity of the $Ti(OPh)_4$ was also highlighted within the framework of a study dealing with the recycling of the polycarbonate (PC) by reactive extrusion. The objective was then to synthesize PC–PCL (poly-ε-caprolactone) copolymers (ideally block copolymers) by the *in situ* polymerization of the caprolactone (CL) in the presence of PC during melt processing. The purpose was to control the length of the copolymer sequences obtained (Figure 7.13). In this case, the bonds type TiOPh generated *in situ* by the exchange reaction between the carbonate groups and the titanium tetrapropoxyde at $280\,°C$ can initiate the CL polymerization.

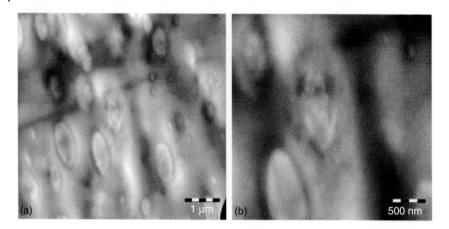

Figure 7.11 Morphology of the blend obtained from the *in situ* CBT polymerization in the presence of (EVA28 800/cBT$_{100}$/Ti(OPh)$_4$). (Bahloul *et al.* 2009 [32]. Reproduced with permission of Elsevier.)

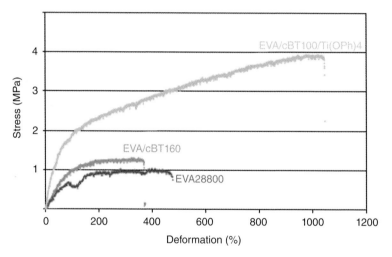

Figure 7.12 Mechanical properties of EVA28 800/cBT$_{100}$/Ti(OPh)$_4$ compared to EVA alone and to the blend obtained with another cyclic monomer (cBT$_{160}$). (Bahloul *et al.* 2009 [32]. Reproduced with permission of Elsevier.)

Typically, the synthesis of PC–PCL copolymer in the extruder is schematized in Figure 7.14. First, the PC granules are introduced through the feeder zone. The titanium tetrapropoxyde is incorporated into the molten PC at $L/D = 12$ so as to generate the new macro-initiator, Ti–O–Ph. Then the CL is injected in order to synthesize the PC/PCL copolymer.

NMR studies clearly highlight the fact that carbonate repetitive units decrease in number according to the proportion of PC/PCL, the rate of initiator, and the time and the temperature of extrusion. Concomitantly, the characteristic signals of the PC–PCL bonds are observed. A major contribution of this study was that it was possibile to control or at least to increase the block character of the

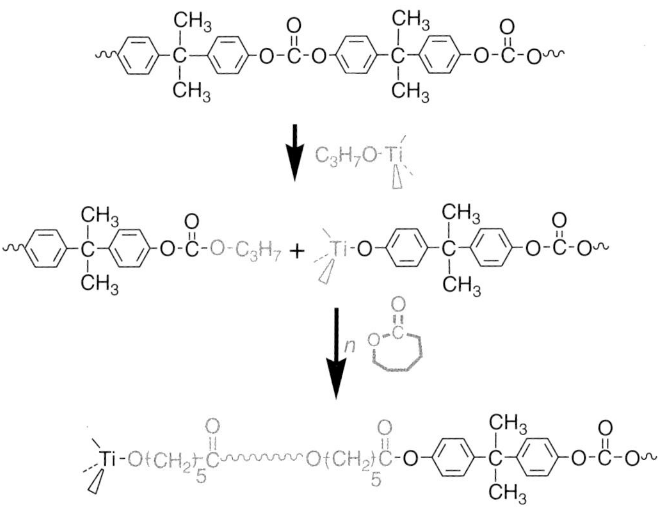

7.3 Polymerization of a Thermoplastic Minor Phase: Toward Blend Nanostructuration

Figure 7.13 Reactional scheme leading to PC–PCL copolymers formation under molten conditions during reactive extrusion.

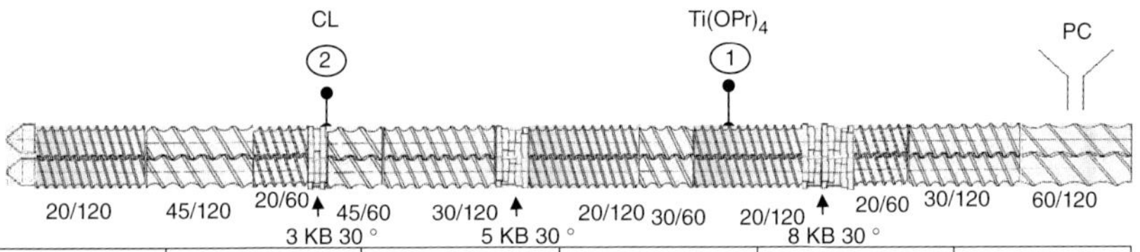

Figure 7.14 Synthesis by reactive extrusion of PC/PCL copolymers. The table shows the different temperature profiles used in this work.

270 °C	270 °C	270 °C	270 °C	280 °C	P
240 °C	270 °C	270 °C	270 °C	280 °C	P1
200 °C	225 °C	250 °C	270 °C	280 °C	P2
185 °C	200 °C	235 °C	270 °C	280 °C	P3
Die					

formed copolymers by modifying the temperature of extrusion. The evolution of the lengths of PC and PCL sequences (L_{PC} and L_{PCL}, respectively) as well as the block character (B) of the synthesized copolymers is presented in Table 7.3. The coefficient (B) is by definition between 0 and 2. It is equal to 0 for homopolymer mixtures, 1 for random copolymers, 1.5 for alternating, and 2 for alternating copolymers

This means that the more copolymers are in blocks, the more B is close to 0.

From these results, the coefficient B is very close to 1 for the first three studied profiles of temperature showing a statistical character of the PC/PCL copolymer. To decrease this factor B, the first method is based on the increase of the k_p/k_t ratio of the polymerization (k_p and k_t are the constant of propagation and transfer rate, respectively) by lowering, for example, the temperatures of extrusion. This is actually possible because the exchange reaction between the titanium tetrapropoxyde and carbonate groups takes place in the first part of the extruder

Table 7.3 Nature of the PC/PCL copolymers (block character B) according to the different temperature profiles (Figure 7.14).

Temperature profile	$(W_{Ti(OPr)4}/W_{PC})_0$	$(W_{PC}/W_{CL})_0$	B	L_{PCL}	L_{PC}
P	1.8/98.2	54/46	0.98	1.85	1.25
P1	1.4/98.6	51/49	0.78	2.64	1.39
P2	1.5/98.5	48/52	0.60	3.81	1.67
P3	1.4/98.6	52/48	0.52	3.95	1.94

leading to chain scissions and consequently to a decrease in the viscosity of the reactional medium. It is then possible after this point ($L/D = 12$) to decrease the barrel temperatures according to the temperature profile P3.

Based on NMR analysis, the synthesized copolymer according to the profile P3 is characterized by a coefficient of randomization equal to 0.52. This result confirms the observation that the choice of the initiator impacts the final morphologies and thus on the properties of the PC/PCL blends. Actually, this study evidences the role of titanium tetraphenoxyde on the grafting rate. This is again a typical illustration of the constraints, such as compatible kinetics, residence time, and temperature, in carrying out reactive extrusion for developing specific initiators.

The synthesis of PE-based grafted copolymers via a melt grafting reaction is always of high interest due to the perpetual developments of polyolefin formulations. Indeed, polyolefins have low surface free energy, strong hydrophobicity, and excellent resistance to chemicals, but the inert and nonpolar character of such materials is a major problem when adhesion to its surface or its mixing is desired. Polyolefins functionalized with MA as a polar monomer are generally used as compatibilizing agents that improve the mechanical properties upon blending with polyamide-6, for example. Regarding the synthesis of polyolefin graft copolymer by reactive extrusion, Badel *et al.* investigated from "grafting from" technique the synthesis of copolymer between poly(ethylene-co-1-octene (Enage copolymer from DOW) and poly (methacrylate) random polymer branches from grafting from technique by reactive extrusion. The *in situ* radical polymerization of methyl methacrylate from the polyolefin backbone was investigated with two different peroxide initiators (dicumyl peroxide (DCP), and 2-ethylhexylcarbonate of tertiobutylhydroperoxide (TBEC)) at 135 and 150 °C, respectively. This grafting of PMMA onto the poly(ethylene-*co*-octene) under molten conditions was done in a co-rotating twin-screw extruder (Leistritz LSM 30–34) at processing temperatures of 135 and 150 °C.

The effect of residence times on MMA conversion and PMMA polymerization degree has been studied. As expected, it was observed that monomer conversion decreases from 53 to 37 mol% by decreasing the residence time from 180 to 60 s. Furthermore, the PMMA molar mass has been kept almost constant irrespective of the residence time and a fine morphology was observed as shown in Figure 7.15.

Figure 7.15 TEM images of the morphology of engage (70%)/PMMA blends by *in situ* synthesis of MMA. (Badel *et al.* 2012 [39]. Reproduced with permission of Wiley.)

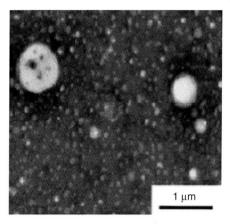

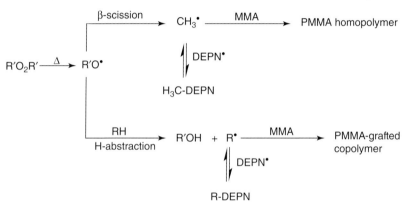

Figure 7.16 Reaction scheme for the poly(ethylene-*co*-1-octene) functionalization with PMMA in the presence of DEPN. (Badel *et al.* 2012 [39]. Reproduced with permission of Wiley.)

The addition of *N-tert*-butyl-*N*-[1-diethylphosphono(2,2-dimethylpropyl)] nitroxide (DEPN) radicals has been carried out to check DEPNs' radical scavenging ability toward methyl radicals (Figure 7.16). As a result, it was observed that monomer conversion decreases from 53 to 26 mol% by increasing DEPN concentration from 0 to 0.8 wt%. Moreover, it is likely that increasing the amount of DEPN results in a decrease of the PMMA average polymerization degree because of the competition that might exist between addition of the radicals formed upon β-scission and H abstraction to engage chains and trapping of radicals by DEPN. Finally, grafting efficiency and grafting degree are not subsequently improved by using DEPN radical scavenger.

In conclusion, the chemical modification of polyolefins under molten conditions by radical chemistry is quite complicated [40]. Actually, it is difficult to ensure the abstraction of hydrogen atoms from the backbone of macromolecular chains at temperatures above 150 °C without any side reactions by using conventional peroxide-based radicals: the free radical process is responsible for degradation (β-scission, cross-linking), and in the case of graftingreactions according to the "grafting from" process, involving a polymer, a monomer, and a peroxide initiator, it results in the formation of both a grafted copolymer and a homopolymer. Thus, it becomes very difficult to control most of the molecular parameters of the chemical species that are produced during these reactions by reactive extrusion. In order to improve the selectivity and to avoid some of the side reactions related to the use of peroxides, tretramethyl piperidino radicals (TEMPO) or indolynoxyl radicals (DPAIO) as radical scavengers at 160 °C have been used [41] or new H-abstracting agents such as N-acetoxy-phthalimide (NAPI), have been for example recently developed to enhance the radical MA grafting onto a PE [42]; however, they are still under investigation for further use in industrial applications.

7.4 Polymerization of a Thermoset Minor Phase Under Shear

The first reference to thermoplastic/thermoset blends (TP/TS) was during the end of the 1960s. Pioneer authors worked on the reinforcement of epoxy thermoset matrices by elastomers of the family of poly(butadiene co-acrylonitrile). They showed that the presence of partially miscible copolymers in the thermoset matrix allows increasing the resistance to cracks propagation appreciably [43]. Then, from the beginning of the 1980s elastomers were replaced by high T_g or high T_m amorphous or semi-crystalline thermoplastics. A multitude of thermoplastics were then tested among which were polyethersulfone (PES), polyetherimide (PEI), poly(phenylene oxidize) (PPO), and polyimides (PI). Mechanical properties were enhanced but the major problem came from the increase of the initial viscosity of the mixtures, which made their processing difficult [44, 45]. Notice that in these studies the thermoset was the matrix.

Later, researchers thought of using the solubility attribute of the epoxy in a variety of polymers in order to facilitate the processing of high glass temperature polymers said to be intractable, that is, very difficult to process because they degrade at the process temperature. The idea was to plasticize these polymers in order to lower their processing temperature [46–48]. As an example, poly(phenylene ether) can be processed at 180 °C instead of 250 °C if 40 wt% of epoxy/amine is added. The concept is the following: the monomers initially act as plasticizers of the high T_g thermoplastic during a mixing stage at moderate temperature in a twin-screw extruder. Then a curing stage out of the extruder is applied to crosslink the epoxy. The epoxy-comonomer play the role of reactive solvents since they are finally polymerized and produce a rigid TS phase.

To summarize, a typical processing scheme is:

1) Mixing the thermoplastic and the monomers in a twin-screw extruder at a temperature selected to limit epoxy reaction to the minimum during this stage. The obtained mixture is homogeneous since monomers are soluble in the thermoplastic polymer.
2) Heating the part at a controlled temperature and time to ensure a complete crosslinking of the epoxy network. Phase separation occurs during this stage leading to a biphasic material.

The final material is a thermoplastic with high T_g crosslinked inclusions or more rarely a thermoset with thermoplastic-rich inclusion. Mechanical and thermomechanical properties are very good but a drawback of this concept is that it implies two steps: mixing and then curing at high temperature for a period that may be relatively long [49, 50].

Based on the above observations, a few authors have extended the TP/TS concept to systems where the thermoplastic is the major phase and the monomers are selected so that they can crosslink very fast. In that particular case, mixing and curing may be achieved in the same operation. Notice that the thermoset is crosslinked inside the processing equipment, under shear. This situation has been scarcely studied.

Examples of works where thermoset precursors are crosslinked inside a thermoplastic matrix, under shear, are presented in the following text. Two situations are encountered. First, the thermoset precursors may be miscible in the thermoplastic, forming initially a homogeneous mixture that will phase separate as polymerization proceeds (Reaction Induced Phase Separation or RIPS). Compositions and properties of both phases evolve during the reaction. Second, the monomers may be immiscible in the thermoplastic; thus, no phase separation is observed and only the dispersed phase properties evolve.

7.4.1 Thermoplastic Polymer/Epoxy-Amine Miscible Blends

Vivier [51] has studied a system with polystyrene (PS) as the major phase and several epoxy/amine couples with different reactivities and degrees of miscibility with PS. Diglycidyl ether of bisphenol A (DGEBA) was the epoxy used, isophorone diamine (IPD) was the curing agent selected because it was miscible with PS. Mixing of PS/DGEBA was realized in an internal mixer at 180 °C and IPD was added subsequently. Crosslinking of DGEBA/IPD proceeded either inside the internal mixer (dynamic curing) or inside a compression press (static curing). Particles obtained in static conditions are spherical with a diameter around 1.4 μm while dynamic curing produces large irregular particles (Figure 7.17).

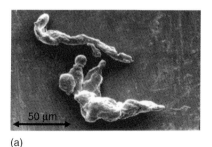

(a)

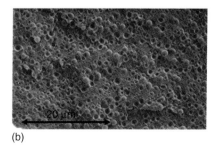

(b)

Figure 7.17 DGEBA/IPD particles after crosslinking of a PS/25% DGEBA + IPD blend (SEM images). (a) The blend was mixed and crosslinked at 180 °C in the internal mixer at 64 rpm. Large irregular shape particles of DGEBA/IPD were obtained. (b) The blend was mixed at 180 °C in an internal mixer and then crosslinked in static conditions in a compression press. 1.4 μm spherical particles were formed [51].

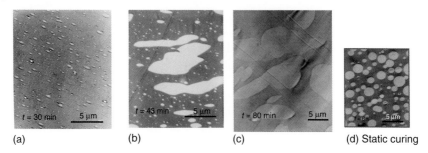

Figure 7.18 PS/40 wt% DGEBA–MDEA blend cured at 177 °C. Transmission electron microscopy images of the blend crosslinked in dynamic conditions inside the internal mixer during (a) $t = 30$ min, (b) $t = 43$ min, (c) $t = 80$ min. (d) Same blend cured in static conditions during 6 h. DGEBA–MDEA appears in light gray. (Meynié et al. 2004 [53]. Reproduced with permission of Elsevier.)

The above observations gives the opportunity to comment on a typical behavior of monomers that crosslink under shear, that is, the morphology of the thermoset coarsens drastically when the monomers are reacting under shear.

Meynié [52, 53] focused also on PS with DGEBA and, after having tested different amines, used 4,4′-methylenebis(2,6-diethylaniline) (MDEA) and a fraction of 40 wt% epoxy–amine as a model system. This amine reacts slower with DGEBA than IPD and a more detailed analysis of the evolution of the system with mixing time was possible. The blend was crosslinked at 177 °C either in static curing (hot press) or in dynamic curing (internal mixer) and extensive characterization was undertaken with measurement of conversion, gel fraction, gel time, T_g of the two phases, dispersed phase morphology, and size distribution. The micrographs show the drastic evolution of the epoxy drops morphology and a final coarse morphology when compared to the blend crosslinked in the press (Figure 7.18). This confirms the findings of Vivier and a more precise insight on the mechanisms involved in coarsening is possible. The blend is initially miscible and homogeneous. As soon as a critical conversion is reached, phase separation occurs, creating a phase of droplets rich in epoxy–amine while the matrix is rich in PS but still contains epoxy–amine (Figure 7.18a). The size of the droplets increases as reaction/phase separation proceeds. However, when the gel point is attained in the drop, breakup mechanism is inhibited because of the high viscosity of the epoxy–amine phase. Coalescence is enhanced and leads to the large irregular crosslinked particles visible in Figure 7.18b. At this intermediate stage, new epoxy–amine droplets are continuously formed by phase separation. They constitute a second population of drops that coexist with the large irregular particles (Figure 7.18b). These new drops rapidly aggregate so that finally only irregular particles remain. The final structure is coarser and more irregular compared to that formed in static curing (compare Figure 7.18c,d). The overall process is summarized in Figure 7.19.

In contrast to morphology, reaction rate and phase compositions are not affected by the curing mode (under shear or in static). Figure 7.20a shows that the evolution of the conversion of epoxy as a function of time is the same when

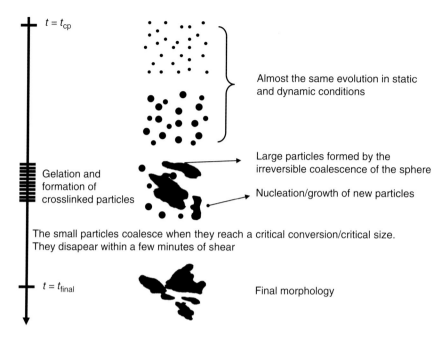

Figure 7.19 Scheme of the proposed mechanism for morphology evolution of initially miscible epoxy/amine crosslinked under shear in a thermoplastic matrix. (Meynié et al. 2004 [53]. Reproduced with permission of Elsevier.)

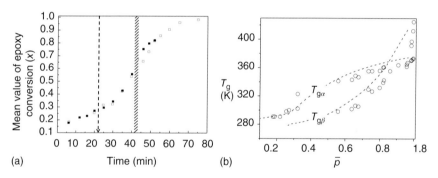

Figure 7.20 PS/40 wt% DGEBA–MDEA blend cured at 177 °C. (a) Epoxy conversion of the blend (□) crosslinked in dynamic condition inside the internal mixer (■) cured in static condition in a hot press [53]. (Reproduced with permission of Elsevier.) (b) T_g data of PS-rich phase ($T_{g\alpha}$) and of the epoxy-rich phase ($T_{g\beta}$) compared to data obtained from simulation as a function of epoxy conversion p. (Riccardi 2004 [54]. Reproduced with permission of Wiley.)

the blend is cured in the mixer or in the press. This confirms the conclusion of Vivier based on his qualitative observations [51].

During the process, PS phase is progressively purifying as can be seen on the $T_{g\alpha}$ value that tend to the neat PS value (Figure 7.20b). $T_{g\beta}$ of the epoxy-rich phase also increases. The figure presents T_g data points for both static and dynamic curing as a function of the average epoxy conversion, p. Notice that although

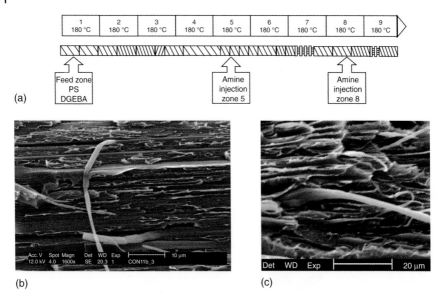

Figure 7.21 (a) Screw and temperature profile of the twin-screw extruder used to prepare PS/DGEBA-AEP oriented blends. The two injection zones of the amine, Z5 and Z8 are indicated. (b) Morphology of PS/DGEBA, no amine, draw ratio = 3 (c) Morphology of PS/DGEBA-AEP, amine injected in zone 8 of the extruder, draw ratio = 3, conversion of the epoxy = 71 wt%. (Fenouillot and Perier-Camby 2004 [56]. Reproduced with permission of Wiley.)

very complex, this system may be modeled. The simulation data calculated by Riccardi et al. [54] are in good agreement with the measured T_g. The rheological evolution of the modulii may also be predicted with a yield model established by Tribut et al. [55].

When applying the concept developed with model TP/epoxy-amine blends to reactive extrusion where residence time is short, one must select a very reactive amine. Fenouillot and Perier-Camby [56] have selected 1-(2-aminoethyl)piperazine (AEP). The objective was to form a fibrillar crosslinked epoxy phase in PS matrix. They used a solid DGEBA that was not miscible in PS while AEP was soluble. By elongating the extrudate at controlled draw ratios at the die exit it was possible to obtain DGEBA fibers in PS (Figure 7.21b). However, when the AEP was injected in the extruder the fibrillar structure coarsened (compare Figure 7.21b,c). Here again, coalescence and also the loss of deformability of epoxy drops was critical when epoxy/amine reacted. The fibrils were preserved only when the amine was injected in zone 5 and the conversion of epoxy was lower but a post-cure was then necessary to increase conversion.

From an applicative point of view, the concepts described above on model blends have been used to formulate a poly(ethylene-*co*-vinyl acetate) (EVA) based material for cable insulation [57]. The thermoset system was a phenol/formaldehyde (Novolac) resin crosslinked with hexamethylene tetramine (HMTA). This thermoset is known to have excellent flame-retardant properties

Table 7.4 Mechanical properties and fire resistance of EVA/novolac-HMTA + fire-retardant fillers prepared and polymerized by reactive extrusion.

Blends	Mechanical properties		Calorimeter cone results		
	Stress at break (MPa)	Elongation at break (%)	Ignition time (s)	Peak heat release (kW m^{-2})	Mean heat release (kW m^{-2})
EVA	25.2	760	40	1470	490
EVA/Novolac 80/20	26.6	630	40	825	175
EVA/Novolac-HMTA 80/20	22.3	660	39	680	132
EVA/Novolac-HMTA 80/20 + MDH 50 phr	7.1	300	52	290	100
EVA/Novolac-HMTA 80/20 + MDH 150 phr	11.2	50	78	164	64
EVA + ATH 150 pph	5.5	170	60	190	87
EVA/Novolac-HMTA 80/20 + ATH 150 phr	6.9	70	88	155	71

Source: Reproduced with permission of Fournier et al. 2011 [57].

and the idea was to incorporate it into polyolefins that are classically used in the cable industry. The EVA/Novolac-HMTA 70/30 and 50/50 blends could be easily processed and polymerized in a twin-screw extruder at moderate temperature with a specific screw profile to limit temperature rise. The resulting TP/TD material was then deposited on the cable by co-extrusion with a single-screw extruder. The Novolac/HMTA crosslinked particles obtained after extrusion had a diameter equal to 1.5 and 3.3 μm for 70/30 and 50/50 blends, respectively.

Mechanical properties were measured by tensile testing and flame retardancy was evaluated with a calorimeter cone at 50 kW m^{-2} (Table 7.4). For optimization, magnesium dihydroxide (MDH) or aluminum trihydroxide (ATH) were added to the blends.

The positive effect of crosslinked phenol-formaldehyde resin on flame retardancy of the material is clear, especially when considering heat release peak and average values. 20 wt% of Novoloc-HMTA decreases the mean heat release drastically, 132 compared to 490 kW m^{-2} with a low impact on ignition time. The addition of only 50 pph of MDH to the composition has a positive effect on ignition time and heat release peak.

When epoxy and amine monomers are immiscible in the thermoplastic, the situation is simpler. The blend is initially heterogeneous as in the PE PE/DGEBA/DDM (4,4'-diaminophenyl methane) blend studied by Vivier [51]. Epoxy and amine react in a common phase and no phase separation is encountered. Vivier succeeded in obtaining a relatively fine dispersion of 5.5 wt% of DGEBA droplets of 3 μm diameter in PE; however, as soon as crosslinking proceeded under shear the final particles were very large (>10 μm). Furthermore, his attempts to compatibilize this blend by using a copolymer failed.

7.4.2 Examples of Stabilization of Thermoplastic Polymer/Epoxy-Amine Blends

The first idea is to stabilize TP/TS blends in the same way classical polymer blends are stabilized. That is, by the addition of a copolymer able to localize at the interface. Meynié *et al.* [58] have tested a polystyrene-poly(methyl methacrylate) (PMMA) block copolymer (P(S-*b*-MMA)) associated with their PS/DGEBA/MDEA model blend. PMMA is miscible in DGEBA and has also proved to be soluble in epoxy networks, and therefore it is expected that the copolymer will be stable at the interface [59]. Indeed, the size of DGEBA–MDEA particles after dynamic curing are two to three times smaller when 15 wt% of PS-*b*-PMMA is added. However, it appeared that the copolymer is expelled from the interface when DGEBA–MDEA phase reaches the gel point. The efficiency of this stabilization is thus limited.

The second idea is to use a reactive copolymer. Vivier has tested a PS functionalized with epoxy or succinic anhydride that is able to react fast with the diamine. The authors used 6.5 wt% of this PS and the final morphology was not enhanced [51]. Meynié *et al.* have used a P(S-*b*-MMA) copolymer containing about 10 epoxy groups on the PMMA block (SMX copolymer). These epoxy groups were expected to react with MDEA and prevent the copolymer from being expelled from the interface. Epoxy-amine particles much smaller than those without copolymer (1 µm diameter) are obtained although they were not spherical (Figure 7.22) [58].

These results prove that a reactive copolymer is able to limit coalescence of the epoxy-amine drops drastically but its amount must be relatively high.

7.4.3 Blends of Thermoplastic Polymer with Monomers Crosslinking via Radical Polymerization

For a thermoset polymerized by radical reactions, the gel point is observed at a much lower conversion than in thermoset polymerized by step polymerization. Thus we may expect that phase separating drops will gel very fast so that

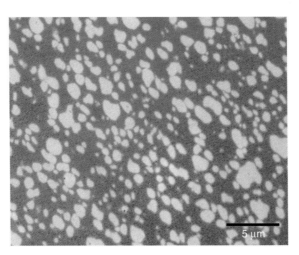

Figure 7.22 PS/40 wt% DGEBA–MDEA with 15 wt% SMX reactive copolymer cured at 177 °C in the internal mixer. Transmission electron microscopy image where DGEBA–MDEA appears in light gray. (Meynié *et al.* 2005 [58]. Reproduced with permission of Wiley.)

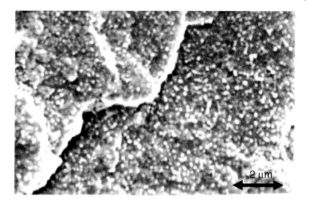

Figure 7.23 PS + 25 wt% trimethylol propane triacrylate + initiator cured at 130 °C in the internal mixer [51].

collision/coalescence events will not have time to agglomerate the TS precursor drops too much. This very interesting idea has been studied by Vivier [51, 60]. He used PS as the matrix and trimethylol propane triacrylate (TRIM) as the monomer polymerized with a thermal initiator. The monomer was totally soluble in PS above 60 °C and served as the plasticizer. The morphology of the blend crosslinked under shear at 130 °C is radically different from what was obtained with PS/DGEBA/IPD. Very small nodules with a diameter around 100–200 nm were obtained (Figure 7.23). Depending on the initiator and temperature, a partial crosslinking of the PS was noted.

Different reasons may explain such a fine morphology: A phase separation that proceeds differently with radical reactions crosslinking and especially an early gelation of the droplets could prevent their coalescence. Also, the formation of PS–TRIM copolymers at the interface, stabilizing the dispersion could have a positive effect on the morphology.

Based on this model blend, several compositions with PMMA, polyvinylidene fluoride (PVDF), or polyamide 12 (PA12) matrix with different network precursors have been patented. PVDF- and PA12-based blends proved to have a much better impact than the virgin thermoplastic [60].

7.5 Conclusion

The objective of this chapter was to describe how new types of polymer materials can be synthesized by reactive extrusion. In the first part, the *in situ* synthesis of an inorganic phase inside a thermoplastic matrix during extrusion processing was investigated. The formation of nanocomposites and organic–inorganic hybrids by sol–gel chemistry was then highlighted. The reactive extrusion is presented as a unique and elegant way to build a nanocomposite by controlling the hydrolysis–condensations reactions of inorganic precursors in a molten polymer. Particular interest in such an approach is to take benefits of shearing and high extrusion temperature to enhance the reactions conversion and fine morphologies. By the same unique reactive step, specific functional inorganic precursors also allow to propose functionalized nanocomposites. In that context, functionalities of these new materials relevant to a number of applications, such as mechanical reinforcement, fire retardancy, and biocide properties were presented and discussed.

In the second part, the polymerization and/or crosslinking of thermoplastic and thermoset minor phase under shear was described. It was then shown that the development and the control of the morphology from nanoscale to microscale can be achieved under specific conditions of processing. The approach of grafting from or elaboration of block copolymers by *in situ* polymerization of a monomer in a molten polymer has been discussed in detail in terms of specific parameters such as molar masses and number of reactive sites. The influence of reactive extrusion conditions and the role of a typical initiator have also been investigated in order to enhance and to control the desired final structure and morphology.

Concerning thermoset precursor polymerizing inside a thermoplastic polymer in an extruder, the control of the final size and shape of the thermoset particles is very tricky for step polymerization monomers. After the gel point is reached, irreversible coalescence dominates the process and fine morphologies cannot be obtained without adding a substantial amount of compatibilizing reactive copolymer. Nevertheless, sometimes the size of the morphology is not critical for obtaining interesting properties. With chain polymerization monomers precursors of a thermoset, the dispersed phase gels at very low conversion and interfacial free radical reactions are possible so that fine structures can be expected.

References

1 Cassagnau, P., Bounor-Legaré, V., and Fenouillot, F. (2007) Reactive processing of thermoplastic polymers: a review of the fundamental aspects. *Int. Polym. Process.*, **22**, 218–258.

2 Kango, S., Kalia, S., Celli, A., Njuguna, J., Habibi, Y., and Kumar, R. (2013) Surface modification of inorganic nanoparticles for development of organic–inorganic nanocomposites—a review. *Prog. Polym. Sci.*, **32**, 1232–1261.

3 Novak, B.M. (1993) Hybrid nanocomposite materials—between inorganic glasses and organic polymers. *Adv. Mater.*, **5**, 422–433.

4 Schmidt, H., Jonschker, G., Goedicke, S., and Mennig, M. (2000) The sol–gel process as a basic technology for nanoparticle-dispersed inorganic–organic composites. *J. Sol-Gel Sci. Technol.*, **19**, 39–51.

5 Brinker, C.J. and Scherrer, G.W. (1990) *Sol–gel Science: The Physics and Chemistry of Sol–gel Processing*, Academic Press, San Diego, Calif, 908 pp.

6 Bounor-Legaré, V. and Cassagnau, P. (2014) *In situ* synthesis of organic–inorganic hybrids or nanocomposites from sol–gel chemistry in molten polymers. *Prog. Polym. Sci.*, **39** (8), 1473–1497.

7 Bahloul, W., Oddes, O., Bounor-Legaré, V., Mélis, F., Cassagnau, P., and Vergnes, B. (2011) Reactive extrusion processing of polypropylene/TiO_2 Nanocomposites by *In situ* synthesis of the nanofillers: experiments and modeling. *AIChE J.*, **57** (8), 2174–2184.

8 Kluenker, E., Faymonville, J., Peter, K., Moeller, M., and Hopmann, C. (2014) Reactive extrusion processing of polypropylene/SiO$_2$ nanocomposites by *in situ* synthesis of the nanofillers: experiments and properties. *Polymer*, **55**, 5370–5380.

9 Oliveira, M. and Machado, A.V. (2014) Nanocomposites prepared by reactive extrusion: effect of the polymer reactive groups. *Macromol. React. Eng.*, **8**, 134–140.

10 Miloskovska, E., Hristova-Bogaerds, D., van Duin, M., and de With, G. (2015) *In situ* silica–EPDM nanocomposites obtained via reactive processing. *Eur. Polym. J.*, **69**, 260–272.

11 Vergnes, B., Della Valle, G., and Delamare, L. (1998) A global computer software for polymer flows in corotating twin screw extruders. *Polym. Eng. Sci.*, **38**, 1781–17929.

12 Cassagnau, P. and Fenouillot, F. (2004) Rheological study of mixing in molten polymers: 1-mixing of low viscous additives. *Polymer*, **45** (23), 8019–8030.

13 Cassagnau, P. and Fenouillot, F. (2004) Rheological study of mixing in molten polymers: 2-mixing of reactive systems. *Polymer*, **45** (23), 8031–8040.

14 Balhoul, W., Melis, F., Bounor-Legaré, V., and Cassagnau, P. (2012) Structural characterisation and antibacterial activity of PP/TiO$_2$ nanocomposites prepared by an *in situ* sol–gel method. *Mater. Chem. Phys.*, **134**, 399–406.

15 Bahloul, W., Bounor-Legaré, V., David, L., and Cassagnau, P. (2010) Morphology and viscoelasticity of PP/TiO$_2$ nanocomposites prepared by *in situ* sol–gel method. *J. Polym. Sci. B Polym. Phys.*, **48** (11), 1213–1222.

16 Kubacka, A., Serrano, C., Ferrer, M., Lunsdorf, H., Bielecki, P., Luisa, M., Fernandez-Garcia, M., and Fernandez-Garcia, M. (2007) High-performance dual-action polymer-TiO$_2$ nanocomposite films via melting processing. *Nano Lett.*, **7**, 2529–2534.

17 Fenoglio, I., Greco, G., Livraghi, S., and Fubini, B. (2009) Non-UV-induced radical reactions at the surface of TiO$_2$ nanoparticles that may trigger toxic responses. *Chem. Eur. J.*, **15**, 4614–4621.

18 Theil-Van Nieuwenhuyse, P., Bounor-Legaré, V., Bardollet, P., Cassagnau, P., Michel, A., David, L., Babonneau, F., and Camino, G. (2013) Phosphorylates Silica/PA6 nanocomposites by *in situ* sol–gel method in molten conditions: impact on the fire-retardancy. *Polym. Degrad. Stab.*, **98**, 2635–2644.

19 Sahyoun, J., Bounor-Legare, V., Ferry, L., Sonnier, R., Da Cruz-Boisson, F., Melis, F., Bonhomme, A., and Cassagnau, P. (2015) Synthesis of a new organophosphorous alkoxysilane precursor and its effect on the thermal and fire behavior of a PA66/PA6 copolymer. *Eur. Polym. J.*, **66**, 352–366.

20 Sahyoun, J., Bounor-Legare, V., Ferry, L., Sonnier, R., Bonhomme, A., and Cassagnau, P. (2015) Influence of organophosphorous silica precursor on the thermal and fire behaviour of a PA66/PA6 copolymer. *Polym. Degrad. Stab.*, **115**, 117–128.

21 Seck, S., Magana, S., Prebe, A., Niepceron, F., Bounor-Legare, V., Bigarre, J., Buvat, P., and Gerard, J.F. (2015) PVDF-HFP/silica-SH nanocomposite synthesis for PEMFC membranes through simultaneous one-step sol–gel reaction and reactive extrusion. *Mater. Chem. Phys.*, **163**, 54–62.

22 Kye, H. and White, J.L. (1996) Continuous polymerization of caprolactam-polyether sulfone solutions in a twin screw extruder to form reactive polyamide-6/polyether sulfone blends and their melt spun fibers. *Int. Polym. Process.*, **11** (4), 310–319.

23 Cassagnau, P., Niestch, T., Bert, M., and Michel, A. (1998) Reactive blending by *in situ* polymerization of the dispersed phase. *Polymer*, **40**, 131–138.

24 Charoensirisomboon, P., Inoue, T., and Weber, M. (2000) Pull-out of copolymer in situ-formed during reactive blending: effect of the copolymer architecture. *Polymer*, **41**, 6907–60912.

25 Ruzette, A.V. and Leibler, L. (2002) Block copolymers in tomorrow's plastics. *Nat. Mater.*, **4**, 19–31.

26 Flat, J.J. (2007) New comb-like nanostructured copolymers: a promising route towards new industrial applications. *Polym. Degrad. Stab.*, **90**, 2278–2286.

27 Pernot, H., Baumert, M., Court, F., and Leibler, L. (2002) Design and properties of co-continuous nanostructured polymers by reactive blending. *Nat. Mater.*, **1**, 54–58.

28 Freluche, M., Iliopoulos, I., Milléquant, M., Flat, J.J., and Leibler, L. (2006) Graft copolymers of poly(methyl methacrylate) and polyamide-6: synthesis by reactive blending and characterization. *Macromolecules*, **39**, 6905–6912.

29 Freluche, M., Iliopoulos, I., Flat, J.J., Ruzette, A.V., and Leibler, L. (2005) Self-organized materials and graft copolymers of polymethylmethacrylate and polyamide-6 obtained by reactive blending. *Polymer*, **46**, 6554–6562.

30 Harrats, C., Thomas, S., and Groeninckx, G. (2006) *Micro- and Nanostructured Multiphase Polymer Blend Systems*, CRC Press LLC.

31 Baker, W., Scott, C., and Hu, G.H. (2001) *Reactive Polymer Blending*, Hanser Publishers, Munich.

32 Bahloul, W., Bounor-Legare, V., Fenouillot, F., and Cassagnau, P. (2009) EVA/PBT nanostructured blends synthesized by *in situ* polymerization of cyclic cBT (cyclic butylene terephthalate) in molten EVA. *Polymer*, **50** (12), 2527–2534.

33 Stevels, M., Bernard, A., Van de Witte, P., Dijstra, P.J., and Feijen, J. (1996) Block copolymers of poly(L-lactide) and poly(ε-caprolactone) or poly(ethylene glycol) prepared by reactive extrusion. *J. Appl. Polym. Sci.*, **62**, 1295–1301.

34 Lee, B.H. and White, J.L. (2002) Formation of a polyetheramide triblock copolymer by reactive extrusion; process and properties. *Polym. Eng. Sci.*, **42** (8), 1710–1723.

35 Zuniga-Martinez, J.A. and Yanez-Flores, J.G. (2004) Synthesis by reactive processing of block copolymers of nylon6/poly(ether-esteramide). *Polym. Bull.*, **53**, 25–34.

36 Madbouly, S.A., Otaigbe, J.U., and Ougizawa, T. (2006) Morphology and properties of novel blends prepared from simultaneous in situ polymerization and compatibilization of macrocyclic carbonates and maleated poly(propylene). *Macromol. Chem. Phys.*, **207**, 1233–1243.

37 Becquart, F., Taha, M., Zerroukhi, A., Chalamet, Y., Kaczun, J., and Llauro, M.F. (2007) Microstructure and properties of poly(vinyl alcohol-*co*-vinyl acetate)-*g*-ε-caprolactone. *Eur. Polym. J.*, **43** (4), 1549–1556.

38 Valle, M. (2006) Synthèse de copolymères greffés EVA ou EMA/poly (ε-caprolactone) à architectures contrôlées par polymérisation de l'ε-caprolactone en présence d'EVA ou d'EMA au cours d'une opération d'extrusion. PhD thesis. University Lyon 1.
39 Badel, T., Beyou, E., Bounor-Legare, V., Chaumont, P., Cassagnau, P., Flat, J.J., and Michel, A. (2012) Synthesis of poly(methyl methacrylate)-grafted poly(ethylene-co-1-octene) copolymers by a "grafting from" melt process. *Macromol. Mater. Eng.*, **297** (7), 702–710.
40 Passaglia, E., Coiai, S., and Augier, S. (2009) Control of macromolecular architecture during the reactive functionalization in the melt of olefin polymers. *Prog. Polym. Sci.*, **34**, 911–947.
41 Belekian, D., Beyou, E., Chaumont, P., Cassagnau, P., Flat, J.J., Quinebeche, S., Guillaneuf, Y., and Gigmes, D. (2015) Effect of nitroxyl-based radicals on the melt radical grafting of maleic anhydride onto polyethylene in presence of a peroxide. *Eur. Polym. J.*, **66**, 342–351.
42 Belekian, D., Cassagnau, P., Flat, J.J., Quinebeche, S., Autissier, L., Bertin, D., Siri, D., Gigmes, D., Guillaneuf, Y., Chaumont, P., and Beyou, E. (2013) N-Acetoxy-phthalimide (NAPI) as a new H-abstracting agent at high temperature: application to the melt functionalization of polyethylene. *Polym. Chem.*, **9**, 2676–2679.
43 Riew, C.K. and Gillham, J.K. (1984) *Rubber-Modified Thermoset Resins*, Advanced Chemical Science, ACS, Washington DC, p. 208.
44 Pascal, T., Mercier, R., and Sillon, B. (1990) New semi-interpenetrating polymeric networks from linear polyimides and thermosetting bismaleimides: 2. Mechanical and thermal properties of the blends. *Polymer*, **31**, 78–83.
45 Pascault, J.P. (2000) Formulation and characterization of thermoset-thermoplastic blends, in *Polymer Blends: Volume I* (eds D.R. Paul and C.B. Bucknall), John Wiley & Sons, New York, pp. 379–415.
46 Pearson, R.A. and Yee, A. (1993) The preparation and morphology of PPO-epoxy blends. *J. Appl. Polym. Sci.*, **48**, 1051–1060.
47 Jansen, B.J.P., Meijer, H.E.H., and Lemstra, P.J. (1999) Processing of intractable polymers using reactive solvents, part 5: morphology control during phase separation. *Polymer*, **40**, 2917–2927.
48 Venderbosch, R.W., Meijer, H.E.H., and Lemstra, P.J. (1995) Processing of intractable polymers using reactive solvents: 2. Poly(2,6-dimethyl-1,4-phenylene ether) as a matrix material for high performance composites. *Polymer*, **36** (6), 1167–1178.
49 Bonnet, A., Pascault, J.P., Sautereau, H. et al. (1999) Epoxy-diamine thermoset/thermoplastic blends. 1. Rates of reaction before and after phase separation. *Macromolecules*, **32** (25), 8517–8523.
50 Venderbosch, R.W., Meijer, H.E.H., and Lemstra, P.J. (1994) *Polymer*, **33** (20), 4349–4357.
51 Vivier, T. (1996) Formation de nodules thermodurcis au sein d'une matrice thermoplastique. PhD thesis 96 STR1 3164. Louis Pasteur University of Strasbourg, France.
52 Meynié, L. (2003) Evolution et contrôle de la morphologie d'un mélange thermoplastique/thermodurcissable polymérisé sous cisaillement. PhD thesis, 03ISAL0066. INSA-Lyon.

53 Meynié, L., Fenouillot, F., and Pascault, J.P. (2004) Polymerization of a thermoset into a thermoplastic matrix: effect of the shear. *Polymer*, **45** (6), 1867–1877.

54 Riccardi, C.C., Borrajo, J., Meynié, L., Fenouillot, F., and Pascault, J.P. (2004) Thermodynamic analysis of the phase separation during the polymerization of a thermoset system into a thermoplastic matrix. Part 2: prediction of phase composition and volume fraction of dispersed phase. *J.Polym. B*, **42** (8), 1361–1368.

55 Tribut, L., Carrot, C., Fenouillot, F., and Pascault, J.-P. (2008) A phenomelogical modification of rheological models for concentrated two-phase sytems: application to a thermoplastic/thermoset blend. *Rheol. Acta*, **47** (4), 459–468.

56 Fenouillot, F. and Perier-Camby, H. (2004) Formation of a fibrillar morphology of crosslinked epoxy in a polystyrene continuous phase by reactive extrusion. *Polym. Eng. Sci.*, **44** (4), 625–637.

57 Fournier, J., Piechaczyk, A., Pinto, O., Pascault, J.-P., Fenouillot, F., and Tribut, L. (2011) Power and/or telecommunications cable having improved fire-retardant properties. US Patent 2,011,301,274 (A1).

58 Meynié, L., Habrard, A., Fenouillot, F., and Pascault, J.P. (2005) Limitation of the coalescence of evolutive droplets by the use of copolymers in a thermoplastic/thermoset blend. *Macromol. Mater. Eng.*, **290** (9), 906–911.

59 Ritzenthaler, S., Girard-Reydet, E., and Pascault, J.P. (2000) Influence of epoxy hardeners on miscibility of blends of PMMA and epoxy networks. *Polymer*, **41** (16), 6375–6386.

60 Bertin, D., Bouilloux, A., Teze, L., and Vivier, T. (2004) Thermoplastic resin compositions comprising a rigid dispersed phase. US Patent 6,462,129 (B1), Atofina patent.

8

Concept of (Reactive) Compatibilizer-Tracer for Emulsification Curve Build-up, Compatibilizer Selection, and Process Optimization of Immiscible Polymer Blends

Cai-Liang Zhang[1], Wei-Yun Ji[1], Lian-Fang Feng[1], and Guo-Hua Hu[2]

[1] *State Key Laboratory of Chemical Engineering, College of Chemical and Biochemical Engineering, Zhejiang University, Hangzhou 310027, China*
[2] *Université de Lorraine, CNRS, Laboratoire Rèactions et Gènie des Procèdès (UMR 7274), 1 rue Grandville, BP 20451, 54001 Nancy, France*

8.1 Introduction

Blending existing polymers is a very attractive and inexpensive way of obtaining new materials without the need to create new molecules [1, 2]. Polymer blending processes use mainly batch mixers and continuous mixers of type-screw extruders, especially twin-screw extruders. The former are mainly used in the laboratory and the latter both in the laboratory and for industrial production.

Since most polymers are mutually immiscible, a compatibilizer of type block or graft copolymer is often required to reduce the interfacial tension, promote the dispersion of one of the polymer components in another, and stabilize the morphology of the resulting polymer blend. A so-called emulsification curve, which is the evolution of the diameter of the dispersed phase domain (DD) of a polymer blend as a function of the compatibilizer concentration (CC), can be used to evaluate the interfacial behavior and compatibilizing efficiency of the compatibilizer [2–8].

When a batch mixer is used, the emulsification curve of a polymer blend can be built up in the following manner. A given composition of the polymer components of the blend as well as that of the compatibilizer is charged to the mixer. After a certain lapse of time (usually 5–10 min), the process reaches a steady state and samples are taken from the mixer. The DD of the blend is measured. This makes up a point on the emulsification curve. The above process is repeated upon varying the CC. An emulsification curve is then built up for the said composition of the blend. The case of a screw extruder is similar to a batch mixer. So-called steady-state experiments can be carried out, namely, a given composition of the polymer components of the blend as well as the compatibilizer is charged to the extruder. After a certain lapse of time (usually more than 10 times the average residence time (RT)), the process reaches a steady state and samples are taken from the die exit for measurement.

Reactive Extrusion: Principles and Applications, First Edition. Edited by Günter Beyer and Christian Hopmann.
© 2018 Wiley-VCH Verlag GmbH & Co. KGaA. Published 2018 by Wiley-VCH Verlag GmbH & Co. KGaA.

Studies reported in literature concerning emulsification curves often use batch mixers and not extruders. The reason is that batch mixers are often more accessible in a laboratory than extruders and they are much easier to operate. More importantly, the amount of the compatibilizer required for building up an emulsification curve in a batch mixer is often much smaller than that required in an extruder. This is especially true for a pilot or industrial-scale screw extruder the throughput of which can vary from a few dozens of kilograms per hour to a few tones per hour. In such a case, the amount of the compatibilizer required to build up an emulsification curve is uneconomically high. However, a compatibilizer which is very efficient in a batch mixer or a laboratory scale extruder may become inefficient in an industrial-scale extruder due to differences in thermo-mechanical conditions. Therefore, the compatibilizing efficiency of a compatibilizer has to be evaluated under the conditions in which it is used.

To overcome problems associated with situations where the amount of the compatibilizer available is too small to build up an emulsification curve, a so-called (reactive) compatibilizer-tracer concept is developed. The idea is based on transient experiments used for measuring residence time distributions (RTD) [8–12]. Unlike a steady-state experiment in which a given composition of the polymer components of the blend and the compatibilizer is charged to the extruder altogether, in a transient experiment the polymer components are first fed to the extruder. When the process reaches its steady state, a given small amount of the compatibilizer is introduced to the extruder as a pulse. Samples are taken at the die exit as a function of RT. Both the evolution of the CC and that of the DD as a function of time can be obtained. The former provides the compatibilizer concentration distribution (CCD) and the latter the diameter of the dispersed phase domain distribution (DDD) of the blend. From these two distributions, an emulsification curve can be easily deduced.

8.2 Emulsification Curves of Immiscible Polymer Blends in a Batch Mixer

Very much like classical emulsions, it is convenient to evaluate the compatibilizing efficiency of a compatibilizer for an immiscible polymer blend based on an emulsification curve [6, 13, 14]. For example, Zhang et al. [8] built up and then used emulsification curves to compare the compatibilizing efficiencies of four graft copolymers (PS-g-PA6) with PS as a backbone and PA6 as grafts in PS/PA6 blends. The matrix PS was a commercial grade of BASF Yangzi Petrochemical Co., Ltd, China. The dispersed phase PA6 was of a commercial grade from UBE Nylon Ltd, Thailand. The molecular structures of four PS-g-PA6 grade copolymers are shown in Table 8.1. The PS and PA6 compositions of PS-g-PA6a were similar to those of PS-g-PA6b. PS-g-PA6a and PS-g-PA6b bore 15.3 and 7.1 PA6 grafts per PS backbone, respectively. PS-g-PA6b and PS-g-PA6c had the same

Table 8.1 Selected characteristics of four PS-g-PA6 graft copolymers [8].

Copolymer designation	Composition of PS-g-PA6		Number of PA6 grafts per PS backbone	M_n (kg mol^{-1})[a]		
	PS backbone	PA6 grafts		PS backbone	Each PA6 graft	PS-g-PA6
PS-g-PA6a	78.4	21.6	15.3	34.3	0.6	26.8
PS-g-PA6b	76.5	23.5	7.1	33.9	1.5	27.3
PS-g-PA6c	66.9	33.1	7.1	33.9	2.4	28.3
PS-g-PA6d	71.2	28.8	3.8	32.0	3.4	33.1

a) Molar masses measured by GPC using PS standards for the calibration and tetrahydrofuran (THF) as the eluent. PA6 was first N-trifluoroaeetylated before the GPC measurement.

PS backbone and the same number of the PA6 grafts per PS backbone but with different PA6 graft lengths. The last graft copolymer, PS-g-PA6d, differed from PS-g-PA6c in that the number of the PA6 grafts per PS backbone was not 7.1 but 3.8. The polymer blending was carried out in a batch mixer of type Brabender torque rheometer with a mixing chamber of 50 ml capacity.

The morphologies of the PS/PA6 blends taken from a batch mixer were observed by a scanning electron microscopy (SEM). Before the SEM observations, samples were first fractured in liquid nitrogen, and then immersed in formic acid at room temperature for 12 h in order to remove the dispersed phase (PA6) domains. They were dried in a vacuum oven at 80 °C for 12 h and then gold sputtered. The voltage for the SEM was 5.0 kV. The DD was measured using a semi-automatic image analysis method. It was characterized by volume average particle diameters, d_v, defined by:

$$d_v = \frac{\sum n_i d_i^4}{\sum n_i d_i^3} \quad (8.1)$$

For each blend, at least 500 particles were counted for statistically meaningful values of d_v.

Figures 8.1 and 8.2 show, respectively, the SEM images and emulsification curves of the PS/PA6 (80/20 by mass) blends with and without PS-g-PA6a, PS-g-PA6b, PS-g-PA6c, and PS-g-PA6d as the compatibilizers. It can be seen from the SEM images that the DD decreases with increasing CC. Moreover, these emulsification curves show that the DD follows the order: PS-g-PA6d < PS-g-PA6c < PS-g-PA6b ≈ PS-g-PA6a. Therefore, from these four PS-g-PA6 compatibilizers it can be concluded that for given PS backbone and number of PA6 grafts per PS backbone, the compatibilizing efficiency is higher when the PA6 grafts are longer (PS-g-PA6b vs PS-g-PA6c); for a given backbone, graft copolymers having fewer and longer grafts are more efficient (PS-g-PA6c vs PS-g-PA6d).

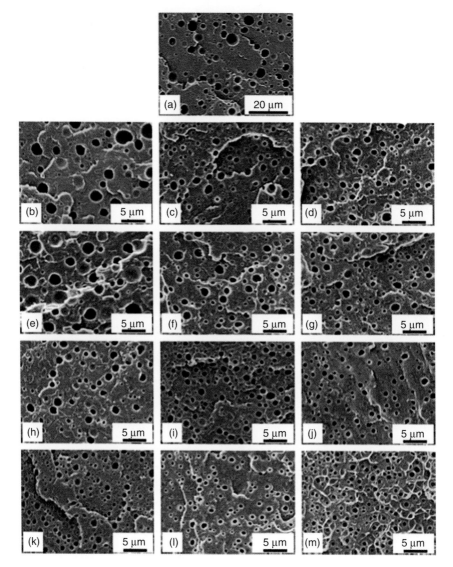

Figure 8.1 SEM images of microtomed surfaces of various PS/PA6 (80/20 by mass) blends: (a) without PS-g-PA6, (b) PS-g-PA6a (5 wt%), (c) PS-g-PA6a (10 wt%), (d) PS-g-PA6a (15 wt%), (e) PS-g-PA6b (5 wt%), (f) PS-g-PA6b (10 wt%), (g) PS-g-PA6b (15 wt%), (h) PS-g-PA6c (5 wt%), (i) PS-g-PA6c (10 wt%), (j) PS-g-PA6c (15 wt%), (k) PS-g-PA6d (5 wt%), (l) PS-g-PA6d (10 wt%), and (m) PS-g-PA6d (15 wt%). PS-g-PA6 concentration is based on the mass of the dispersed phase. Mixing temperature: 230 °C; mixing time: 8.5 min and rotation speed: 65 rpm. (Zhang et al. 2010 [8]. Reproduced with permission of Wiley.)

Figure 8.2 Effect of the molecular architecture of PS-*g*-PA6 as a compatibilizer on the emulsification curve of the PS/PA6 (80/20 by mass) blend. The PS-*g*-PA6 concentration is based on the mass of the dispersed phase. Mixing temperature: 230 °C; mixing time: 8.5 min; rotation speed: 65 rpm. Symbols: experimental data; curves: trends. (Zhang et al. 2010 [8]. Reproduced with permission of Wiley.)

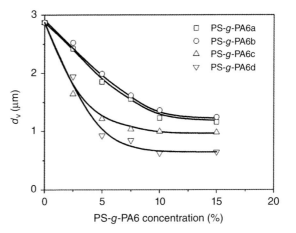

8.3 Emulsification Curves of Immiscible Polymer Blends in a Twin-Screw Extruder Using the Concept of (Reactive) Compatibilizer

8.3.1 Synthesis of (Reactive) Compatibilizer-Tracers

A compatibilizer-tracer is a compatibilizer (a block or graft copolymer), which bears moieties capable of being detected in a very sensitive manner, even if their content is very small. Moieties such as anthracene are highly sensitive to fluorescence [15, 16]. An example of a compatibilizer-tracer is a graft copolymer PS-*g*-PA6 containing 9-(methylaminomethyl) anthracene (MAMA) with fluorescent properties, which is designated as PS-*g*-PA6-MAMA, for compatibilizng PS/PA6 blends. Its backbone is PS and bears PA6 as grafts and anthracene moieties as tracers (see Figure 8.3) [17–19].

The PS-*g*-PA6-MAMA can be synthesized in the following three steps. The first one is to prepare a compatibilizer-tracer precursor, a copolymer of styrene and 3-isopropenyl-α,α′-dimethybenzyl isocyanate (TMI) denoted as PS-*co*-TMI.

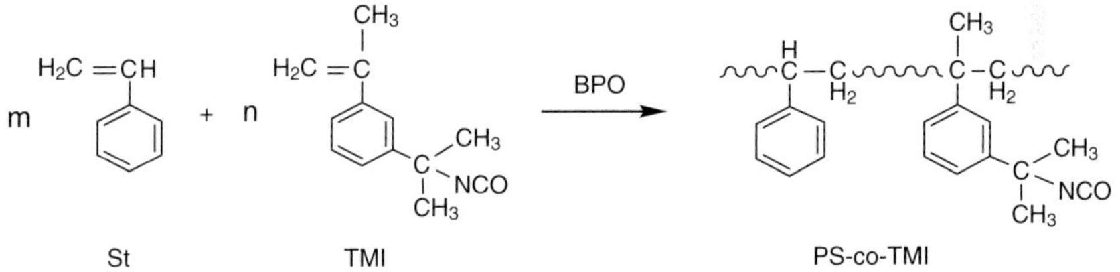

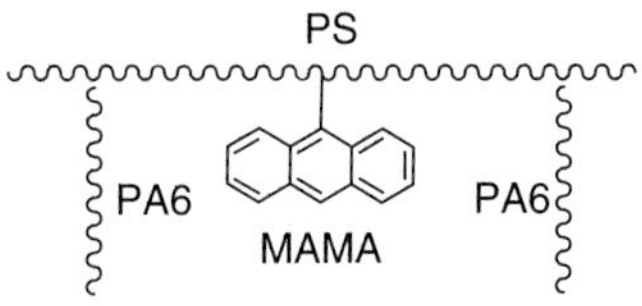

Figure 8.3 Schematic of the structure of PS-*g*-PA6-MAMA compatibilizer-tracer.

The second step is to incorporate anthracene moieties by reacting MAMA with a fraction of the isocyanate moieties of the PS-*co*-TMI.

The last step is to use the remaining intact isocyanate moieties of the PS-*co*-TMI-MAMA as activating centers from which polyamide grafts grow in the presence of ε-caprolactam (CL) and a catalyst like ε-caprolactam sodium (NaCL). A PS-*g*-PA6-MAMA graft copolymer is obtained.

PS-*co*-TMI-MAMA obtained in step 2 can also be used as a reactive compatibilizer-tracer for compatibilizing PS/PA6 blends. A reactive compatibilizer-tracer bears reactive groups capable of reacting with at least one of the polymer components of the blend to form a block or a graft copolymer *in situ* during the blending process, and moieties capable of being detected in a very sensitive manner, even if their content is very small. The PS-*co*-TMI-MAMA bears anthracene moieties capable of being detected on the order of ppm and isocyanate groups capable of reacting with the terminal amine group of the PA6 leading to the formation of PS-*g*-PA6-MAMA.

8.3.2 Development of an In-line Fluorescence Measuring Device

An in-line fluorescence measuring device has been developed to detect the concentration of anthracene bearing (reactive) compatibilizer-tracers which will be called the compatibilizer-tracer concentration (CC). As shown in Figure 8.4, it is composed of the following three main parts: fluorescent light generation, in-line fluorescent light detection device, and signal processing. The source of the fluorescent light is an ultraviolet high-pressure mercury lamp (125 W) and is divided into two beams so that the in-line device can simultaneously detect the fluorescent signal at two different locations along the extruder. Each of them successively passes through its own coupler and bifurcated optical fiber before it irradiates the anthracene tracer containing polymer flow stream in a screw extruder. The light

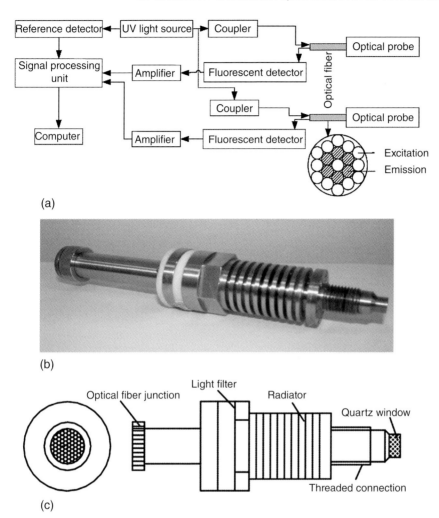

Figure 8.4 (a) Diagram of the in-line fluorescence measuring device involving three main parts: a fluorescent light generating source, fluorescent light detection and signal processing; (b) a photo of a bifurcated optical probe, which has the same dimensions as a pressure transducer mounted on a screw extruder; (c) a schematic of a bifurcated optical probe. (Zhang et al. 2006 [20]. Reproduced with permission of Wiley.)

with a specific wavelength emitted from the tracer is subsequently transmitted to a fluorescent detector (a photomultiplier) through the bifurcated optical fiber probe and is then amplified through an amplifier. Finally, the amplified optical signals coming from the two fluorescent detectors reach the signal processing unit. The latter converts them to two analogs. They are then collected by a computer system and are displayed in real time on the screen [17–21]. The sensitivity of the measuring system with respect to different amounts of the tracer can be regulated by a knob. Signals are collected once every second.

Table 8.2 Molar masses, PA6, and MAMA contents of two compatibilizer-tracers.

Sample	M_n[a] (kg mol^{-1})	M_w[a] (kg mol^{-1})	MAMA content[b] (wt%)	PA6 content[c] (wt%)	M_n of each PA6 graft (kg mol^{-1})
PS-g-PA6-MAMA-1	37.8	137.0	0.10	35.3	3.0
PS-g-PA6-MAMA-2	32.3	88.2	0.23	13.2	1.1

a) Molar masses of N-trifluoroacetylated PS-g-PA6 are measured by SEC using a refractometer detector and PS standards.
b) UV at 367 nm.
c) SEC using dual UV detection at 238 and 260 nm.

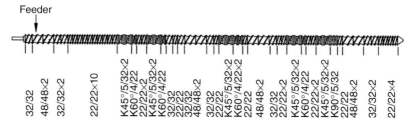

Figure 8.5 Screw configuration of the twin-screw extruder [17–19]. (Zhang et al. 2012 [18]. Reproduced with permission of Wiley.)

8.3.3 Experimental Procedure for Emulsification Curve Build-up

A co-rotating twin-screw extruder with a diameter of 35 mm and a length to diameter ratio of 48 from Nanjing Ruiya Extrusion System Limited, China, is used to show the working principle and potential applications of the concept of compatibilizer-tracer for building up emulsification curves of immiscible polymer blends. PS and PA6 are used as polymer components of the blend and PS-g-PA6-MAMA-1 as the compatibilizer-tracer whose molecular architecture is shown in Table 8.2 of Section 8.3.2. Figure 8.5 shows the screw configuration. It is basically composed of right-handed screw elements and kneading blocks. The head of the extruder is equipped with a strip die of 30 mm in length and 5 mm in width. The die exit is cuboid and has a hole on one side allowing for installing a bifurcated optical probe which is in direct contact with the flow stream and detects the fluorescent signal (Figure 8.6) [17–21].

After the extrusion process of the PS/PA6 blend runs steadily, a given amount (a few grams maximum) of the PS-g-PA6-MAMA-1 compatibilizer-tracer is added to it from the hopper as a pulse. At the same time, the in-line fluorescent measuring device starts to record the fluorescent signal. Before the fluorescent signal begins to increase, a sample of the extrudate is taken from the die and is used as a blank reference. Subsequently, samples are taken once every 10 s and are quenched immediately in liquid nitrogen until the fluorescent signal reaches its baseline value.

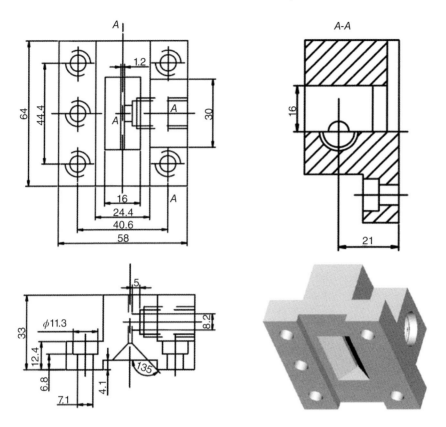

Figure 8.6 Geometry of the extrusion die used to install a bifurcated optical probe in the side hole (unit: mm). (Zhang et al. 2012 [17]. Reproduced with permission of Wiley.)

With the assumption that the value of the analog fluorescent signal (voltage) is proportional to the CC, the RTD function, $E(t)$, can be calculated according to the following expression:

$$E(t_i) = \frac{V_i(t)}{\sum_{i=1}^{n} V(t_i)(t_i - t_{i-1})} \tag{8.2}$$

where V_i is the voltage at time t_i ($i = 1, 2, \ldots, n$).

Morphologies of the PS/PA6 blends taken at the die exit are characterized by SEM, following the procedure described in Section 8.2. Figure 8.7 shows the CCD, namely, CC versus. RT, for three different initial amounts of the PS-g-PA6-MAMA-1. The composition of the PS/PA6 is 80/20 by mass. It can be seen that for a given amount of the compatibilizer-tracer introduced to the extruder as a pulse, a minimum lapse of time is necessary for the compatibilizer-tracer to exit the die. This minimum lapse of time or the minimum RT of the compatibilizer-tracer, decreases as its initial amount increases.

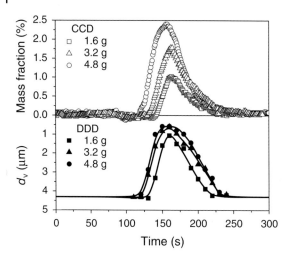

Figure 8.7 Effect of the amount of the compatibilizer-tracer on the CCD and the corresponding DDD of the PS/PA6 (80/20 by mass) blend. Feed rate: 13 kg h^{-1}; screw speed: 100 rpm. The amount of the PS-g-PA6-MAMA-1 compatibilizer-tracer is obtained from the anthracene concentration by the in-line fluorescent measuring device. The initial mass of the compatibilizer-tracer is 1.6, 3.2, or 4.8 g. (Zhang et al. 2012 [19]. Reproduced with permission of Wiley.)

This may be because an increase in the initial amount of the compatibilizer-tracer brings about an increase in the local throughput. After the minimum lapse of time, the compatibilizer-tracer mass fraction increases with increasing time, reaches a maximum and then decreases with a further increase in time. After a sufficiently long period of time, all the initial amount of the compatibilizer-tracer is washed out and the compatibilizer-tracer mass becomes zero again. From Figure 8.7, the DDD follows an opposite trend of the CCD, namely, the DD decreases with increasing CC.

The DDD and CCD shown in Figure 8.7 can be converted to the emulsification curves of the PS-g-PA6-MAMA-1 compatibilized PS/PA6 (80/20) blend corresponding to three initial masses of the compatibilizer-tracer (1.6, 3.2, and 4.8 g). These emulsification curves are shown in Figure 8.8. It is interesting to note that they are not single curves but loops. The two parts of a loop correspond, respectively, to the short-time and long time domains demarcated by the maximum of the corresponding CCD. In the short time domain, the CC increases with increasing RT, while in the long time domain it decreases with increasing RT. Moreover, both the short time domain and long time domain curves shift upward

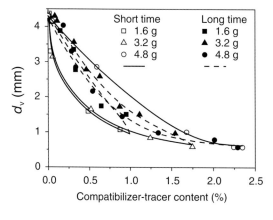

Figure 8.8 Emulsification curves obtained from the CCD and DDD curves illustrated in Figure 8.7 for the PS/PA6 (80/20 by mass) blend. (Zhang et al. 2012 [19]. Reproduced with permission of Wiley.)

with increasing initial amount of the compatibilizer-tracer. This implies that the compatibilizing efficiency of the compatibilizer-tracer decreases when its initial amount is increased. This phenomenon could be related to the fact that it would be more difficult to disperse a larger amount of the compatibilizer-tracer so that it could totally diffuse to the interfaces. A compatibilizer-tracer acts as a compatibilizer only when it is located at the interfaces.

It is also seen that when the initial amount of the compatibilizer-tracer is 1.6 or 3.2 g, the emulsification curve in the long time domain is above the corresponding one in the short time domain. By contrast, the trend is the opposite when the initial amount is 4.8 g. Closer observation of Figure 8.7 reveals that when the initial amount is 1.6 and 3.2 g, the DDD curves exhibit obvious maxima, and when it is 4.8 g the maximum is no longer a single point but a plateau. This indicates that the amount of the tracer-compatibilizer might be in excess in the plateau region so that micelles would be formed and a longer time would have been needed to disperse them. As a result, the emulsification curve in the short time domain is located above the corresponding one in the long time domain.

The above results show that the concept of compatibilizer-tracer together with transit experiments can allow building up an emulsification curve in a pilot and/or industrial scale extruder by using a very small amount of compatibilizer-tracer and by conducting one single experiment. In what follows, examples will be given to show potential applications of this methodology for compatibilizer selection and process optimization.

8.3.4 Compatibilizer Selection Using the Concept of Compatibilizer-Tracer

Consider two compatibilizer-tracers, PS-g-PA6-MAMA-1 and PS-g-PA6-MAMA-2, whose characteristics are shown in Table 8.2. They have a similar PS backbone length and the same number of PA6 grafts. They differ in that the length of PA6 grafts of the former one is almost three times that of the latter one.

Figure 8.9 shows the CCD and DDD curves of the PS/PA6 (80/20 by mass) blend compatibilized by PS-g-PA6-MAMA-1 and PS-g-PA6-MAMA-2, respectively. Their emulsification curves are shown in Figure 8.10. It is seen that

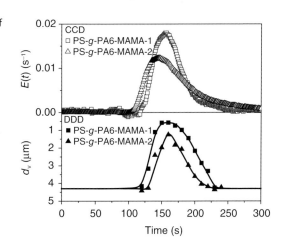

Figure 8.9 CCD and DDD curves of the PS/PA6 (80/20 by mass) blend compatibilized by PS-g-PA6-MAMA-1 and PS-g-PA6-MAMA-2, respectively. Die width: 5 mm; feed rate: 13 kg h^{-1}; mass of the compatibilizer-tracer: 4.8 g; screw speed: 100 rpm. (Zhang et al. 2012 [19]. Reproduced with permission of Wiley.)

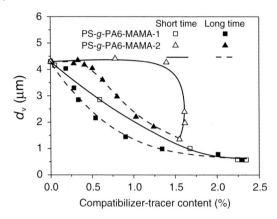

Figure 8.10 Effect of the molecular architecture of the compatibilizer-tracer on the emulsification curve of the PS/PA6 (80/20 by mass) blend. Die width: 5 mm; feed rate: 13 kg h^{-1}; mass of the compatibilizer-tracer: 4.8 g; screw speed: 100 rpm. (Zhang et al. 2012 [19]. Reproduced with permission of Wiley.)

the emulsification loop of PS-*g*-PA6-MAMA-2 is located well above that of PS-*g*-PA6-MAMA-1, indicating that the former is much less efficient than the latter. Moreover, in the case of PS-*g*-PA6-MAMA-2, the DD in the short-time domain does not decrease with increasing CC until the latter exceeds 1.4 wt%. In other words, PS-*g*-PA6-MAMA-2 does not have any compatibilizing efficiency when its concentration is below 1.4 wt%. This is likely because its RT is not long enough for it to reach the interfaces between the PS and PA6. Indeed, PS-*g*-PA6-MAMA-2 is expected to have much higher miscibility with the PS than with PA6 because its PA6 grafts are short. It tends to stay in the PS matrix. This may explain the reason for its compatibilizing efficiency becoming much higher in the long-time domain (see closed symbols in Figure 8.10).

8.3.5 Process Optimization Using the Concept of Compatibilizer-Tracer

8.3.5.1 Effect of Screw Speed

Screw speed is one of the most important processing parameters for polymer blending in a twin-screw extruder. Figure 8.11 compares the CCD and DDD

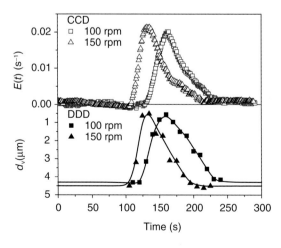

Figure 8.11 Effect of screw speed on the CCD and DDD curves of the PS/PA6 (80/20 by mass) blend. Die width: 5 mm; feed rate: 13 kg h^{-1}; mass of the compatibilizer-tracer PS-*g*-PA6-MAMA-1: 3.2 g. (Zhang et al. 2009 [18]. Reproduced with permission of Wiley.)

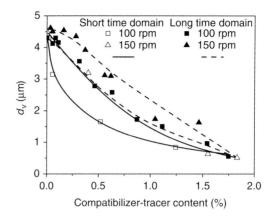

Figure 8.12 Effect of screw speed on the emulsification curves of the PS/PA6 (80/20 by mass) blend. Die width: 5 mm; feed rate: 13 kg h^{-1}; mass of the compatibilizer-tracer PS-g-PA6-MAMA-1: 3.2 g. (Zhang et al. 2009 [18]. Reproduced with permission of Wiley.)

curves for two screw speeds, 100 and 150 rpm. An increase in screw speed does not modify much the shape of the curve but shifts it to the short-time domain.

Figure 8.12 shows the emulsification curves (loops) for these two screw speeds. The one at 100 rpm is below that at 150 rpm, indicating that the compatibilizer-tracer has higher compatibilizing efficiency at 100 rpm. In other words, under the specified conditions, increasing screw speed does not favor the localization of the compatibilizer-tracer at the interfaces. As the screw speed increases, the mixing intensity increases. However, the RT of the compatibilizer-tracer in the extruder decreases accordingly. The compatibilizer-tracer needs a certain amount of time to reach the PS/PA6 interfaces to reduce the interfacial tension and prevent the dispersed phase domains from coalescing. At 150 rpm, it does not have enough time to migrate to the interface despite more intensive mixing.

8.3.5.2 Effects of the Type of Mixer

Different types of mixers may provide different thermo-mechanical and temporal conditions. Therefore, the compatibilizing efficiency of a compatibilizer in one type of mixer may be different from that in another type of mixer. Figure 8.13 compares the emulsification curves of PS-g-PA6-MAMA-1 compatibilized PS/PA6 blends obtained from the twin-screw extruder and a Haake rheometer type batch mixer. Irrespective of the PS/PA6 mass ratio, the emulsification curves obtained from the twin-screw extruder are significantly below those obtained from the batch mixer, indicating that the twin-screw extruder has a much higher mixing capacity than the batch mixer.

8.3.6 Section Summary

The above results show that the concept of the compatibilizer-tracer together with transient experiments offers a very simple, convenient, and reliable method to obtain emulsification curves in a screw extruder, especially in pilot and industrial ones, which would not be possible otherwise due to the need to use large amounts of compatibilizers. Emulsification curves provide guidance for compatibilizer selection, process scale up, and optimization.

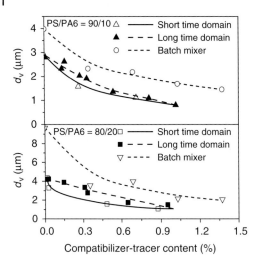

Figure 8.13 Effect of the type of mixer on the emulsification curves of the PS/PA6 blends. The data for the short time domain and long time domain are obtained from the twin-screw extruder, whereas those for a batch mixer are obtained from a Haake torque rheometer. Blending conditions in the twin-screw extruder are: die width: 5 mm; feed rate: 13 kg mol^{-1}; mass of PS-g-PA6-MAMA-1: 1.6 g for PS/PA6 (90/10 by mass) and 3.2 g for PS/PA6 (80/20 by mass); screw speed: 100 rpm. Blending conditions in the batch mixer are: screw speed: 65 rpm; mixing time: 8 min. (Zhang et al. 2012 [17]. Reproduced with permission of Wiley.)

8.4 Emulsification Curves of Reactive Immiscible Polymer Blends in a Twin-Screw Exturder

A compatibilizer can be premade or synthesized *in situ* during the blending process. The latter case is called reactive polymer blending, reactive compatibilization, or *in situ* compatibilization. In this case, either both polymer components of the blend or fractions of them bear reactive groups that can react with each other at the interfaces leading to *in situ* formation of a block or graft copolymer during the blending process. This *in situ*-formed block or graft copolymer compatibilizes the blend. In cases where one of the polymer components does not bear any reactive groups, a third polymer component, which is identical to or miscible with it and which bears reactive groups capable of reacting with the reactive polymer component, has to be added. This reactive polymer is called reactive compatibilizer. Its compatibilizing efficiency depends, among others, on the reaction kinetics between reactive groups, which in turn depend on the types of reactive groups [22] and interfacial mixing [23–25].

8.4.1 Reaction Kinetics between Reactive Functional Groups

A (reactive) polymer blending process typically takes less than 10 min in a batch mixer and less than 3 min in a screw extruder. Therefore, the interfacial reaction kinetics should be compatible with the process time. Common interfacial reactions between functional groups are shown in Figure 8.14.

Table 8.3 shows the conversion at 2 min of the reactions among a number of polystyrene types with different reactive groups at 180 °C. The reactivity of the functional group pair is in the following order: amine/isocyanate > amine/anhydride > carboxylic/anhydride > acid/epoxy > acid/oxazoline > amine/epoxy. Although the reaction between amine and isocyanate is the fastest, it has not been applied to commercial blends due to isocyanate's sensitivity to water and instability under high temperature. Consequently, the amine/anhydride is the most common reactive pair that is frequently used for reactive compatibilization

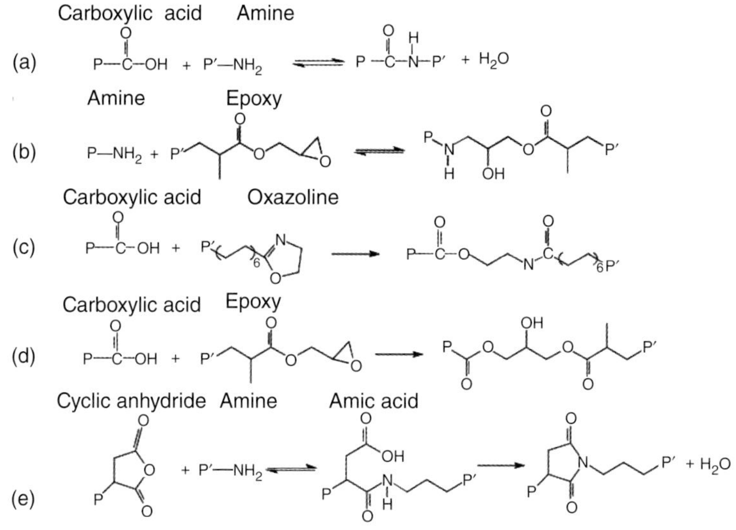

Figure 8.14 Common chemical reactions between functional groups: (a) acid/amine reaction to form an amide; (b) amine/epoxy reactions; (c) carboxylic acid/oxazoline; (d) carboxylic acid/epoxy; (e) two-step reaction to form an imide from amines and cyclic anhydrides [22].

Table 8.3 Comparison of conversion at 2 min and 180 °C for reactive pairs [26].

Group 1	Group 2	Conversion (%)	k (kg mol^{-1} min^{-1})
Carboxylic acid	Aliphatic amine	0	—
Aromatic amine	Aliphatic epoxy	0.6	0.1
Aromatic amine	GMA epoxy	0.7	0.15
Aliphatic amine	Aliphatic epoxy	1.1	0.28
Aliphatic amine	GMA epoxy	1.8	0.34
Carboxylic acid	oxazoline	2.1	0.92
Carboxylic acid	GMA epoxy	9.0	2.1
Aromatic amine	Cyclic anhydride	12.5	3.3
Aliphatic amine	Cyclic anhydride	99	$\sim 10^3$

of commercial polymer blends. Nevertheless, it should be pointed out that mixing may significantly increase the interfacial reaction between two immiscible reactive polymers [25]. In other words, contrary to one's intuition, under mixing, reactions between two functional groups attached to two immiscible polymer backbones may proceed even much faster than those between two small molecule analogs.

8.4.2 (Non-reactive) Compatibilizers Versus Reactive Compatibilizers

Table 8.4 summarizes several characteristics of non-reactive compatibilizers and reactive ones.

Table 8.4 Characteristics of non-reactive compatibilizers and reactive ones.

	Non-reactive compatibilizers	Reactive compatibilizers
Synthesis	Synthesized before blending	Synthesized during blending
Molecular structure	Fixed and can be characterized before blending	Evolving[a] and unknown during blending and could be characterized after blending
Amount	Fixed and known before blending	Evolving[a] and unknown during blending and could be determined after blending
Location	Not all compatibilizers could reach and stay at the interfaces; real amount of the compatibilizer at the interfaces unknown	Not all reactive compatibilizers could reach and react at the interfaces; the blend remains potentially reactive after blending; all copolymers (compatibilizer) are formed at the interfaces; not all of the *in situ* formed copolymers can stay at the interfaces; real amount of the copolymers at the interfaces unknown
Deformulation	Much easier	Much more difficult

a) In case the reactive compatibilizer or one of the two polymer components bears two or more functional groups.

8.4.3 An Example of Reactive Compatibilizer-Tracer

A typical reactive compatibilizer-tracer bears both reactive groups capable of reacting with its counterparts to form a copolymer *in situ* for compatibilization of a polymer blend, and moieties which allow accurate determination of the amount of the *in situ* formed compatibilizer even if its amount is small.

Figure 8.15 shows an example of a reactive compatibilizer-tracer for PS/PA6 blends. It is denoted as PS-*co*-TMI-MAMA and bears isocyanate (NCO) groups along the PS backbone, which are able to react with the terminal amine of the PA6, forming a graft copolymer of PS and PA6 as shown in Figure 8.16. Unlike classical isocyanate groups, the one attached to TMI is relatively inert to moisture. The amount of the *in situ* formed graft copolymer PS-*g*-PA6-MAMA

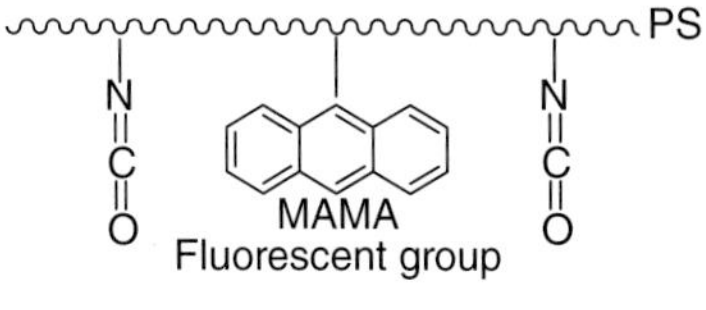

Figure 8.15 Schematic of the structure of PS-*co*-TMI-MAMA reactive compatibilizer-tracer.

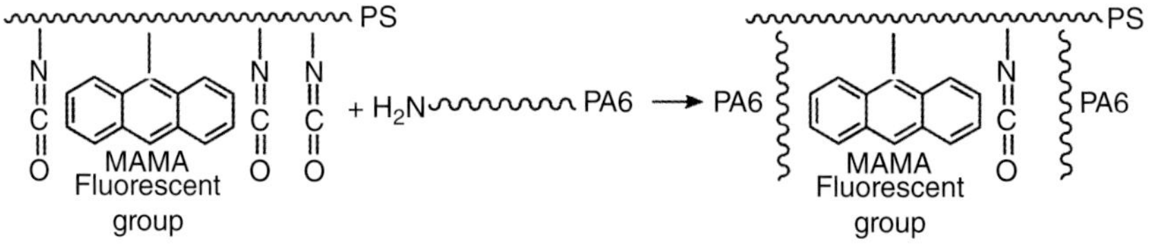

Figure 8.16 Formation of the PS-*g*-PA6-MAMA graft copolymer by the interfacial reaction between the PS-*co*-TMI-MAMA and PA6.

can be determined in an accurate manner using size exclusion chromatography (SEC) with an ultraviolet detector because MAMA moieties have very strong UV-absorption at 367 nm while PA6 and PS do not have any absorption at this wavelength. PS-co-TMI-MAMA can be obtained by reacting a fraction of the isocyanate groups of PS-co-TMI with MAMA, as described in Section 8.3.1.

8.4.4 Assessment of the Morphology Development of Reactive Immiscible Polymer Blends Using the Concept of Reactive Compatibilizer

The PS-TMI-MAMA is used as a reactive compatibilizer-tracer to compatiblize PS/PA6 blends in a batch mixer and to study their morphology development [27]. Two mixing modes are adopted, as shown in Figure 8.17. One is continuous mixing, namely, the components of the blend are mixed at 100 rpm and 230 °C for various time intervals ranging from 1 to 10 min. The other mixing mode is stepwise mixing, that is, after the components of the blend are mixed at 100 rpm and 230 °C for a prescribed time interval, the rotors stopped rotating, that is, 0 rpm or no mixing, and then the resulting blend is annealed under the quiescent condition at 230 °C for a given period of time. Thereafter, the mixing is resumed at 100 rpm and 230 °C.

Figure 8.18 presents the evolution of the percentage of the PS-TMI-MAMA that has reacted with at least one PA6 chain and that of the DD of the PS/PA6/PS-TMI-MAMA blends as a function of time under the continuous mixing conditions for two compositions: 95/5/1.5 and 80/20/1.5 by mass. As expected, irrespective of the blend composition, the percentage of the reacted PS-TMI-MAMA first increases rapidly with increasing mixing time, and then slows down with a further increase in mixing time (Figure 8.18a). On the other hand, the DD first decreases slightly and then starts to drastically increase after about 4 min of mixing, reaching a value close to the one without the reactive compatibilizer in less than 1 min (Figure 8.18b). Figure 8.18c shows the evolution of d_v as a function of the percentage of the reacted PS-TMI-MAMA. For both compositions, d_v starts to drastically increase when the percentage of the reacted PS-TMI-MAMA reaches a critical value of about 50% for the 95/5/1.5 by mass and about 66% for the 80/20/1.5 by mass.

Figure 8.19a compares the evolution of the percentage of the reacted PS-TMI-MAMA of the PS/PA6/PS-TMI-MAMA (80/20/1.5 by mass) reactive

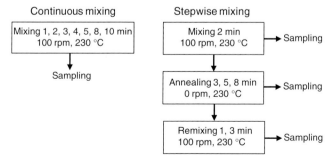

Figure 8.17 Two mixing modes for reactive blending of PS and PA6 using PS-co-TMI-MAMA as a reactive compatibilizer-tracer: continuous mixing and stepwise mixing.

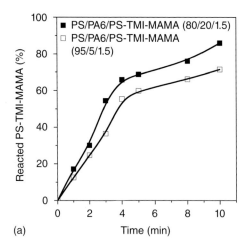
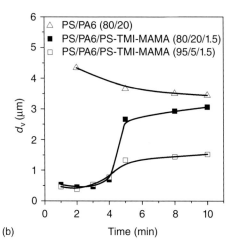
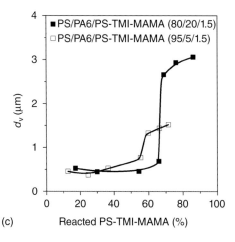

Figure 8.18 Evolution of the percentage of the reacted PS-TMI-MAMA (a) and that of the DD of the PA6 (b) of the PS/PA6/PS-TMI-MAMA reactive blends as a function of mixing time for the continuous mixing case. The PS/PA6 (80/20 by mass) non-reactive blend in terms of d_v is shown as a reference. (c) Evolution of the DD of the PA6 of the PS/PA6/PS-TMI-MAMA reactive blends as a function of the percentage of the reacted PS-TMI-MAMA for the continuous mixing case. (Ji 2016 [27]. Reproduced with permission of Wiley.)

Figure 8.19 (a) Evolution of the percentage of the reacted PS-TMI-MAMA (a) and that of the DD (b) of the PS/PA6/PS-TMI-MAMA (80/20/1.5 by mass) reactive blends as a function of mixing time for the stepwise mixing case. The continuous mixing case and the PS/PA6 (80/20 by mass) non-reactive blend are shown as references. (c) Comparison between the continuous and stepwise mixing cases for the PS/PA6/PS-TMI-MAMA (80/20/1.5 by mass) reactive blend in terms of d_v as a function of the percentage of the reacted PS-TMI-MAMA. (Ji 2016 [27]. Reproduced with permission of Wiley.)

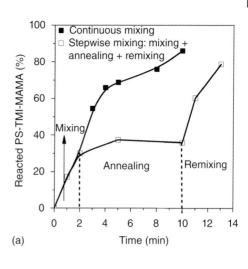

(a)

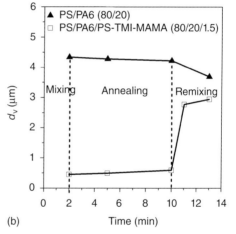

(b)

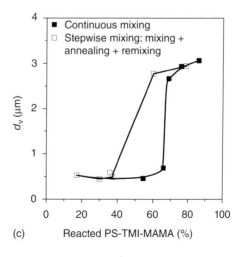

(c)

blend as a function of time under the continuous mixing and stepwise mixing, respectively. It can be seen that when mixing stops after 2 min, the interfacial reaction almost stops as well. When it is resumed after 8 min of quiescent annealing, the interfacial reaction also starts to proceed immediately. Similar phenomena are found in the literature [25, 28]. The evolution of the DD with time (Figure 8.19b) shows that during the quiescent annealing from 2 to 10 min, the dispersed phase domains have undergone little coarsening. When mixing is resumed, they suddenly increase in diameter and reach a value close to that of the PS/PA6 (80/20 by mass) without the reactive compatibilizer-tracer (PS-TMI-MAMA). This is very similar to the continuous mixing case.

The evolutions of the DD as a function of the percentage of the reacted PS-TMI-MAMA for the PS/PA6/PS-TMI-MAMA (80/20/1.5 by mass) reactive blend under the continuous and stepwise mixing are shown in Figure 8.19c. It is clear that in the stepwise mixing case, the DD also increases suddenly when the amount of the *in situ* formed copolymer reaches a critical value of about 40%. This critical value is significantly smaller than that for the continuous mixing case, namely 66%. This indicates that the quiescent annealing tends to decrease the threshold beyond which the catastrophic coarsening occurs.

From the above experimental results, it is obvious that the reactive compatibilizer-tracer is very efficient at reducing the DD at the beginning of the reactive blending process. Moreover, there exists a critical threshold for the amount of the *in situ* formed compatibilizer beyond which the DD sharply increases and tends to reach a value close to the corresponding PS/PA6 non-reactive blend, as if the whole amount of the *in situ* formed copolymer had completely lost its compatibilizing efficiency. This phenomenon is undesirable for a reactive blending process. It should be kept in mind that an important feature of a typical reactive blending system is that both the amount and the molecular architecture of the *in situ* formed copolymer keep evolving during blending (see Table 8.4). Therefore, the interfacial driven change in the molecular architecture of the *in situ* formed graft copolymer may be responsible for the sudden deterioration of its compatibilizing efficiency. At the beginning of mixing, the *in situ* formed PS-*g*-PA6-MAMA graft copolymer may only have one or two PA6 grafts per PS-TMI-MAMA backbone and is therefore stable at the interface preventing the dispersed phase domains from coalescence. As mixing time increases, the interfacial reaction proceeds further and more PA6 grafts are attached to the PS-*co*-TMI-MAMA backbone. As a result, the resulting graft copolymer may no longer be stable at the interface and may be pulled out of the interface by mixing, forming copolymer micelles in the PA6 domains. This can be confirmed by the images of the confocal spectroscopy of the PS/PA6/PS-*co*-TMI-MAMA (80/20/3 by mass) reactive blend after 2 (Figure 8.20a) and 10 min (Figure 8.20b) of the continuous mixing. It is obvious that there are much more micelles (corresponding to turquoise dots) in the blend with 10 min of mixing than in the one with 2 min of mixing. This supports the statement that the detachment of the *in situ* formed PS-g-PA6-MAMA graft copolymer from the interfaces is responsible for the loss of its compatibilizing efficiency.

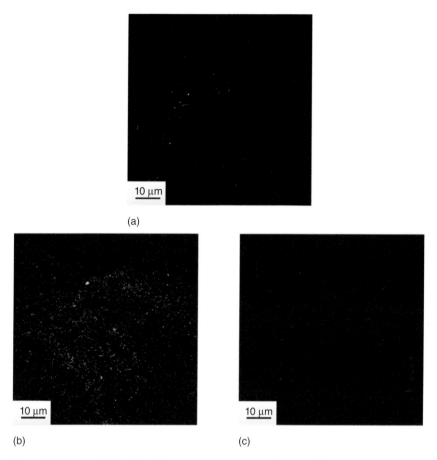

Figure 8.20 (a) and (b) confocal fluorescent spectroscopy images of the PS/PA6/PS-co-TMI-MAMA (80/20/3 by mass) reactive blend after 2 and 10 min of continuous mixing, respectively. (c) Confocal spectroscopy image of the PS/PS-TMI-MAMA (80/3) control system after 10 min of continuous mixing. Turquoise dots correspond to the PS-g-PA6-MAMA graft copolymer micelles whose d_n and d_v are about 200 and 300 nm, respectively. (Ji 2016 [27]. Reproduced with permission of Wiley.)

The above results show that a reactive compatibilizer-tracer does allow determining very small amounts of *in situ* formed graft copolymer during the blending process. Moreover, it has helped reveal and understand an unexpected phenomenon that a reactive compatibilizer which is very efficient under certain thermo-mechanical and/or temporal conditions may become inefficient under others.

8.4.5 Emulsification Curve Build-up in a Twin-Screw Extruder Using the Concept of Reactive Compatibilizer-Tracer

Similar to the concept of compatibilizer-tracer, the concept of reactive compatibilizer-tracer allows building up emulsification curves of reactive

Figure 8.21 Screw configuration of the twin-screw extruder used. (Ji et al. 2015 [29]. Reproduced with permission of American Chemical Society.)

immiscible polymer blends in a screw extruder with small amounts of reactive compatibilizer-tracers. Figure 8.21 shows the screw configuration of the twin-screw extruder. A kneading zone composed of 11 kneading blocks is located in the last part of the screw extruder. Each kneading block is composed of seven kneading elements with a total length of 32 mm and an angle of 30° between two adjacent kneading elements, unless specified otherwise. The PS-co-TMI-MAMA reactive compatibilizer-tracer can be introduced into the screw extruder either from tracer port 1 or tracer port 2.

After the extrusion process of the PS/PA6 (80/20 by mass) blend runs steadily, the reactive compatibilizer-tracer (PS-co-TMI-MAMA) is introduced to the twin-screw extruder from port 1 as a pulse (3.2 g). The in-line fluorescent probe is placed in the die to record the signal of the MAMA concentration of the PS-co-TMI-MAMA. It should be kept in mind that unlike a non-reactive compatibilizer-tracer such as PS-g-PA6-MAMA, which is a premade graft copolymer, a reactive compatibilizer-tracer such as PS-co-TMI-MAMA is a graft copolymer precursor that forms a graft copolymer upon reacting with a reactive polymer counterpart *in situ* at the interfaces during the blending process. Therefore, besides the CCD and DDD, the use of a reactive compatibilizer-tracer can obtain one more curve, the reacted reactive compatibilizer-tracer concentration distribution (RCCD). Figure 8.22 shows the CCD, DDD, and RCCD curves for the PS/PA6 (80/20 by mass) using the PS-co-TMI-MAMA as a reactive compatibilizer-tracer. Both CCD and DDD curves follow the same trend in the sense that the higher the reactive Compatibilizer-tracer Concentration (CC), the smaller the DD. They also show that, as expected, the reactive compatibilizer-tracer works in a way that is very similar to that of a non-reactive compatibilizer-tracer. The RCCD follows more or less the trends of the CCD and DDD.

The CCD and DDD shown in Figure 8.22 can be converted to an emulsification curve. As shown in Figure 8.23, it is a loop instead of a single curve, which is similar to that for a non-reactive compatibilizer-tracer [17–19]. Moreover, the upper and lower part of the loop correspond to the short time domain and the long time domain, respectively. This may imply that under the specified conditions, the compatibilizing efficiency of the reactive compatibilizer-tracer is more efficient in the long time domain than in the short time domain. It is expected that in the long time domain, PS-co-TMI-MAMA stays longer in the extruder and has more opportunities to react with the PA6, resulting in a higher reacted reactive compatibilizer-tracer concentration (RCC) or a higher PS-g-PA6 graft

Figure 8.22 CCD, DDD, and RCCD of the PS/PA6 (80/20 by mass) blend with PS-TMI-MAMA as a reactive compatibilizer-tracer, which is injected as a pulse (3.2 g) at port 1. Feed rate: 13 kg h^{-1}; screw speed: 100 rpm. (Ji et al. 2015 [29]. Reproduced with permission of American Chemical Society.)

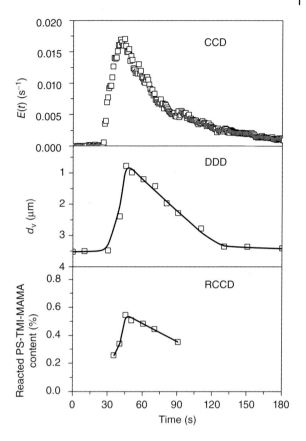

Figure 8.23 Emulsification curve of the PS/PA6 (80/20 by mass) blend using PS-TMI-MAMA as the reactive compatibilizer-tracer (data from Figure 8.21). (Ji et al. 2015 [29]. Reproduced with permission of American Chemical Society.)

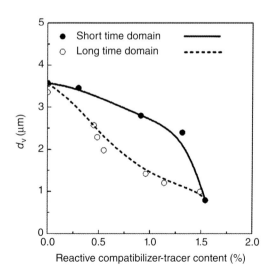

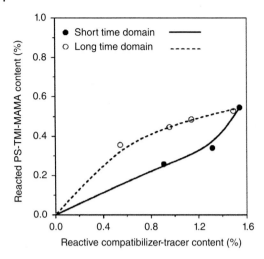

Figure 8.24 RCC versus CC of the PS-TMI-MAMA compatibilized PS/PA6 (80/20 by mass) blend (data from Figure 8.23). (Ji et al. 2015 [29]. Reproduced with permission of American Chemical Society.)

copolymer concentration. This can be confirmed by the RCC as a function of the CC, as shown in Figure 8.24. Like the emulsification curve, a loop is also obtained for the RCC as a function of the CC. Moreover, the values of the loop in the long time domain are indeed higher than those in the short time domain.

Based on the RCC versus CC, an effective emulsification curve, namely, the DD as a function of the RCC instead of the CC can be built up, as shown in Figure 8.25. Interestingly, the curves corresponding to the short and long time domains superimpose. This indicates that under the specified conditions, it is the RCC and not the CC in the blend that determines the diameter of the dispersed domains. This is expected, as it is the *in situ* formed PS-*g*-PA6-MAMA graft copolymer and not the PS-*co*-TMI-MAMA reactive compatibilizer-tracer itself that is the real compatibilizer for the PS/PA6 blend.

The above results show that the use of PS-*co*-TMI-MAMA as a reactive compatibilizer-tracer together with transit experiments can provide information

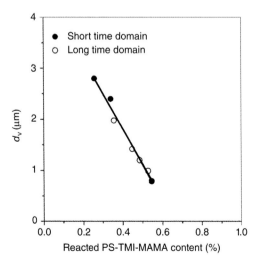

Figure 8.25 Effective emulsification curve of the PS/PA6 (80/20) blend using PS-*co*-TMI-MAMA as the reactive compatibilizer-tracer (data from Figure 8.23). (Ji et al. 2015 [29]. Reproduced with permission of American Chemical Society.)

8.4 Emulsification Curves of Reactive Immiscible Polymer Blends in a Twin-Screw Extruder

on the evolutions of the DD, RCC, the emulsification curve, and the effective emulsification curve. All these pieces of information are very useful for in-depth understanding and optimization of reactive polymer blending processes in screw extruders, especially those used on the industrial scale.

In what follows, the concept of the reactive compatibilizer-tracer will be used to assess the effects of processing parameters on the DD, RCC, the emulsification curve, and the effective emulsification curve. The processing parameters include the reactive compatibilizer-tracer injection location, the composition of the PS/PA6 blend, and the geometry of screw elements.

8.4.6 Assessment of the Effects of Processing Parameters Using the Concept of Reactive Compatibilizer-Tracer

8.4.6.1 Effect of the Reactive Compatibilizer-Tracer Injection Location

The reactive compatibilizer-tracer (PS-*co*-TMI-MAMA) is added into the extruder as a pulse either at port 1 or 2. Figure 8.26 shows the CCD, DDD, and RCCD curves for these two cases. Since port 1 is closer to the die exit than port 2, its curves are all shifted to the short time domain with respect to those of port 2.

Figure 8.27 presents their emulsification curves (a), RCC versus CC (b), and effective emulsification curves (c). In the case of port 2, the emulsification curves

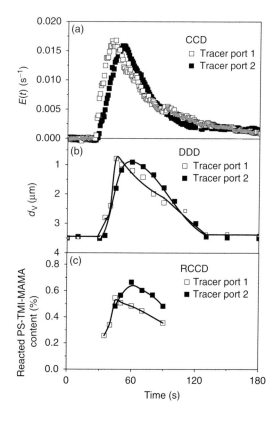

Figure 8.26 Effects of the reactive compatibilizer-tracer injection location (port 1 or port 2) on the CCD (a), DDD (b), and RCCD (c) for the PS/PA6 (80/20 by mass) blend. Feed rate: 13 kg h^{-1}; screw speed: 100 rpm; amount of PS-*co*-TMI-MAMA reactive compatibilizer-tracer: 3.2 g. (Ji et al. 2015 [29]. Reproduced with permission of American Chemical Society.)

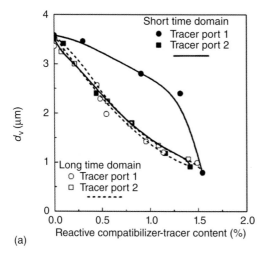
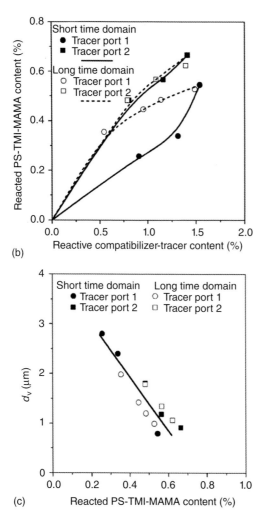

Figure 8.27 Effect of the reactive compatibilizer-tracer injection location on the emulsification curve (a), RCC as a function of CC (b), and the effective emulsification curve (c) of PS/PA6 (80/20 by mass) system. Feed rate: 13 kg h^{-1}; screw speed: 100 rpm; amount of the reactive compatibilizer-tracer PS-co-TMI-MAMA: 3.2 g. (Ji et al. 2015 [29]. Reproduced with permission of American Chemical Society.)

of the short and long time domains almost superimpose, indicating that the mixing performance in the long time domain is still poor and is similar to that in the short time domain. However, in the case of port 1, they do not superimpose any more. The one in the long time domain is significantly below that in the short time domain. This may indicate that in the case of port 1, the long time domain provides a sufficient mixing or reaction time. As a result, the RCC in the long time domain is much higher than that in the short time domain, as show in Figure 8.27b.

The effective emulsification curves are shown in Figure 8.27c. The short and long time domain data all superimpose on a single line within experimental errors, irrespective of the location at which the reactive compatibilizer-tracer is introduced. This indicates that in these cases, the RCC dictates the morphology of the blend.

8.4.6.2 Effect of the Blend Composition

Figure 8.28 compares the CCD, DDD, and RCCD curves between two PS/PA6 (80/20 and 95/5 by mass) blends. The reactive compatibilizer-tracer is introduced at port 2 or 1. Figure 8.29 compares their emulsification curves, the RCC versus CC, and the effective emulsification curves. From the emulsification curves, when there is no reactive compatibilizer-tracer, the DD of the PS/PA6 (95/5 by mass) blend is smaller than that of the PS/PA6 (80/20 by mass), as expected. However, it is surprising that the former is larger than the latter at high reactive compatibilizer-tracer concentration. This is true for both blend compositions and the two reactive compatibilizer-tracer injection locations, especially port 1.

In a reactive polymer blending process, the reactive compatibilizer-tracer (PS-*co*-TMI-MAMA) molecules have to migrate, collide, and react with the PA6. When the PA6 volume fraction of the PS/PA6 blend is very low, the probability of collisions between the reactive compatibilizer-tracer and PA6 is low, resulting in a low RCC (or graft copolymer concentration). This is confirmed by the RCC versus CC curves shown in Figure 8.29c,d. The RCC of the PS/PA6 (95/5 by mass) blend is indeed significantly lower than that of the PS/PA6 (80/20 by mass) blend, which further validates the above hypothesis that the interfacial reaction between the reactive compatibilizer-tracer and PA6 for the PS/PA6 (95/5 by mass) blend proceeds more slowly than for the PS/PA6 (80/20 by mass) blend because of the lower PA6 volume fraction. Thus, it is not difficult to understand that in the presence of a reactive compatibilizer-tracer, the DD of the PS/PA6 (95/5 by mass) blend can be even larger than that of the PS/PA6 (80/20 by mass) blend.

From the effective emulsification curves (Figures 8.29e,f), it is seen that as the interfacial reaction increases, the difference in DD between the two PS/PA6 (95/5 and 80/20 by mass) blends gradually narrows, indicating that for both reactive compatiblizer-tracer injection locations, the effect of the blend composition on the morphology decreases as the *in situ* formed graft copolymer concentration increases, due to reduced interfacial tension and coalescence.

Figure 8.30 depicts a simplified scheme of the development of the interfacial reaction and morphology of the PS/PA6 (80/20 by mass) and PS/PA6 (95/5 by mass) blends using a reactive compatibilizer-tracer such as PS-*co*-TMI-MAMA. Generally speaking, when the PA6 volume fraction is lower, the distance between

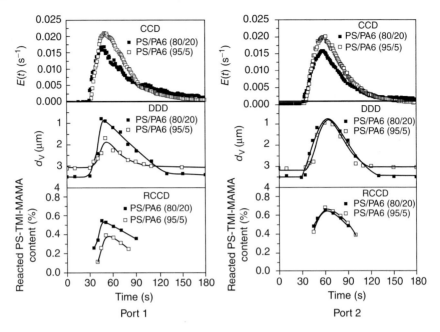

Figure 8.28 Effect of the PS/PA6 blend composition on the CCD, DDD, and RCCD using PS-*co*-TMI-MAMA as a reactive compatibilizer-tracer. Feed rate: 13 kg h^{-1}; screw speed: 100 rpm; amount of the PS-TMI-MAMA reactive compatibilizer-tracer: 3.2 g. The reactive compatibilizer-tracer is injected at port 1 or port 2. (Ji *et al*. 2015 [29]. Reproduced with permission of American Chemical Society.)

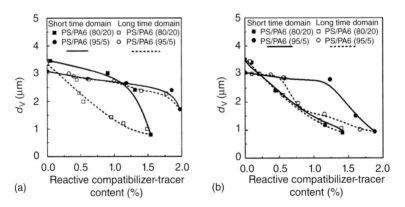

Figure 8.29 Effect of the PS/PA6 blend composition on the emulsification curves (a,b), the RCC versus CC (c,d), and the effective emulsification curves (e,f) using PS-*co*-TMI-MAMA as a reactive compatibilizer-tracer. Feed rate: 13 kg h^{-1}; screw speed: 100 rpm; amount of tracer-compatibilizer: 3.2 g. (Ji *et al*. 2015 [29]. Reproduced with permission of American Chemical Society.)

8.4 Emulsification Curves of Reactive Immiscible Polymer Blends in a Twin-Screw Exturder | 237

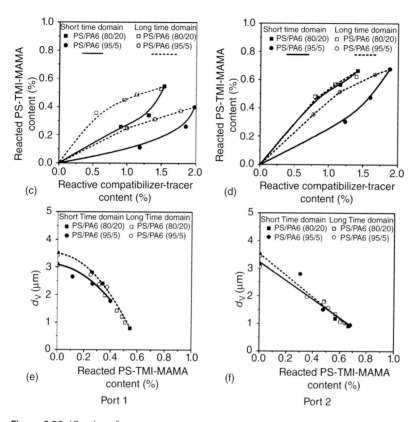

Figure 8.29 (Continued)

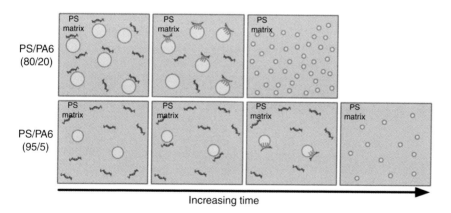

Figure 8.30 A schematic of the development of the interfacial reaction and morphology of the PS/PA6 (80/20 by mass) and PS/PA6 (95/5 by mass) blends using a reactive compatibilizer-tracer of type PS-TMI-MAMA. (Ji et al. 2015 [29]. Reproduced with permission of American Chemical Society.)

the PA6 domains and the reactive compatibilizer-tracer is longer, requiring a longer mixing time (or a stronger mixing intensity) for them to collide and react with each other at the interfaces, leading to the *in situ* formation of the graft copolymer. Therefore, it is possible that the DD of the PS/PA6 blend containing a lower PA6 volume fraction may be even temporally larger than that of the PS/PA6 blend containing a higher PA6 volume fraction. In other words, a polymer blend with a lower dispersed polymer concentration may require more mixing (a longer mixing time, a longer screw length and/or higher mixing intensity) than one with a higher dispersed polymer concentration.

The above scheme could help understand further the data in Figure 8.29 in order to shed more light on processing parameters that dictate the relationship between mixing, interfacial reaction, and morphology development in reactive blending processes. For example, in the case of the PS/PA6 (80/20 by mass) blend, the area of the loop of the DD versus CC narrows considerably when the reactive compatibilizer-tracer injection location is shifted from port 1 to port 2 (Figure 8.29b). This narrowing results when the reactive compatibilizer-tracer injection location is shifted from port 1 to port 2; however, the DD in the long time domain is not much affected by the shift because in both cases the mixing time is long enough. However, the DD in the short time domain is significantly decreased because the shift of the injection location from port 1 to port 2 increases the mixing time significantly. On the other hand, in the case of the PS/PA6 (95/5 by mass) blend, the area of the loop of the DD versus CC enlarges considerably when the reactive compatibilizer-tracer is shifted from port 1 to port 2 (Figure 8.29a). This is mainly because irrespective of the reactive compatibilizer-tracer injection location (port 2 or 1), when the PA6 volume fraction is low, the time is too short to create enough collisions between the reactive compatibilizer-tracer and PA6. Thus, the **DD** remains large.

8.4.6.3 Effect of the Geometry of Screw Elements

The geometry of screw elements is one of the most important processing parameters for a twin-screw extruder. However, there are few data to quantify its effect

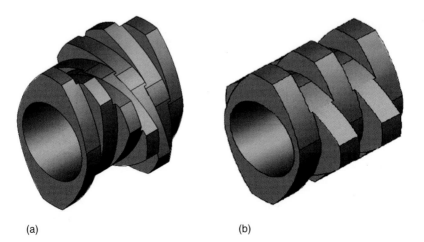

(a) (b)

Figure 8.31 Screw elements used in the experiments. (a) 30° kneading block and (b) 90° kneading block.

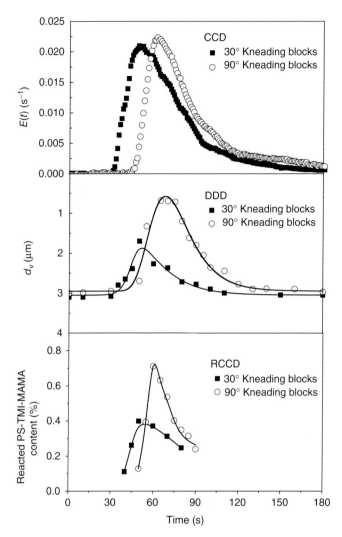

Figure 8.32 Effects of the geometry of the kneading block (30° vs 90°) on the CCD, DDD, and RCCD of the PS/PA6 (95/5 by mass) blend using the PS-co-TMI-MAMA as a reactive compatibilizer-tracer. Feed rate: 13 kg h^{-1}; screw speed: 100 rpm; amount of compatibilizer-tracer: 3.2 g; reactive compatibilizer-tracer injection location: port 1. (Ji et al. 2015 [29]. Reproduced with permission of American Chemical Society.)

due to experimental difficulties. The concept of reactive compatibilizer-tracer allows overcoming these difficulties.

The mixing performance of two types of kneading blocks: 30° and 90° is compared. Their geometries are shown in Figure 8.31. The PS/PA6 blend composition is 95/5 by mass and the reactive compatibilizer-tracer is injected at port 1.

Figure 8.32 shows CCD, DDD, and RCCD curves for these two different kneading blocks. Compared with that of the 30° kneading block, the CCD of the 90° kneading block is shifted toward the long time domain by about 15 s. Its DDD and RCCD curves are all shifted to the long time domain accordingly.

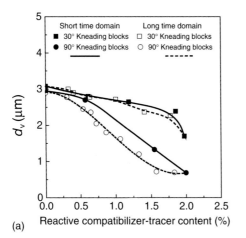
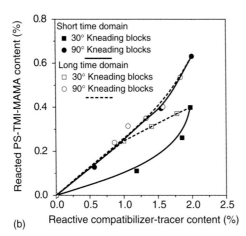
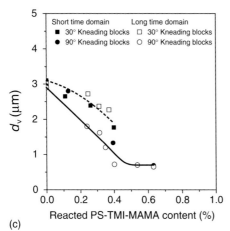

Figure 8.33 Effect of the geometry of the kneading zone on the emulsification curve (a), RCC versus CC (b), and effective emulsification curve (c) using PS-co-TMI-MAMA as a reactive compatibilizer-tracer of the PS/PA6 (95/5 by mass) blend. Feed rate: 13 kg h^{-1}; screw speed: 100 rpm; amount of tracer-compatibilizer: 3.2 g. The reactive compatibilizer-tracer is injected at port 1. (Ji et al. 2015 [29]. Reproduced with permission of American Chemical Society.)

Figure 8.33 compares the 30° and 90° kneading blocks in terms of the emulsification curve (a), RCC versus CC (b), and the effective emulsification curve (c). The emulsification curve of the 90° kneading block is significantly below that of the 30° kneading block. Moreover, its RCC versus CC curve is significantly above that of the 30° kneading block. The fact that the RCC versus CC of the 90° kneading block is significantly higher than that of the 30° kneading block indicates that the former provides much better distributive and dispersive mixing than the latter. It is also noted that the effective emulsification curve of the 90° kneading block is significantly below that of the 30° kneading block. This further confirms that the 90° kneading block has a better dispersive mixing performance than the 30° kneading block.

The above results show that the concept of reactive compatibilizer-tracer is a powerful tool for assessing the mixing performance of different screw elements.

8.5 Conclusion

A compatibilizer-tracer is a compatibilizer (a block or graft copolymer) that bears moieties capable of being detected in a very sensitive manner, even if their content is very small. A reactive compatibilizer-tracer is a compatibilizer-precursor that bears both reactive groups capable of reacting with its counterpart to form a copolymer *in situ* as a compatibilizer, and moieties that allow accurate determination of the amount of the *in situ* formed compatibilizer even if its amount is small.

The concepts of compatibilizer-tracer and reactive compatibilizer-tracer with transient experiments allow building up emulsification curves with very small amounts of compatibilizers. They also allow probing interfacial phenomena of (reactive) polymer blending processes. They are a powerful tool for understanding and optimizing polymer blending processes in terms of the molecular architecture of (reactive) compatibilizers, screw profile, process conditions, and so on.

References

1 Hu, G.H. and Kadri, I. (1998) Modeling reactive blending: an experimental approach. *J. Polym. Sci., Part B: Polym. Phys.*, **36** (12), 2153–2163.
2 Ryan, A.J. (2002) Polymer science: designer polymer blends. *Nat. Mater.*, **1** (1), 8–10.
3 Matos, M., Favis, B.D., and Lomellini, P. (1995) Interfacial modification of polymer blends-the emulsification curve: 1. Influence of molecular weight and chemical composition of the interfacial modifier. *Polymer*, **36**, 3899–3907.
4 Cigana, P. and Favis, B.D. (1998) The relative efficacy of diblock and triblock copolymers for a polystyrene/ethylene-propylene rubber interface. *Polymer*, **39** (15), 3373–3378.

5 Chio, W.M., Park, O.O., and Lim, J.G. (2004) Effect of diblock copolymers on morphology and mechanical properties for syndiotactic polystyrene/ethylene-propylene copolymer blends. *J. Appl. Polym. Sci.*, **91** (6), 3618–3626.

6 Harrats, C., Fayt, R., and Jérôme, R. (2002) Effect of block copolymers of various molecular architecture on the phase morphology and tensile properties of LDPE rich (LDPE/PS) blends. *Polymer*, **43** (3), 863–873.

7 Zhang, C.L., Feng, L.F., Gu, X.P., Hoppe, S., and Hu, G.H. (2007) Efficiency of graft copolymers as compatibilizers for immiscible polymer blends. *Polymer*, **48** (20), 5940–5949.

8 Zhang, C.L., Feng, L.F., Gu, X.P., Hoppe, S., and Hu, G.H. (2010) Blend composition dependence of the compatibilizing efficiency of graft copolymers for immiscible polymer blends. *Polym. Eng. Sci.*, **50** (11), 2243–2251.

9 Nauman, E.B. and Buffham, B.A. (1983) *Mixing in Continuous Flow System*, John Wiley & Sons Inc, New York.

10 Chen, L.Q., Pan, Z.Q., and Hu, G.H. (1993) Residence time distribution in screw extruders. *AIChE J.*, **39** (9), 1455–1464.

11 Chen, L.Q. and Hu, G.H. (1993) Application of a statistical theory to residence time distribution. *AIChE J.*, **39** (9), 1558–1562.

12 Chen, L., Hu, G.H., and Lindt, J.T. (1995) Residence time distribution in non-intermeshing counter-rotating twin screw extruders. *Polym. Eng. Sci.*, **35** (7), 598–603.

13 Lacasse, C. and Favis, B.D. (1999) Interface/morphology/property relationships in polyamide-6/ABS blends. *Adv. Polym. Technol.*, **18** (3), 255–265.

14 Cigana, P., Favis, B.D., and Jérôme, R. (1996) Diblock copolymers as emulsifying agents in polymer blends: influence of molecular weight, architecture, and chemical composition. *J. Polym. Sci., Part B: Polym. Phys.*, **34** (9), 1691–1700.

15 Hu, G.H. and Kadri, I. (1999) Preparation of macromolecular tracers and their use for studying the residence time distribution of polymeric systems. *Polym. Eng. Sci.*, **39** (2), 299–311.

16 Hu, G.H., Kadri, I., and Picot, C. (1999) On-line measurement of the residence time distribution in screw extruders. *Polym. Eng. Sci.*, **39** (5), 930–939.

17 Zhang, C.L., Feng, L.F., Hoope, S., and Hu, G.H. (2012) Compatibilizer-tracer: a powerful concept for polymer-blending processes. *AIChE J.*, **58** (6), 1921–1928.

18 Zhang, C.L., Feng, L.F., Hoope, S., and Hu, G.H. (2009) Residence time distribution: an old concept in chemical engineering and a new application in polymer processing. *AIChE J.*, **55** (1), 279–283.

19 Zhang, C.L., Feng, L.F., Gu, X.P., Hoope, S., and Hu, G.H. (2012) Tracer-compatibilizer: synthesis and applications in polymer blending processes. *Polym. Eng. Sci.*, **52** (2), 300–308.

20 Zhang, X.M., Xu, Z.B., Feng, L.F., Song, X.B., and Hu, G.H. (2006) Assessing local residence time distributions in screw extruders through a new in-line measurement instrument. *Polym. Eng. Sci.*, **46** (4), 510–519.

21 Zhang, X.M., Feng, L.F., Hoppe, S., and Hu, G.H. (2008) Local residence time, residence revolution, and residence volume distributions in twin-screw extruders. *Polym. Eng. Sci.*, **48** (1), 19–28.

22 Macosko, C.W., Jeon, H.K., and Hoye, T.R. (2005) Reactions at polymer–polymer interfaces for blend compatibilization. *Prog. Polym. Sci.*, **30** (8), 939–947.

23 Xie, F., Zhou, C., and Yu, W. (2008) Effects of small-amplitude oscillatory shear on polymeric reaction. *Polym. Compos.*, **29** (1), 72–76.

24 Zhang, J., Ji, S., Song, J., Lodge, T.P., and Macosko, C.W. (2010) Flow accelerates interfacial coupling reactions. *Macromolecules*, **43** (18), 7617–7624.

25 Feng, L.F. and Hu, G.H. (2004) Reaction kinetics of multiphase polymer systems under flow. *AIChE J.*, **50** (10), 2604–2612.

26 Orr, C.A., Cernohous, J.J., Guegan, P., Hirao, A., Jeon, H.K., and Macosko, C.W. (2001) Homogeneous reactive coupling of terminally functional polymers. *Polymer*, **42** (19), 8171–8178.

27 Ji, W.Y., Feng, L.F., Zhang, C.L., Hoope, S., Hu, G.H., and Dumas, D. (2016) A concept of reactive compatibilizer-tracer for studying reactive polymer blending processes. *AIChE J.*, **62** (2), 359–366.

28 Hu, G.H. and Kadri, I. (1998) Modeling reactive blending: an experimental approach. *J. Polym. Sci., Part B: Polym. Phys.*, **36** (12), 2153–2163.

29 Ji, W.Y., Feng, L.F., Zhang, C.L., Hoope, S., and Hu, G.H. (2015) Development of a reactive compatibilizer-tracer for studying reactive polymer blends in a twin-screw extruder. *Ind. Eng. Chem. Res.*, **54** (43), 10698–10706.

Part V

Selected Examples of Synthesis

# 9

Nano-structuring of Polymer Blends by *in situ* Polymerization and *in situ* Compatibilization Processes

Cai-Liang Zhang[1], Lian-Fang Feng[1], and Guo-Hua Hu[2]

[1] State Key Laboratory of Chemical Engineering, College of Chemical and Biochemical Engineering, Zhejiang University, Hangzhou 310027, China
[2] Université de Lorraine, CNRS, Laboratoire Rèactions et Gènie des Procèdès (UMR 7274), 1 rue Grandville, BP 20451, 54001 Nancy, France

9.1 Introduction

Nowadays, it has been common practice to blend existing polymers to create new polymer materials, instead of searching for new monomers because it is often more costly and time-consuming to create new monomers [1]. The final properties of such a polymer blend depend on the state of dispersion or its morphology. However, since most polymer pairs are thermodynamically immiscible, it may not be easy to disperse one polymer component in the other in a sufficiently fine manner and therefore simply blending them may likely yield materials with poor properties.

To overcome these problems associated with the immiscibility between polymers, block or graft copolymers, whose segments are chemically identical or similar to those of the polymer components, are often used as compatibilizers (also called interfacial modifiers or emulsifiers) to reduce the interfacial tension, promote the dispersion of one polymer component in the other, and stabilize the morphology of the blend [2]. Such copolymers can be premade and then added to a polymer blend, or generated *in situ* during the blending process. The first method has the advantage of better controlling the molecular architecture of the copolymer. However, many chemical routes cannot be followed easily or in an economically feasible fashion to synthesize block or graft copolymers for industrial compatibilization purposes. More importantly, it is not always easy for the copolymer to migrate and locate at the interfaces between polymer components. The second method is called reactive blending, reactive compatibilization, or *in situ* compatibilization. In this method, the copolymer is directly formed *in situ* at the interfaces during polymer blending. Therefore, the issue of copolymer migration to the interfaces does not exist anymore. For this reason, *in situ* compatibilization has been considered as the most efficient compatibilization method to stabilize immiscible polymer blends [3, 4].

It should be pointed out that it is virtually impossible to structure immiscible polymer blends at a nanometer scale, namely, the diameter of the dispersed phase

Reactive Extrusion: Principles and Applications, First Edition. Edited by Günter Beyer and Christian Hopmann.
© 2018 Wiley-VCH Verlag GmbH & Co. KGaA. Published 2018 by Wiley-VCH Verlag GmbH & Co. KGaA.

domains is less than 100 nm, irrespective of the compatibilization method used. This is true even in the case of *in situ* compatibilization because the amount of copolymer formed is often small, unless at least one of the reactive polymer components has a low molar mass.

This chapter aims at presenting an *in situ* polymerization and *in situ* compatibilization methodology to nano-structure immiscible polymer blends. To that end, it first describes classical immiscible polymer blending processes in order to better understand why they rarely lead to nano-blends, namely, polymer blends structured at a nanometer scale. It then presents the principles and examples of nano-structuring processes of immiscible polymer blends by *in situ* polymerization and *in situ* compatibilization. It gives a summary at the end.

9.2 Morphology Development of Classical Immiscible Polymer Blending Processes

Polymer blends are often obtained by melting and mixing polymers in a batch mixer or a continuous screw extruder. Their morphology may not necessarily be in a thermodynamic equilibrium state but may be altered by thermomechanical conditions.

A typical morphology development of an immiscible polymer blend is depicted in Figure 9.1 [5]. A solid pellet (S) melts so that the morphology can develop subsequently. The molten polymer layers (M) stripped from the solid pellet undergo a transient affine deformation to form sheets or threads (T). Driven by interfacial tension, the latter break up into small particles (P). These small particles may undergo coalescence because of interfacial tension and form larger particles (C). Obviously, the morphology of an immiscible polymer blend results from complex interactions between flow and events occurring at droplet-length scales: melting, deformation, breakup, coalescence, and hydrodynamic interactions, as pointed out by Ottino *et al.* [6].

Despite its complexity, the morphology development of polymer blends can be divided into two stages: the solid–liquid transition stage or melting stage where at least one of the polymer components is still undergoing melting; the liquid–liquid stage or melt flow stage where both polymer components have become totally molten [5]. Controversy remains over which stage determines the final morphology. One school of thought is that the ultimate morphology is formed almost as soon as the transition of polymer components from solid to liquid is over. This deduction is based on the fact that the most significant changes of morphology are accomplished in the very early stages of mixing, and further mixing often has

Figure 9.1 Different steps involved in the morphology development of an immiscible polymer blend starting from solid pellets: melting of solid pellets (S); stretching/deformation of the molten polymer (M) to slender threads (T); breakup of slender threads to small particles (P); and coalescence of small particles to larger ones (C). (Li and Hu 2001 [5]. Reproduced with permission of Wiley.)

little or no influence on the morphology. Another school of thought is that the final morphology is critically determined in the liquid–liquid stage, and not in the solid–liquid transition stage. Therefore, one has to understand its morphology change in those two stages in order to evaluate quantitatively the ultimate size of the dispersed phase domains.

9.2.1 Solid–Liquid Transition Stage

In the solid–liquid transition stage, the solid pellets of polymers are molten and softened to form polymer melts. A mechanism for morphology development at short mixing times for a polymer blend proposed by Scott and Macosko [7] is shown in Figure 9.2. It involves the formation of sheets or ribbons of the dispersed phase in the matrix, which are drawn out of a large mass of the dispersed phase. Similar phenomena were also observed by Lindt and Ghosh [9], Sundararaj *et al.* [8], and Willemse *et al.* [10]. The sheets are unstable and holes begin to form in them because of the effects of flow stresses and interfacial stress. The holes are filled with the material. When they attain a sufficient size and concentration, a fragile lace structure is formed and then breaks apart into irregularly shaped pieces that further continue to break down until all of the particles become nearly spherical droplets.

Generally speaking, the change of morphology from sheets or ribbons to spherical particles in the solid–liquid stage can lead to a drastic reduction in the size of the dispersed phase domains [11]. As shown in Figure 9.3, the volume average diameter of the dispersed phase domains for polyamide (PA)/maleic anhydride

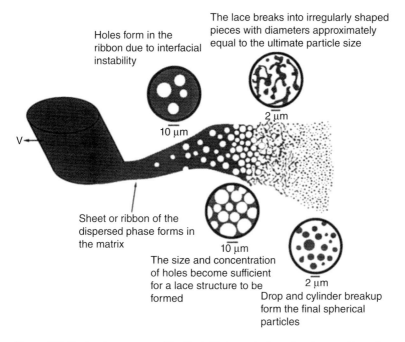

Figure 9.2 Mechanism proposed for the initial morphology development in polymer blending. (Sundararaj *et al.* 1995 [8]. Reproduced with permission of Elsevier.)

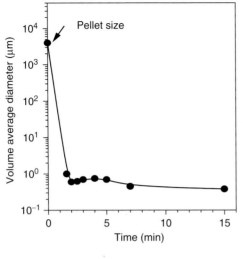

Figure 9.3 Morphology development for a polyamide (PA)/maleic anhydride modified ethylene–propylene rubber (EP-MA) blend system with 20 wt% EP-MA in a batch intensive mixer. The rotation speed of the rotors is fixed at 50 rpm. Volume average diameter can be statistically characterized by the following equation: $d_v = \dfrac{\sum n_i d_i^4}{\sum n_i d_i^3}$. (Scott and Macosko 1994 [12]. Reproduced with permission of Elsevier.)

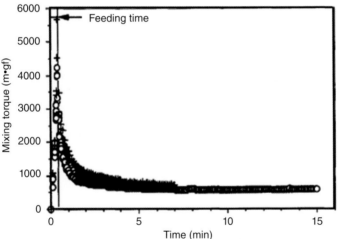

Figure 9.4 Torque as a function of mixing time for: (○) polystyrene (PS)/polyamide66 (PA66) blends (data from seven runs are overlaid); (+) vinyl oxazoline-modified polystyrene (PS-Ox)/PA66 blend (data from two runs are overlaid). (Scott and Macosko 1995 [7]. Reproduced with permission of Elsevier.)

modified ethylene-propylene rubber (EP-MA) blend system decreases from a millimeter scale to a micrometer scale in the solid–liquid stage [12]. The main reason for this drastic reduction in the dispersed phase domain size is that stresses applied to the materials are greatest in the whole polymer blend process, which can be confirmed by the evolution of the torque as a function of mixing time as shown in Figure 9.4 [7]. Despite the fact that the rotor speed remains the same during the entire mixing period, the torque applied to the material in the solid–liquid transition stage is far higher than that in the following melt flow stage because the viscosity of the material is highest at this time.

What is the size of the droplets obtained from breakup of threads? According to Tomotika's analysis [13], the diameter of the droplets ($d_{droplet}$) resulting from breakup of an infinitely long and viscous thread embedded in a quiescent matrix of an immiscible and viscous melt is related to its initial diameter, d_0, by the following equation:

$$d_{droplet} = \left(\frac{3\pi}{2X_m}\right)^{1/3} d_0 \qquad (9.1)$$

where X_m is the dominant wave number of the thread leading to breakup, which is a function of viscosity ratio between the thread and the matrix. Usually, fine dispersion is facilitated for polymer components with similar viscosity. When the viscosity ratio equals to 1, and X_m equals to 0.56, Eq. (9.1) then becomes $d_{droplet} = 2d_0$ [14].

For the diameter of the droplets to be smaller than 100 nm, the diameter of the thread to breakup has to be smaller than 50 nm. Although this is theoretically under certain circumstances, it has not been feasible in practice because of agglomerates whose droplets are bound by van der Waals forces. As a matter of fact, the average number of droplets in final fragments $\langle i \rangle$ depends greatly on the dissipation energy imposed by flow, E, according to the following equations [15]:

$$\text{Simple shear flow}: \langle i \rangle = 5.57 \times 10^5 E^{-0.468} \qquad (9.2)$$

$$\text{Simple elongation flow}: \langle i \rangle = 3.20 \times 10^7 E^{-0.798} \qquad (9.3)$$

It is evident that nano-structure of droplets is possible only when agglomerates are composed of nano-droplets and when $\langle i \rangle$ is unity. For simple shear flow, in order for $\langle i \rangle$ to be unity, E has to be 1.89×10^{12} W m^{-3}. This is simply not attainable in a batch mixer or an extruder. In fact, E ranges from 10^4 to 10^9 W m^{-3} for extruders. Even in the upper limit of E for extruders, $\langle i \rangle$ is 34. Under this circumstance, the final droplets will be 1.7 µm even if the diameter of the particles is 50 nm [14]. Obviously, the above analyses show that it is very difficult, if not possible at all, to reach nano-dispersion starting from polymer pellets/powders.

9.2.2 Melt Flow Stage

For a polymer blending process in a batch mixer or a screw extruder, there is a flow field. The stress of the flow field can deform polymer droplets and then may break them into smaller domains. At the same time, the flow filed can also promote collisions between the dispersed phase domains, leading to coalescence. Thus, the morphology development of polymer blends in a melt flow stage results from the competition between the breakup and coalescence of the dispersed phase domains. Whether or not a dispersed phase domain breaks up depends on the capillary number, Ca, which is the ratio of deforming viscous stress to resisting interfacial stress (Eq. (9.4)). Similarly, whether or not coalescence takes place depends on the fragmentation number, Fr, which is the ratio of deforming viscous stress to resisting cohesive strength of the agglomerates (Eq. (9.5)).

$$Ca = \frac{\text{Deforming viscous stress}}{\text{Resisting interfacial stress}} = \frac{\eta_m \dot{\gamma}}{2\sigma/d} \qquad (9.4)$$

$$Fr = \frac{\text{Deforming viscous stress}}{\text{Cohesives strength of agglomerates}} = \frac{\eta_m \dot{\gamma}}{T} \qquad (9.5)$$

where η_m is viscosity of the continuous phase, $\dot{\gamma}$ is the shear rate, σ is the interfacial tension, d is the diameter of the dispersed phase domain, and T is the cohesive strength of the agglomerate that corresponds to inter-particle bonds due to electrostatic charges and van der Waals forces.

On the other hand, the flow stress deforms and breaks the dispersed phase domains and the interfacial tension resists deformation. When the flow stress and interfacial stress reach a balance, and the viscosity ratio of the dispersed phase domain and the matrix (η_r) is below 2.5, the diameter of the dispersed phase domain can be calculated using Eq. (9.6) proposed by Taylor [16, 17]:

$$d = \frac{4\sigma(\eta_r + 1)}{\dot{\gamma}\eta_m \left[\left(\frac{19}{4}\right)\eta_r + 4\right]} \tag{9.6}$$

When η_r is close to 1, dispersed phase domains break most easily. Additionally, if η_r is greater than 5, the breakup of droplets does not occur regardless of how high the capillary number is.

It should be noticed that the Taylor equation is based on the assumption that the dispersed phase is dispersed in infinitely dilute Newtonian systems as spheres. However, polymer melts are mostly viscoelastic and non-Newtonian fluids. Under a shear stress, the elasticity of the dispersed phase can resist its deformation so that it hardly breaks up. Based on blends composed of PA6 and copolymer of ethylene and propylene rubber (EPR), Wu [18] modified the Taylor equation to the following one:

$$d = \frac{4\sigma\eta_r^{\pm 0.84}}{\dot{\gamma}\eta_m} \tag{9.7}$$

where the plus sign is used for $\eta_r > 1$ and minus sign for $\eta_r < 1$. However, it ignores the influence of blend composition on the morphology. As a matter of fact, there are many empirical equations based on different blend systems. They all have different limitations. Moreover, irrespective of the empirical equation, under normal thermodynamic and mechanical conditions, the predicted particle size cannot reach a nanometer scale.

On the one hand, for breakup of the dispersed phase domain to occur, the capillary number must exceed a critical value (Ca_{crit}). Thus, this critical value of the capillary number gives the maximum value of the dispersed phase domain size that can survive in a given flow in the absence of coalescence. For two immiscible polymers under mixing, the maximum diameter of the dispersed phase domain, denoted as d_{max}, can be expressed by the following equation:

$$d_{max} = \frac{2\sigma}{\eta_m \dot{\gamma}} Ca_{crit} \tag{9.8}$$

If polymer components are of equal viscosity, the value of Ca_{crit} is close to unity. Taking the values of $\sigma, \eta_m, \dot{\gamma}$ as 1×10^{-2} N m^{-2}, 100 Pa s, 100 s^{-1}, respectively, the corresponding value of d_{max} is equal to 2 µm. This implies that under such typical processing conditions, dispersed phase domains whose diameters are greater

Figure 9.5 Evolution of the radius of PA6 particles in the PP matrix for PP/PA6 (100/0.5 by mass) blend with and without a graft copolymer of PP and PA6 as a compatibilizer as a function of mixing time in an internal batch mixer of type Haake Rheocord. (Li and Hu 2001 [5]. Reproduced with permission of Wiley.)

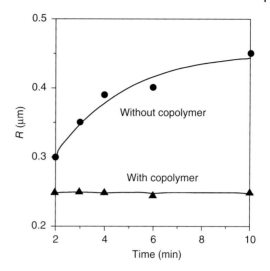

than 2 μm cannot be stable and will be broken up into smaller ones. Those whose diameters are smaller than 2 μm will be stable, if no coalescence occurs [14].

However, the rate of coalescence cannot be zero without a compatibilizer, even if the fraction of the dispersed phase is very small. For example, for a polyamide 6 (PA6)/polypropylene (PP) blending containing 0.5 wt% PA6, the radius of the PA6 domains without a compatibilizer is still above 0.3 μm. And it increases constantly with mixing time due to coalescence, as shown in Figure 9.5.

From the above analysis on the mechanism of morphology development of polymer blends during melt blending, the reduction of interfacial tension and coalescence is very crucial for obtaining fine dispersed particles.

9.2.3 Effect of Compatibilizer

The above discussion is based on uncompatibilized polymer blends. The addition of a compatibilizer, whether it is a premade copolymer or formed *in situ* during the melt blending process, has a significant influence on the morphology development. It is well known that the presence of a compatibilizer can accelerate the morphology development, enhance the interfacial adhesion, stabilize the morphology, and reduce the disperse phase domain size. Early researchers attributed these changes to a lower interfacial tension. However, Lyu *et al.* [19] found that the reduction in interfacial tension alone is not reason enough to explain the decrease of the dispersed phase domain size, and that the suppression of coalescence may be the main reason.

There are two mechanisms for the suppression of coalescence, which are shown in Figure 9.6. One is caused by Marangoni force induced by a gradient of copolymer concentration at interfaces, which was proposed by Milner and Xi [20]. As shown in Figure 9.6a, when two droplets approach each other, the matrix escaping

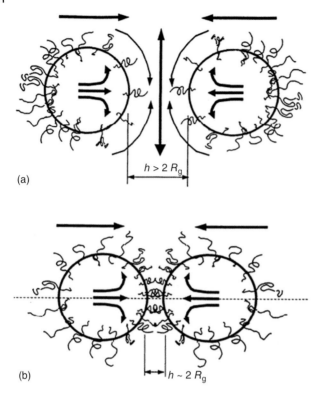

Figure 9.6 Two mechanisms proposed for the suppression of coalescence by copolymers: (a) surface tension gradient (Marangoni) force and (b) steric repulsion. (Lyu et al. 2002 [19]. Reproduced with permission of American Chemical Society.)

from the gap between them sets up a recirculating fountain flow that carries the copolymer at the interface out of the gap, forming a concentration gradient.

The other is a steric interaction between droplets resulting from the compression between the block copolymer layers that are attached to the interfaces, as proposed by Sundararaj and Macosko [21]. Usually, the interface is surrounded either by a single layer or by multi-layers of copolymers. When two droplets approach each other, the copolymer at the interface hinders these two droplets from meeting and colliding, as shown in Figure 9.6b

However, many experiments [5, 21–25] have indicated that the diameter of the dispersed phase domains of immiscible polymer blends is rarely below 100 nm even if there is no coalescence. As shown in Figure 9.5, the radius of the dispersed phase domain for a PP/PA6 (100/0.5 by mass) blend remains constant at about 0.25 μm during the entire blending process in the presence of a graft copolymer of PP and PA6 as a compatibilizer.

To sum up, the above discussions show that it is very difficult, if not impossible, to achieve nano-structuring of immiscible polymer blends by melt processing in a batch mixer or a screw extruder. Therefore, it needs a new technique and/or strategy to generate dispersed phase domains at a nanometer scale.

9.3 *In situ* Polymerization and *in situ* Compatibilization of Polymer Blends

9.3.1 Principles

An *in situ* polymerization and *in situ* compatibilization process consists of polymerizing the monomer (MB) of a polymer component B in the presence of the other polymer component A. A fraction of polymer component A chains bears initiating sites either at the chain end(s) or along the chains, denoted as A′, from which MB can polymerize to form a graft or block copolymer of polymer A and polymer B. Thus, there are four phenomena involved in this process as shown in Figure 9.7: (1) polymerization of MB in the presence of polymer A, which produces polymer B; (2) phase separation between polymers A and B as a result of the *in situ* polymerization; (3) polymerization of MB from polymer A leading to the formation of a copolymer; and (4) compatibilization of the *in situ* polymerized blend by the *in situ* formed copolymer.

Conceptually, the *in situ* polymerization and *in situ* compatibilization method can allow obtaining a variety of morphologies including nano-dispersion of one polymer component in the other by controlling the kinetics of *in situ* polymerization, phase separation, and *in situ* copolymer formation. In order to acquire a nano-blend, both the amount and the rate of formation of the copolymer should be sufficiently high compared with the rates of polymerization and phase separation so that the morphology can be stabilized from the early stages of the phase separation [26].

9.3.2 Classical Polymer Blending Versus *in situ* Polymerization and *in situ* Compatibilization

Figure 9.8a,b depicts a classical polymer blending process and an *in situ* polymerization and *in situ* compatibilization process, respectively. When a mixture

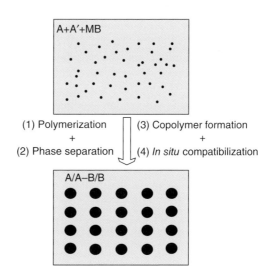

Figure 9.7 Schematic description of the *in situ* polymerization and *in situ* compatibilization methodology for preparing (nano-)blends of polymer A and polymer B. A′: polymer A chain bears initiating sites either at the chain end(s) or along the chain; MB: the monomer of polymer B; A–B: copolymer of polymer A and polymer B.

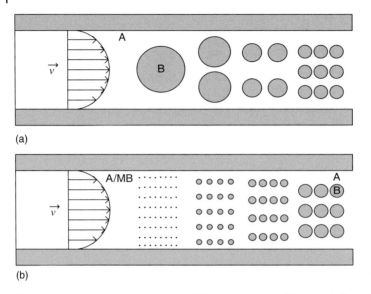

Figure 9.8 Schematic representation of the mechanisms of the morphology development of two different systems subjected to a flow field. (a) An immiscible mixture of polymer A and polymer B; (b) a homogeneous mixture of polymer A and a polymerizable monomer of polymer B (MB). (Hu and Cartier 1999 [27]. Reproduced with permission of American Chemical Society.)

of two polymers is subjected to a flow field, the dispersed phase domains reduce in size along the flow stream. As their size decreases, they become less and less deformable. A threshold will be reached when a dynamic equilibrium is established between thermomechanical and thermodynamic forces. At this point, the dispersed phase domain can no longer be deformed and their breakup will become impossible. As discussed earlier, the equilibrium size can rarely be smaller than 100 nm, irrespective of the compatibilization method used (the compatibilizer is premade and then added to the blend or is generated *in situ* during blending). Thus, classical polymer blending in the melt is a size-reduction process, from millimeter to micrometer.

As for an *in situ* polymerization and *in situ* compatibilization process, for the sake of simplicity consider a homogeneous mixture of polymer A and a polymerizable monomer (MB), which is subjected to a flow filed. Prior to the polymerization, the system is monophasic. As the polymerization of MB proceeds, polymer A and the resulting polymer B will phase separate if they are mutually immiscible. At the same time, the size of the dispersed phase (polymer B) domains increases. If there is no interfacial compatibilizer present in the blend system, the resulting blend will not be very different from the one obtained by directly blending premade polymers A and B whose morphology is gross and unstable [27]. However, it should be noted that this process is a particle-growth process, from nanometer to micrometer. This particle-growth process can be reduced or stopped by the *in situ* formation of a copolymer. Thus the final size of the dispersed polymer phase domains is a result of the rates of particle-growth by coalescence and *in situ* copolymer formation. If the latter is significantly faster than the former, the final size of the dispersed polymer phase domains can be at a nanometer scale.

A classical polymer blending process and an *in situ* polymerization and *in situ* compatibilization process also differ in the mechanism of copolymer formation. In the first case, as reactive polymer components can only meet and react at the interfaces, the amount of the resulting copolymer cannot be large because the total interfacial volume is small. However, if the molar mass of at least one of the reactive polymer components is small, the amount of the copolymer can be high. This is because in this case the interfacial volume can be larger. More importantly, the molecular architecture of the resulting copolymer is so asymmetrical that it may leave the interfaces where they are formed, sparing the interfaces for further reaction.

In the case of *in situ* polymerization and *in situ* compatibilization, the copolymer is formed by polymerization from the initiating sites of a polymer chain. Therefore, the issue of interface does not exist anymore if the system is homogeneous and the amount of the copolymer attainable can be very high. If the system is not homogeneous, the amount of copolymer is still expected to be higher than that for a classical polymer blending system. This is because in the case of *in situ* polymerization and *in situ* compatibilization, the copolymer formation is between a polymer and a monomer, whereas in the case of a classical reactive polymer blending, it is between two immiscible reactive polymers.

9.3.3 Examples of Nano-structured Polymer Blends by *in situ* Polymerization and *in situ* Compatibilization

In what follows, three examples will be given to illustrate how to achieve nano-structuration of polymer blends using the above mentioned "from nanometer to nanometer" methodology, on the one hand, and to provide answers to the question of what properties nano-structuration can bring about.

9.3.3.1 PP/PA6 Nano-blends

Underlying Chemistry and Strategy PA6 can be prepared via an activated anionic polymerization of ε-caprolactam (CL) in the presence of a catalyst such as sodium caprolactam (NaCL) and an activator such as an isocyanate-bearing compound. The reaction mechanism of this activated anionic polymerization is composed of activation, initiation, and propagation steps, as shown in Figure 9.9. Basically, R—N=C=O first activates CL to form an acyl caprolactam. This acyl caprolactam then reacts readily with NaCL forming a new reactive sodium salt. This new salt initiates the polymerization of CL; at the same time, the catalyst (NaCL) is restored. Finally, the repetition of the propagation reaction leads to a high molar mass PA6 [27]. More importantly, the overall polymerization rate is very fast and can be accomplished in a few minutes or less, which is compatible with the residence time of a reactive extrusion process.

In addition to synthesizing PA6, the chemical process can be explored for the preparation of pure copolymers containing polyamide segments when isocyanate (N=C=O) moieties are attached onto a polymer chain, namely, R of R—N=C=O is a polymer chain [26, 28–30]. For example, if PP bears isocyanate moieties along

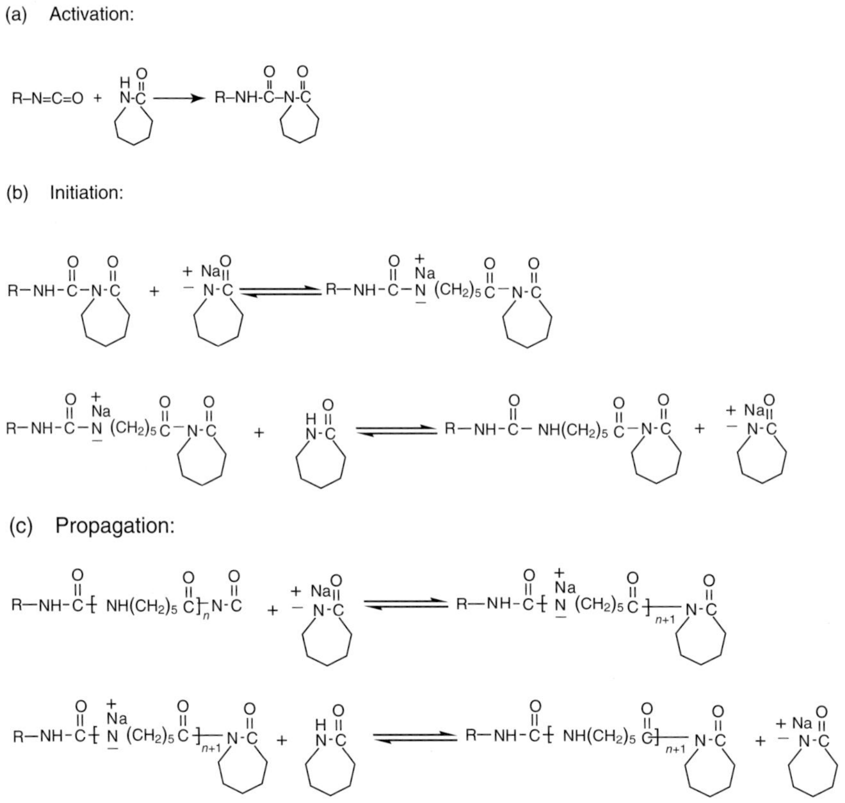

(a) Activation:

(b) Initiation:

(c) Propagation:

Figure 9.9 Schematic description of the activated anionic polymerization of ε-caprolactam (CL).

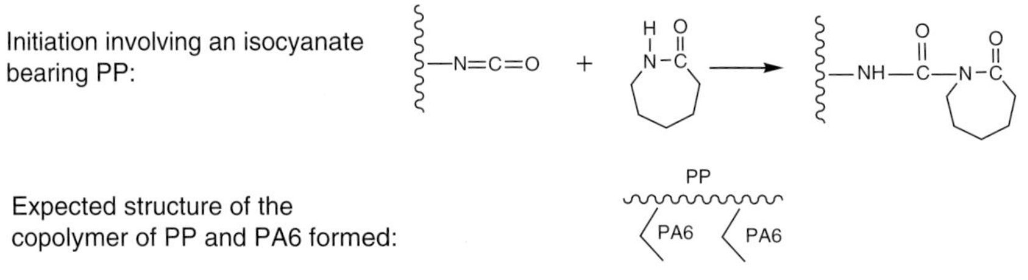

Initiation involving an isocyanate bearing PP:

Expected structure of the copolymer of PP and PA6 formed:

Figure 9.10 Mechanism of formation of a graft copolymer of PP and PA6, denoted as PP-g-PA6, using an isocyanate-bearing PP as a macromolecular activator [27].

its chain, a pure graft copolymer of PP and PA6 will be obtained. The underlying chemical mechanism is shown in Figure 9.10.

Hu et al. [25, 30, 31] grafted 3-isopropenyl-α, α-dimethylbenzene isocyanate (TMI) onto PP by a free radical mechanism in the presence of 2,5-dimethyl-2,5-di(*tert*-butylperoxy)-hexane (DHBP) as a free radical initiator. Figure 9.11 shows the molecular structure of TMI. The TMI-modified PP is denoted as PP-g-TMI. CL was polymerized using PP-g-TMI as a macro-activator, CL blocked hexamethylene diisocyanate as a micro-activator, and NaCL as a

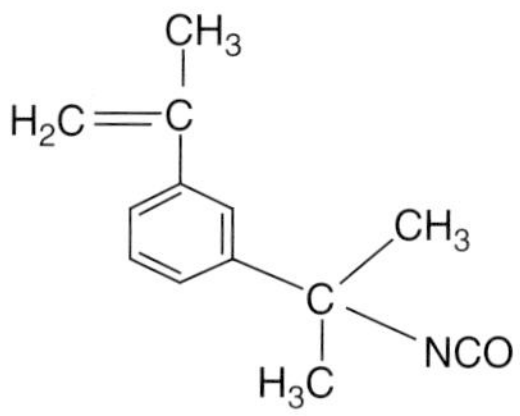

Figure 9.11 Molecular structure of TMI [27, 32].

catalyst. The PP-g-TMI macro-activator and the micro-activator led to the *in situ* formation of PP-g-PA6 graft copolymer and PA6, respectively. The morphology of the resulting PA6/PP blend could be well stabilized and the size of the dispersed phase (PA6) domains could be below 100 nm, depending on the rate of the PS-g-PA6 graft copolymer formation with respect to that of the PA6 formation.

Teng et al. [33] used a maleic anhydride grafted PP as a macro-initiator to obtain nano-blend by *in situ* polymerization and *in situ* compatibilization.

Activated Anionic Polymerization of CL in the Presence of PP, PP-g-TMI or PP/PP-g-TMI
Unlike classical isocyanate groups, those in PP-g-TMI are stable enough under process conditions: high temperature and exposure to moisture. Their stability is due to the protective steric effect of the two adjacent methyl groups of TMI. At the same time, they are reactive enough to activate the polymerization of CL. To activate the polymerization of CL in a rapid manner, the isocyanate moieties should react with CL rapidly. Hu and coworkers [30] evaluated the reactivity of PP-g-TMI toward CL under real polymerization conditions in a Haake torque rheometer in terms of the disappearance of the isocyanate group of the PP-g-TMI as a function of mixing time at 180 and 200 °C, respectively. As shown in Figure 9.12, it is seen that the reaction between PP-g-TMI and CL was

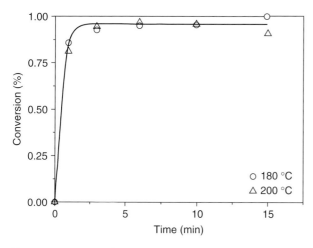

Figure 9.12 Reaction between PP-g-TMI (75 wt%) and CL (25 wt%) at 180 and 200 °C. The TMI content of the PP-g-TMI is 0.4 wt%. (Zhang et al. 2011 [30]. Reproduced with permission of Wiley.)

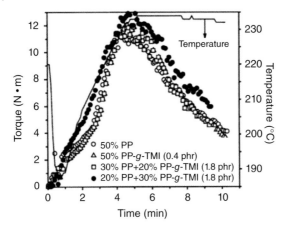

Figure 9.13 Evolution of the torque and temperature during the *in situ* polymerization of CL in a batch mixer in the presence of PP or PP-*g*-TMI. NaCL/micro-activator = 3.0/100 g of CL/PP or CL/PP/PP-*g*-TMI. All the polymerizing systems have almost the same temperature profiles, as shown by the single temperature–time curve. (Zhang *et al.* 2011 [30]. Reproduced with permission of Wiley.)

Table 9.1 Four different *in situ* polymerization systems.

Polymerizing system	Final material
CL/PP/NaCL/micro-activator (50/50/3/3 by mass)	Uncompatibilized PP/PA6 blend
CL/PP-*g*-TMI/NaCL/micro-activator (50/50/3/3 by mass). TMI% = 0.4 wt%	Compatibilized PP/PA6 blend
CL/PP + PP-*g*-TMI/NaCL/micro-activator (50/20 + 30/3/3 by mass). TMI% = 1.8 wt%	Better compatibilized PP/PA6 blend
CL/PP + PP-*g*-TMI/NaCL/micro-activator (50/20 + 20/3/3 by mass). TMI% = 1.8 wt%	Best compatibilized PP/PA6 blend

Source: Zhang *et al.* 2011 [30]. Reproduced with permission of Wiley.

almost completed in less than 3 min (this time included the time necessary for melting PP-*g*-TMI and mixing it with CL). This indicates that the reactivity of the isocyanate group of the PP-*g*-TMI toward CL is indeed very high.

Figure 9.13 compares the evolutions of temperature and torque of four polymerization systems (Table 9.1) as a function of time. Overall, they are good signatures of the CL polymerization process in the batch mixer. The viscosity and therefore the torque of the polymerization system increase as CL is polymerized. The torques of all the four polymerization systems first increase sharply, reach a maximum at about 4–5 min, and then decrease in a drastic way. This indicates that for all the four polymerization systems, CL starts to polymerize immediately after the reactants are introduced to the mixing chamber, and the polymerization has achieved completion in 5 min. A slightly higher torque is observed for the CL/PP/PP-*g*-TMI/NaCL/activator (50/30/20/3/3 by mass) system, likely because a higher concentration of PP-*g*-PA6 copolymer is formed, resulting in a slightly higher viscosity with respect to the other polymerization systems. The conversions of CL to PA6 and PP-*g*-PA6 are about 96% for these four systems.

The amounts of the PA6 homopolymer in polymerized materials can be determined by solvent extraction in formic acid at room temperature. For the PP-*g*-TMI/CL/NaCL/micro-activator (50/50/3/3 by mass) system, the amount

Table 9.2 Effect of the composition of the PP/PP-g-TMI/ε-CL/NaCL/ micro-activator polymerization system on the PP-g-PA6 graft copolymer content obtained after successive extraction in hot xylene and then in formic acid.

PP/PP-g-TMI/CL/ NaCL/micro- activator (by mass)	PP-g-TMI/ micro-activator (NCO/NCO)	PP-g-PA6 graft copolymer (wt%)
50/0/50/3/3	0	0
48/2/50/3/3	0.03	2.4
30/20/50/3/3	0.30	12.1
0/50/50/3/3	0.75	16.1

The PP-g-TMI contains 1.8 wt.% TMI. The percentage of the PP-g-PA6 graft copolymer is calculated on the total amount of the PP/PA6 blends. For example, 16.1% means that 100 g of the PP/PA6 blend contains 16.1 g of PS-g-PA6 graft copolymer.
Source: Zhang et al. 2011 [30]. Reproduced with permission of Wiley.

of the PA6 homopolymer is 29.3 wt%, which implies that the amount of PA6 grafts in the PP-g-PA6 graft copolymer is 19.7 wt%. In order to further confirm that PP-g-PA6 graft copolymer is formed, successive extraction in hot xylene and formic acid is carried out since PP can be insoluble in hot xylene and PA6 in formic acid. For an uncompatibilized PP/PA6 blend, polymerized materials are solubilized completely after successive extraction in hot xylene and formic acid. For a compatibilized PP/PA6 blend, a certain amount of the material remains insoluble after successive extraction in hot xylene and formic acid. This insoluble material is not crosslinked due to its thermoplastic behavior. Thus, it has to be the PP-g-PA6 graft copolymer. Table 9.1 shows the amounts of the insoluble material for the PP/PP-g-TMI/ε-CL/NaCL/micro-activator polymerization system with four different PP-g-TMI contents. It is obvious that when the PP-g-TMI content is high, the insoluble material (PP-g-PA6) content is also high (Table 9.2).

Morphologies of in situ *Polymerized and* in situ *Compatibilized PP/PA6 Blends*
Figure 9.14 compares the morphologies of PP/PA6 blends from the polymerization of the PP/CL/NaCL/micro-activator (50/50/3/3 by mass) with those from the polymerization of the PP-g-TMI/CL/NaCL/micro-activator (50/50/3/3 by mass) systems. The morphology of the former blend is very gross and its dispersed phase domain sizes are as big as 20 μm. This is typical of an uncompatibilized immiscible polymer blend whose composition is close to 50/50. By contrast, the dispersed phase domains of the PP-g-TMI/CL/NaCL/micro-activator (50/50/3/3 by mass) system after polymerization are extremely fine and their sizes are as small as 80 nm. Moreover, each PA6 particle is covered by a layer of the PP-g-PA6 graft copolymer, as shown by the black layer in Figure 9.14b.

Figure 9.15 compares the DSC curves between the PA6 homopolymer, the uncompatibilized PP/PA6 blend, and the compatibilized PP/PA6 blend obtained from the CL/NaCL/micro-activator (100/3/3 by mass), the

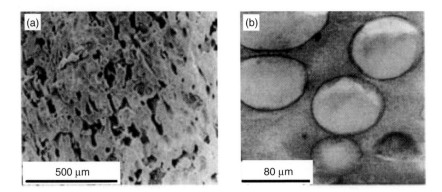

Figure 9.14 Morphologies of the *in situ* polymerized blends from the PP/CL/NaCL/micro-activator (50/50/3/3 by mass) system (a) and the PP-*g*-TMI/CL/NaCL/micro-activator (50/50/3/3 by mass) system (b), respectively. (Hu and Cartier 1999 [27]. Reproduced with permission of American Chemical Society.)

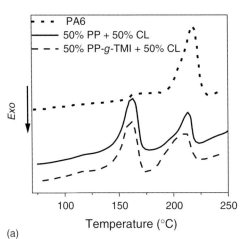

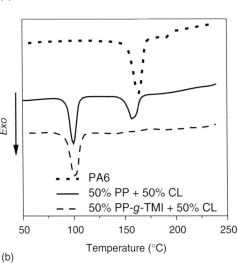

Figure 9.15 Comparison of the DSC thermograms of the polymerized material from the CL/NaCL/micro-activator (100/3/3 by mass), PP/CL/NaCL/micro-activator (50/50/3/3 by mass), and PP-*g*-TMI/CL/NaCL/micro-activator (50/50/3/3 by mass) during heating (a) and cooling (b). Heating and cooling rates are 10 °C min^{-1}. (Zhang *et al.* 2011 [30]. Reproduced with permission of Wiley.)

PP/CL/NaCL/micro-activator (50/50/3/3 by mass), and the PP-g-TMI/CL/NaCL/micro-activator (50/50/3/3 by mass), respectively. It can be observed from Figure 9.15a that both PP/PA6 blends display two distinct peaks at 161 and 212 °C, corresponding to the melting temperatures of the PP and PA6, respectively. In the cooling process (Figure 9.15b), the uncompatibilized PP/PA6 blend from the PP/CL/NaCL/micro-activator (50/50/3/3 by mass) polymerized system shows two peaks at 101 and 160 °C, corresponding to the recrystallization temperatures of the PP and PA6, respectively. However, the compatibilized PP/PA6 blend from the PP-g-TMI/CL/NaCL/micro-activator (50/50/3/3 by mass) polymerized system yields only one peak at 101 °C. This phenomenon is typical for a compatibilized PP/PA6 blend.

Mechanical Properties of PP/PA6 Blends Obtained by in situ *Polymerization and* in situ *Compatibilization* As shown in Figure 9.16, the PP/PA6 blend from the PP/CL/NaCL/micro-activator (50/50/3/3 by mass) polymerized system is very brittle and its elongation at break is less than 10% for a tensile speed of 50 mm min^{-1}. By contrast, the one from the PP-g-TMI/CL/NaCL/micro-activator (50/50/3/3 by mass) system exhibits a very ductile behavior at the same test speed. In fact, it does not break down within the span of the tensile machine used (250%). This is again the case when the testing speed is increased from 50 to 150 mm min^{-1}. The elongation at break still reaches 142% when the tensile speed is as high as 500 mm min^{-1}. This is truly remarkable for an immiscible polymer blend, especially for a blend composition close to 50/50 by mass.

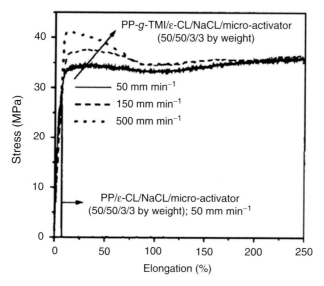

Figure 9.16 Stress–strain traces of the PP/PA6 blends prepared from the PP-g-TMI/CL/NaCL/micro-activator (50/50/3/3 by mass) system at different testing speeds. Tensile speed = 50 (solid line), 150 (dashed line), and 500 mm min^{-1} (dotted line). The stress–strain curve for the blend prepared from the PP/CL/NaCL/micro-activator (50/50/3/3 by mass) system is also shown for comparison; tensile speed = 50 mm min^{-1} (solid line). (Zhang et al. 2011 [30]. Reproduced with permission of Wiley.)

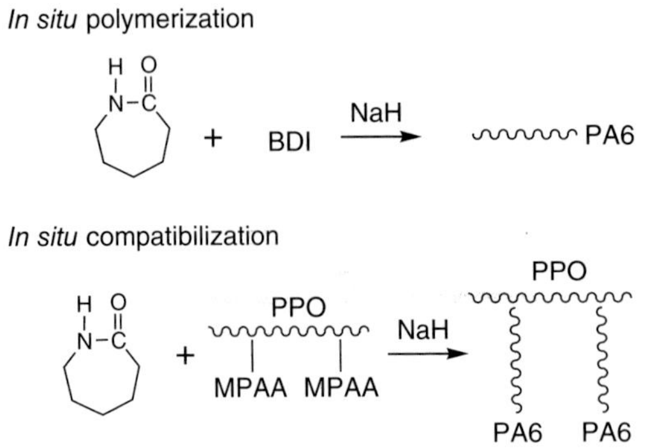

In situ polymerization

In situ compatibilization

Figure 9.17 Reaction mechanism of *in situ* polymerization and *in situ* compatibilization to prepare a compatibilized PPO/PA6 blend [34].

9.3.3.2 PPO/PA6 Nano-blends

Poly(2,6-dimetyl-1,4-phenylene oxide) (PPO) and PA6 blends are of interest as high performance materials. Ji *et al.* [34] adopted successfully the above *in situ* polymerization and *in situ* compatibilization method to compatibilize PPO/PA6 blends. As shown in Figure 9.17, in the presence of a micro-acitvator (2,4-toluene diisocyanate, denotes as BDI), a macro-activator (4-methoxyphenylacrylate (MPAA) modified PPO, denotes as PPO-*g*-MPAA), and a catalyst (NaH), a fraction of the CL is polymerized into PA6 homopolymer while the other fraction grows from the PPO-*g*-MPAA into a graft copolymer of PPO and PA6 (PPO-*g*-PA6). The latter acts as a compatibilizer to control the morphology of the resulting PPO/PA6 blend.

Figure 9.18 shows the effect of the PPO-*g*-MPAA content on the morphology of the PPO/PA6 blend obtained from the PPO/PPO-*g*-MPAA/CL system. In the absence of PPO-*g*-MPAA, the morphology is very gross and the dispersed phase (PPO) domain diameters are larger than 5 μm (Figure 9.18a). The presence of 15 wt% PPO-*g*-MPAA reduces their diameters to about 2.5 μm (Figure 9.18b). An increase in the PPO-*g*-MPAA content to 30 wt% further reduces their sizes and the interfaces become more blurring. When the PPO-*g*-MPAA reaches 40 wt%, the PPO/PA6 blend seems to be nano-structured (Figure 9.18d).

9.3.3.3 PA6/Core–Shell Blends

PA6 as an important thermoplastic has a wide range of engineering applications. However, PA6 is brittle under severe conditions such as high strain rates, low temperatures, and the presence of a notch. The toughness of PA6 can be improved by adding low modulus rubbers. However, the enhancement of the toughness is accompanied by a drastic drop in modulus. To overcome this issue, a core–shell modifier, which contains a rigid core and a soft shell, has been used to impart a substantial toughness to the matrix while keeping a high rigidity. In general, the toughened PA6 with a core–shell modifier can be obtained by melt blending PA6 with rubber and a stiff polymer, or with a premade functional copolymer. However, it is difficult to form a core–shell structure under such conditions because of a low diffusion rate of the viscous rubber in the blend and the micellization of copolymer in the copolymer-toughened system.

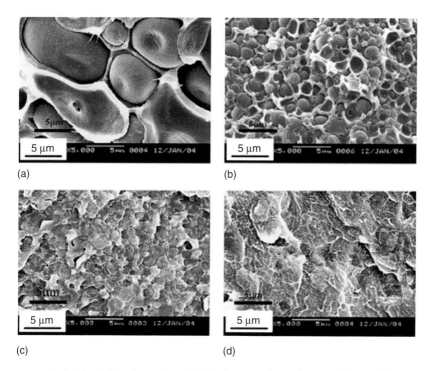

Figure 9.18 Effect of the PPO/PPO-*g*-MPAA/CL mass ratio on the morphology of the polymerized PPO/PPO-*g*-MPAA/CL system. (a) 30/0/70 by mass, (b) 15/15/70 by mass, (c) 0/30/70 by mass, and (d) 0/40/60 by mass. (Ji *et al.* 2005 [34]. Reproduced with permission of Elsevier.)

Yan *et al.* [35] adopt the above *in situ* polymerization and *in situ* compatibilization method to prepare tough PA6/core–shell blends. Figure 9.19 shows a schematic diagram of the formation of PA6/core–shell blends. First, maleic anhydride-modified styrene-ethylene/butylene-styrene block copolymer (SEBS-*g*-MA) and polystyrene (PS) are dissolved in molten CL. Second, an anionic ring-opening polymerization of CL is carried out in the presence of a catalyst and an activator in a twin-screw extruder to produce tough PA6/core–shell blends.

Figure 9.20 shows the morphologies of polymerized CL and CL/SEBS-*g*-MA/PS systems. The SEBS and PS phases of the polymerized CL/SEBS-*g*-MA/PS system are in the form of spherical domains in the PA6 matrix, and the PEB block of the SEBS-*g*-MA is located at the interface between the PA6 and PS phases. Moreover, an increase in the SEBS-*g*-MA content reduces the dispersed phase domain size. Yan *et al.* [35] attributed this to the activation of maleic anhydride groups of the SEBS-*g*-MA, which initiate PA6 chain growth to form EBS-*g*-PA6 graft copolymer as a compatibilizer during the reactive extrusion process. However, this explanation does not seem reasonable. Anhydride groups may not be able to initiate the anionic polymerization of CL but they react with the terminal amine group of the resulting PA6.

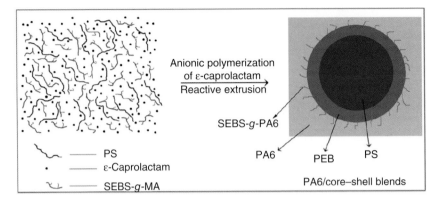

Figure 9.19 Schematic diagram of the formation of PA6/core–shell blends. (Yan *et al*. 2013 [35]. Reproduced with permission of Wiley.)

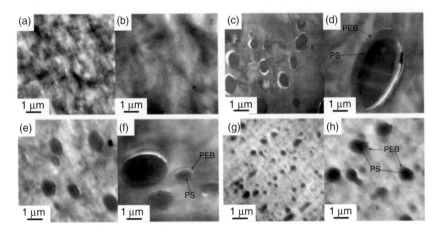

Figure 9.20 TEM images of PA6 (a and b), PA6/SEBS-*g*-MA/PS (85/5/10 by mass) blends (c and d), PA6/SEBS-*g*-MA/PS (85/7.5/7.5 by mass) blends (e and f), and PA6/SEBS-*g*-MA/PS (85/10/5 by mass) blends (g and h) prepared by reactive extrusion of PA6. (Yan *et al*. 2013 [35]. Reproduced with permission of Wiley.)

The PA6 grafts of the SEBS-*g*-PA6 graft copolymer are located on the PEB backbone to which maleic anhydride is grafted. Therefore, the SEBS-*g*-PA6 graft copolymer improves the interfacial adhesion between the PA6 and the PEB block of the SEBS. The poor compatibility between the PS and the PA6 separates them apart. Thus, the PEB block is located at the interface between the PA6 and PS, as shown in Figure 9.20d,f,h.

When the SEBS-*g*-MA content is below 10 wt%, that is, 5 and 7.5 wt%, the PS domains are not completely encapsulated by the PEB block and the interfacial adhesion between the PA6 and PS is weak. Moreover, the size of the dispersed PS phase domains is large. This morphology leads to poor mechanical properties, as shown in Table 9.3. As the SEBS-*g*-MA content reaches 10 wt%, a perfect core–shell structure with the PS as the core and PEB as the shell is formed,

Table 9.3 Mechanical properties of the PA6 and PA6 blends prepared by reactive extrusion.

Samples	Tensile strength (MPa)	Flexural strength (MPa)	Elongation at break (%)	Notched Izod impact strength (KJ m^{-2})
PA6	68	88	50	9.6
PA6/SEBS (85/15 by mass)	50	53	350	26.0
PA6/SEBS/PS (85/5/10 by mass)	55	65	78	6
PA6/SEBS/PS (85/7.5/7.5 by mass)	60	70	80	11
PA6/SEBS/PS (85/10/5 by mass)	65	75	350	22.0

Source: Yan et al. 2013 [35]. Reproduced with permission of Wiley.

as shown in Figure 9.20g,h. Moreover, the diameter and the distance between two core–shell particles are below 1 µm, which not only increases the toughness but also retains the rigidity of the PA6 matrix. As shown in Table 9.3, the elongation at break and notched izod impact strength of the PA6/SEBS/PS (85/10/5 by mass) blends are 7-fold and 2.3-fold higher than those of the pure PA6, respectively.

9.4 Summary

This chapter first shows theoretically that it is very difficult, if not impossible, to obtain nano-structured immiscible polymer blends (the size of the dispersed polymer phase domains is below 100 nm) by melt processing, irrespective of the compatibilization method used (the compatibilizer is premade or generated *in situ* during the melt processing), except that the amount of the compatibilizer is very high, which in practice is very unlikely. When the compatibilizer is formed *in situ*, a large amount of compatibilizer can be formed *in situ* only when the molar mass of at least one of the reactive polymer components is small enough so that the molecular architecture of the resulting copolymer is highly asymmetrical and tends to leave the interfaces, sparing interfaces for further interfacial reaction between reactive polymer components.

It then presents the principles of nano-structuring of immiscible polymer blends by *in situ* polymerization and *in situ* compatibilization processes. Unlike a classical compatibilization process in which two or more existing immiscible polymer components are blended in the melt and an appropriate copolymer is added or produced *in situ* during blending, an *in situ* polymerization and *in situ* compatibilization process consists of polymerizing a monomer of polymer B in the presence of polymer A. A fraction of polymer A chains bear activated sites from which polymer B chain can grow. In the blending process, both polymerization of monomer B and formation of a copolymer of A and B occur simultaneously, leading to an *in situ* compatibilized polymer blend. This way, a very large amount of copolymers as compatibilizers can be formed, allowing nano-structuring of immiscible polymer blends. Nano-blends may exhibit properties conventional polymer blends may not possess.

References

1. Manas-Zloczower, I. (2009) *Mixing and Compounding of Polymers: Theory and Practice*, Hanser.
2. Baker, W., Scott, C., and Hu, G.H. (2001) *Reactive Polymer Blending*, Hanser.
3. Tsou, A.H., Favis, B.D., Hara, Y., Bhadane, P.A., and Kirino, Y. (2009) Reactive compatibilization in brominated poly(isobutylene-*co-p*-methylstyrene) and polyamide blends. *Macromol. Chem. Phys.*, **210** (5), 340–348.
4. Shashidhara, G.M., Biswas, D., Pai, B.S., Kadiyala, A.K., Feroze, G.W., and Ganesh, M. (2009) Effect of PP-*g*-MAH compatibilizer content in polypropylene/nylon-6 blends. *Polym. Bull.*, **63** (1), 147–157.
5. Li, H. and Hu, G.H. (2001) The early stage of the morphology development of immiscible polymer blends during melt blending: compatibilized vs. uncompatibilized blends. *J. Polym. Sci., Part B: Polym. Phys.*, **39** (5), 601–610.
6. Ottino, J.M., DeRoussel, P., Hansen, S., and Khakhar, D.V. (1999) Mixing and dispersion of viscous liquids and powdered solids. *Adv. Chem. Eng.*, **25**, 105–205.
7. Scott, C.E. and Macosko, C.W. (1995) Morphology development during the initial stages of polymer–polymer blending. *Polymer*, **36** (3), 461–470.
8. Sundararaj, U., Dori, Y., and Macosko, C.W. (1995) Sheet formation in immiscible polymer blends: model experiments on initial blend morphology. *Polymer*, **36** (10), 1957–1968.
9. Lindt, J.T. and Ghosh, A.K. (1992) Fluid mechanics of the formation of polymer blends. Part I: formation of lamellar structures. *Polym. Eng. Sci.*, **32** (24), 1802–1813.
10. Willemse, R.C., Ramaker, E.J.J., Van Dam, J., and De Boer, A.P. (1999) Morphology development in immiscible polymer blends: initial blend morphology and phase dimensions. *Polymer*, **40** (24), 6651–6659.
11. Sundararaj, U., Macosko, C.W., Rolando, R.J., and Chan, H.T. (1992) Morphology development in polymer blends. *Polym. Eng. Sci.*, **32** (24), 1814–1823.
12. Scott, C.E. and Macosko, C.W. (1994) Morphology development during reactive and non-reactive blending of an ethylene-propylene rubber with two thermoplastic matrices. *Polymer*, **35** (25), 5422–5433.
13. Tomotika, S. (1935) On the instability of a cylindrical thread of a viscous liquid surrounded by another viscous fluid. *Proc. R. Soc. Ser. A*, **150** (870), 322–337.
14. Hu, G.H. and Feng, L.F. (2003) Extruder processing for nanoblends and nanocomposites. *Macromol. Symp.*, **195** (1), 303–308.
15. Higashitani, K., Iimura, K., and Sanda, H. (2001) Simulation of deformation and breakup of large aggregates in flows of viscous fluids. *Chem. Eng. Sci.*, **56** (9), 2927–2938.
16. Taylor, G.I. (1932) The viscosity of a fluid containing small drops of another fluid. *Proc. R. Soc. Ser. A*, **138** (834), 41–48.
17. Taylor, G.I. (1934) The formation of emulsions in definable fields of flow. *Proc. R. Soc. Ser. A*, **146** (858), 501–523.

18 Wu, S. (1987) Formation of dispersed phase in incompatible polymer blends: interfacial and rheological effects. *Polym. Eng. Sci.*, **27** (5), 335–343.
19 Lyu, S., Jones, T.D., Bates, F.S., and Macosko, C.W. (2002) Role of block copolymers on suppression of droplet coalescence. *Macromolecules*, **35** (20), 7845–7855.
20 Milner, S.T. and Xi, H. (1996) How copolymers promote mixing of immiscible homopolymers. *J. Rheol.*, **40** (4), 663–687.
21 Sundararaj, U. and Macosko, C.W. (1995) Drop breakup and coalescence in polymer blends: the effects of concentration and compatibilization. *Macromolecules*, **28** (8), 2647–2657.
22 Cartier, H. and Hu, G.H. (1999) Morphology development of *in situ* compatibilized semicrystalline polymer blends in a co-rotating twin-screw extruder. *Polym. Eng. Sci.*, **39** (6), 996–1013.
23 Zhang, C.L., Feng, L.F., Zhao, J., Huang, H., Hoppe, S., and Hu, G.H. (2008) Efficiency of graft copolymers at stabilizing co-continuous polymer blends during quiescent annealing. *Polymer*, **49** (16), 3462–3469.
24 Zhang, C.L., Zhang, T., and Feng, L.F. (2014) In situ control of co-continuous phase morphology for PS/PS-*co*-TMI/PA6 blend. *J. Appl. Polym. Sci.*, **131** (6), 39972–39979.
25 Ji, W.Y., Feng, L.F., Zhang, C.L., Hoope, S., and Hu, G.H. (2015) Development of a reactive compatibilizer-tracer for studying reactive polymer blends in a twin-screw extruder. *Ind. Eng. Chem. Res.*, **54** (43), 10698–10706.
26 Hu, G.H., Li, H., and Feng, L.F. (2002) A two-step reactive extrusion process for the synthesis of graft copolymers with polyamides as grafts. *Macromolecules*, **35** (22), 8247–8250.
27 Hu, G.H. and Cartier, H. (1999) Reactive extrusion: toward nanoblends. *Macromolecules*, **32** (14), 4713–4718.
28 Zhang, C.L., Feng, L.F., Hoope, S., and Hu, G.H. (2008) Grafting of polyamide 6 by the anionic polymerization of ε-caprolactam from an isocyanate bearing polystyrene backbone. *J. Polym. Sci., Part A: Polym. Chem.*, **46** (14), 4766–4776.
29 Zhang, C.L., Feng, L.F., Gu, X., Hoppe, S., and Hu, G.H. (2007) Determination of the molar mass of polyamide block/graft copolymers by size-exclusion chromatography at room temperature. *Polym. Test.*, **26** (6), 793–802.
30 Zhang, C.L., Feng, L.F., Gu, X.P., Hoope, S., and Hu, G.H. (2011) Kinetics of the anionic polymerization of ε-caprolactam from an isocyanate bearing polystyrene. *Polym. Eng. Sci.*, **51** (11), 2261–2272.
31 Hu, G.H., Cartier, H., Feng, L.F., and Li, B.G. (2004) Kinetics of the *in situ* polymerization and *in situ* compatibilization of poly(propylene) and polyamide 6 blends. *J. Appl. Polym. Sci.*, **91** (3), 1498–1504.
32 Cartier, H. and Hu, G.H. (2001) A novel reactive extrusion process for compatibilizing immiscible polymer blends. *Polymer*, **42** (21), 8807–8816.
33 Teng, J., Otaigbe, J.U., and Taylor, E.P. (2004) Reactive blending of functionalized polypropylene and polyamide 6: *in situ* polymerization and *in situ* compatibilization. *Polym. Eng. Sci.*, **44** (4), 648–659.

34 Ji, Y., Ma, J., and Liang, B. (2005) In situ polymerization and *in situ* compatibilization of polymer blends of poly(2,6-dimethyl-1,4-phenylene oxide) and polymer 6. *Mater. Lett.*, **59** (16), 1997–2000.

35 Yan, D., Li, G., Huang, M., and Wang, C. (2013) Tough polyamide 6/core–shell blends prepared via *in situ* anionic polymerization of ε-caprolactam by reactive extrusion. *Polym. Eng. Sci.*, **53** (12), 2705–2710.

10

Reactive Comb Compatibilizers for Immiscible Polymer Blends

Yongjin Li, Wenyong Dong, and Hengti Wang

Hangzhou Normal University, College of Material, Chemistry and Chemical Engineering, No. 16 Xuelin Road, Hangzhou 310036, P.R. China

10.1 Introduction

Polymer blending is a convenient and economic pathway to develop novel polymer products with advantageous combinations of useful properties based on their individual components [1, 2]. This technique is accomplished by the addition of pre-made polymers (block, graft, etc.) or *in situ* formed reactive compatibilizers, in order to lower the interfacial tension of immiscible polymers, decrease the dispersed phase size, suppress the coalescence of the dispersed phase, and improve the interfacial adhesion.

Reactive compatibilization is more often used in industry [3–6]. For the formation of block or graft polymers that act as compatibilizers in this technique, the precursors are required to have a backbone that is miscible with one phase and some functional groups that are either distributed along or present at the chain end of the precursor's backbone [3–8]. During melt blending, the functional groups can react with the complementary groups of the other phase and the graft or block compatibilizers are *in situ* formed at the interface. The *in situ* formed graft polymers are more appealing than block ones for industrial applications due to the convenience of synthesis by copolymerization or graft reaction.

As for the linear precursors (defined as reactive linear (RL) polymers hereafter), which have functional groups distributed randomly along the backbone (Figure 10.1a), it is obvious that as the grafting reaction proceeds, the number of the grafted chains increases and the molecular weight between two grafting sites in the backbone decreases correspondingly. These factors significantly influence the balance of the interaction between the backbone with one phase and the grafted side chains with the other phase in the graft polymers. Thus, those graft polymers, which do not have balanced molecular structures, are prone to be "pulled in" or "pulled out" from the interface under shear condition to form micelles (or inverse micelles) in one phase or the other (Figure 10.1a).

Although some authors have shown that the stability of the graft polymers (premade or *in situ* formed) at the interface is significantly influenced by the molecular structures, which finally determines the microstructure morphology and

Reactive Extrusion: Principles and Applications, First Edition. Edited by Günter Beyer and Christian Hopmann.
© 2018 Wiley-VCH Verlag GmbH & Co. KGaA. Published 2018 by Wiley-VCH Verlag GmbH & Co. KGaA.

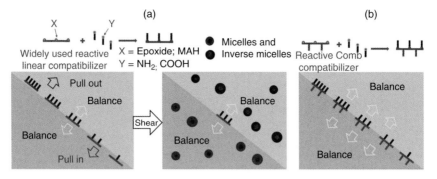

Figure 10.1 Compatibilization by Reactive Linear polymers and the formation of micelles (a); by Reactive Comb polymers (b).

compatibilization efficiency, only a limited number of theoretical and experimental works are devoted to this topic [9–12]. Zhang et al. compared the efficiencies of a series of graft copolymers with PS (polystyrene) as the backbone and PA6 (polyamide 6) as the side chains on stabilizing the co-continuous morphology of PS/PA 6 blends during quiescent annealing; they found that for the graft copolymers with the same number of side chains, the longer the side chains, the higher their compatibilizing and stabilizing efficiencies [10]. Xu et al. applied a graft copolymer with a PE (polyethylene) backbone and some PMMA (poly(methyl methacrylate)) side chains as compatibilizer in PE/PMMA blends and found that the graft copolymers with the shortest side chains were more efficient at reducing the size of the dispersed phase and improving the interfacial adhesion between PE and PMMA. They deduced that the graft copolymers with the shortest side chains could immigrate to the interface more quickly than those with long side chains [11]. Gersappe et al. theoretically indicated that at a fixed volume ratio of the side chains to backbones, graft copolymers with fewer and longer side chains were better compatibilizers than those with multiple, short side chains [12].

To enlarge the stability window of the graft polymers at the interface, a series of comb-like polymers (defined as *Reactive Comb* hereafter) were synthesized and applied as compatibilizers. As shown in Figure 10.1, compared to the RL polymers, Reactive Comb polymers have some side chains that are distributed along the backbone. It is anticipated that the side chains can counterbalance the weakening of the backbone with one phase due to graft reaction as in RL polymers and improve the stability of the RC polymers at the interface.

10.2 Synthesis of Reactive Comb Polymers

The molecular structures of the Reactive Comb polymers are depicted in Figure 10.2. At least four parameters affect the compatibilization efficiency and are defined hereafter as the length of the backbone (M), the content of the reactive groups (E), the distance between the side chains (D), and the length of the side chains (L). We synthesized two kinds of Reactive Comb polymers were synthesized and the backbone of one was made of PMMA, while that of the other was made of PS. Both of them had some PMMA side chains and a few

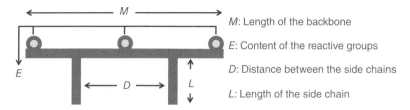

Figure 10.2 Schematic diagram of the molecular structure of Reactive Comb polymers.

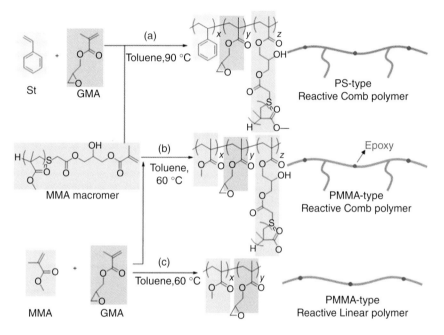

Scheme 10.1 Synthesis of the Reactive Comb and Reactive Linear polymers.

epoxide groups that are distributed randomly along the backbone. As shown in Scheme 10.1, the first step was the synthesis of the MMA (methyl methacrylate) macromers as side chains in the RC polymers. Carboxyl-terminated PMMA was first synthesized by telomerization of MMA with TAC (TGA) as the chain transfer agent, then the carboxyl groups were transformed into double bonds by esterifying with GMA (glycidylmethylacrylate). The second step was copolymerizing MMA macromer, GMA and St (Styrene) for the synthesis of PS-type RC polymers (Path a); the PMMA-type RC or RL polymers were synthesized by copolymerizing MMA and GMA with MMA macromer (Path b) or without MMA macromer (Path c). As shown in Table 10.1, a series of RL or RC polymers with different molecular structures have been successfully synthesized and in the following text we will investigate the influence of the parameters of molecular structures (i.e., M, E, D, L) will be investigated on their compatibilizing efficiencies [13–16].

Table 10.1 Molecular characterization of RL and RC polymers.

Sample ID	Composition (wt%)				M_n (g mol^{-1})	M_w/M_n
	Macromer	GMA	MMA	St		
L-M-0-1-9[a]	0	10	90	0	22 000	2.0
L-M-0-2-8	0	20	80	0	21 000	1.9
C-M-1-1-8(S-2400)[b]	10	10	80	0	21 000	1.9
C-M-1-2-7(S-2400)	10	20	70	0	20 000	2.1
C-M-1-2-7(S-4800)	10	20	70	0	20 000	2.0
C-M-1-2-7(S-6300)	10	20	70	0	19 000	2.0
C-M-2-2-6(S-4800)	20	20	60	0	38 000	1.9
C-St-1-2-7(S-6300)[c]	10	20	0	70	19 000	4.3

a) L denotes Linear, M denotes PMMA-type, and the numbers connected by dash is the weight ratio of MMA macromer/GMA/MMA.
b) C denotes Comb, M denotes PMMA-type, S in brackets denotes side chain, and the number in bracket is the number average molecular weight of the MMA macromer.
c) C denotes Comb, St denotes PS-type, S in brackets denotes side chain, the number in bracket is the number average molecular weight of the MMA macromer and the numbers connected by dash is the weight ratio of MMA macromer/GMA/St.

10.3 Reactive Compatibilization of Immiscible Polymer Blends by Reactive Comb Polymers

In this section, two typical immiscible systems, PLLA/PVDF (poly(L-lactic acid)/poly(vinylidene fluoride)) and PLLA/ABS (acrylonitrile-butadiene-styrene) are selected as model systems. It is well known that the epoxide groups are ready to react with various engineering plastics that have reactive —COOH, —OH, —NH$_2$ groups [17, 18]. Therefore, the RL and RC polymers can react with polyesters and polyamides. For example, the epoxide groups can react readily with the terminal carboxyl groups of PLLA during melt blending. On the other hand, PMMA is also reported to be miscible with some polymers, similarly to PVDF and SAN (styrene-acrylonitrile) [19–22].

10.3.1 PLLA/PVDF Blends Compatibilized by Reactive Comb Polymers

10.3.1.1 Comparison of the Compatibilization Efficiency of Reactive Linear and Reactive Comb Polymers

The neat PLLA/PVDF system (50/50) displayed a very broad size distribution of PVDF particles in PLLA matrix (Figure 10.3a). TEM investigation showed a poor adhesion at the interface of black PVDF dispersed phase and white PLLA matrix (Figure 10.3b). The PMMA-type RL and RC polymers, miscible with PVDF and immiscible with PLLA, were applied as compatibilizers in the immiscible PLLA/PVDF system. The influence of both the PMMA side chains and the compatibilizer's content on the morphology of PLLA/PVDF blends was investigated in detail. Figure 10.4 displays the morphology evolution relative to the increase in content of RL (L-M-0-1-9) or RC (C-M-1-1-8(S-2400)) polymers

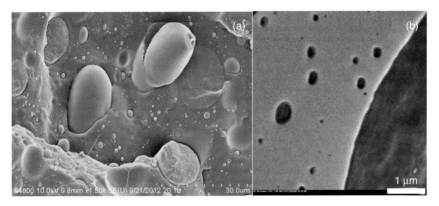

Figure 10.3 SEM (a) and TEM (b) images of uncompatibilized PLLA/PVDF (50/50, w/w) blend (marrix: PLLA, particle: PVDF). All blends were prepared by melt mixing at 190 °C for 10 min using a batch mixer (Haake Polylab QC), with a rotation speed of 50 rpm.

in the PLLA/PVDF system. As shown in Table 10.1, it should be noted that the epoxy groups concentration and the molecular weight of RL and RC polymers were almost the same. Although both the size and the size distribution of the PVDF particles decreased with the addition of RL or RC copolymers, it was found that the particles are distributed more uniformly in RC-compatibilized blends as compared to those compatibilized by RL. Furthermore, it is rather remarkable that the size of the PVDF particles is reduced more sharply with the addition of only 1 wt% of RC and the and that there was no obvious decrease in size even when the weight percentage was increased further, which means that the RC polymers at the interface of matrix and dispersed phase were close to saturation [23, 24]. These results implied that the RC polymers were prone to stay at the interface and inhibited the coalescence of the dispersed phase more efficiently [25].

It is well known that PLLA is a brittle polymer and PVDF is ductile. The mechanical properties of the uncompatibilized PLLA/PVDF blend were even worse than the neat components due to the poor interfacial interaction between the phases. The stress–strain curves and the tensile properties of PLLA/PVDF blends compatibilized by RL or RC polymers are shown in Figure 10.5. It should be noted that with the addition of the RL or RC polymers, the ductility of the PLLA/PVDF blends was significantly improved and, at the same wt% addition, the fracture strain of the blends compatibilized by RC was at least three times higher than those compatibilized by RL. An interesting difference between the blends compatibilized by RL or RC polymers was that the stress–strain curves showed a strain hardening when the strain was over 200% for the RC-compatibilized blends.

Figure 10.6 displays the TEM images of the PLLA/PVDF blends compatibilized by RL (L-M-0-1-9) or by RC (C-M-1-1-8(S-2400)) polymers. The black region was a PVDF dispersed phase, which was stained by OsO_4. It was found that the PVDF particles did not exhibit spherical shape, but deformed into ellipsoidal shape; the extent of deformation was greater when the PLLA/PVDF blends were compatibilized by RC than those compatibilized by RL. This phenomenon was most

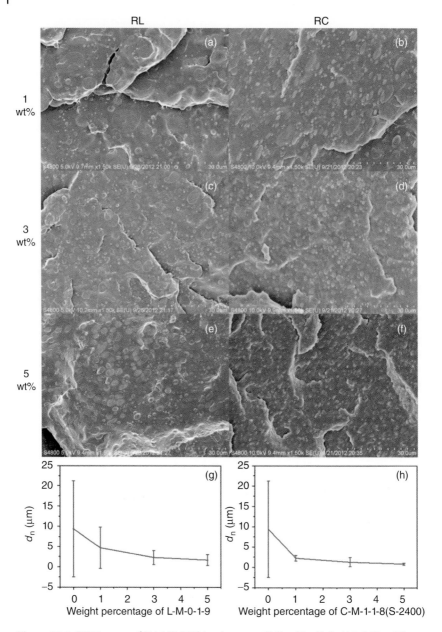

Figure 10.4 SEM images of PLLA/PVDF blends compatibilized by RL (L-M-0-1-9) or RC (C-M-1-1-8(S-2400)) at different weight ratios; (a) PLLA/PVDF/RL (50/50/1), (b) PLLA/PVDF/RC (50/50/1), (c) PLLA/PVDF/RL (50/50/3), (d) PLLA/PVDF/RC (50/50/3), (e) PLLA/PVDF/RL (50/50/5), and (f) PLLA/PVDF/RC (50/50/5). The number average diameter (d_n) of the dispersed phase in PLLA/PVDF (50/50) blends versus wt% of L-M-0-1-9 (g) and C-M-1-1-8(S-2400) (h).

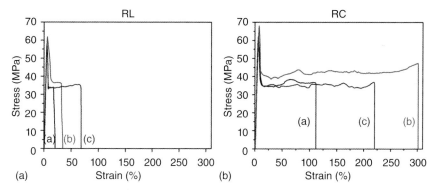

Figure 10.5 Stress–strain curves of PLLA/PVDF blends compatibilized by RL (L-M-0-1-9) (A) or RC(C-M-1-1-8(S-2400)) (B) polymers. The weight ratio of PLLA/PVDF/RL or RC was (a) 50/50/1, (b) 50/50/3 and (c) 50/50/5.

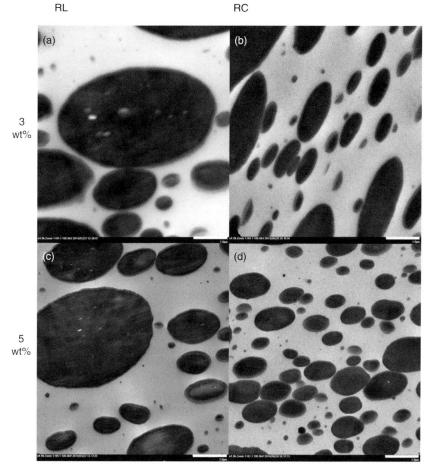

Figure 10.6 TEM images of PLLA/PVDF blends compatibilized by RL (L-M-0-1-9) or RC (C-M-1-1-8(S-2400)) at different weight ratios; (a) PLLA/PVDF/RL (50/50/3), (b) PLLA/PVDF/RC (50/50/3), (c) PLLA/PVDF/RL (50/50/5), and (d) PLLA/PVDF/RC (50/50/5). The scale bar was 1 μm.

obvious in Figure 10.6a,b, when the blends were compatibilized by 3 wt% of RL or RC polymers. The greater deformation implied a lower interfacial tension and, furthermore, a better emulsifying effect of the *in situ* formed PLLA grafted RC polymers at the interface [26–31].

10.3.1.2 Effects of the Molecular Structures on the Compatibilization Efficiency of Reactive Comb Polymers

As mention earlier, the PMMA-type RC polymers were synthesized by copolymerizing MMA, GMA, and MMA macromer. Herein, the weight ratio of these three monomers was further optimized and at the same time, the M_n of the MMA monomer was varied. Thus, a series of RC polymers with different molecular structures were obtained. Then the relationships between the molecular structures of the PMMA-type compatibilizers and the toughness of the compatibilized PLLA/PVDF blends are investigated.

As shown in Table 10.1, the wt% of GMA in both RL and RC was increased from 10 to 20 wt% by changing the feed ratio firstly. Then three kinds of MMA macromers with different molecular weights are selected and a series of RC polymers with different lengths of side chains and the same length of the backbone are thus synthesized. As shown in Figure 10.7, the dark phase seen in the atomic force microscopy (AFM) images corresponded to PLLA and the brighter phase to PVDF. It was found that, firstly, the size of the dispersed phase compatibilized by 3wt% of RC (C-M-1-2-7(S-2400), C-M-1-2-7(S-4800), and C-M-1-2-7(S-6300)) polymers was significantly reduced (Figure 10.7c–f) as compared to 3wt% of RL (L-M-0-2-8) polymers seen in Figure 10.7b. Secondly, as the length of the side chains increased from 2400 to 6300 gmol^{-1} since the RCs were copolymerized with various M_n of macromers, the morphology evolved from slightly irregular big particles, ellipsoidal fibers and finally to homogeneous small particles. Macosko *et al.* once proposed a lacing/sheeting mechanism for the morphology development of polymer blends. The minor PVDF phase was first stretched out by the major PLLA phase to form sheets; under shear condition, holes began to form in the PVDF sheets; the sheets were torn up and stretched into fibers, which finally broke up into small particles. The *in situ* formed compatibilizers at the interface could facilitate the tearing of PVDF sheets and inhibit the coalescence of the particles [32–34]. Another explanation for the morphology evolution could be the capillary number (*Ca*), which was important for the deformation and breakup of drops in shear field. *Ca* was defined as the ratio of the shear force to the interfacial force:

$$Ca = \frac{\eta \dot{\gamma} R}{\sigma} \tag{10.1}$$

in which η was the matrix viscosity, $\dot{\gamma}$ the shear rate and σ the equilibrium interfacial tension. The *in situ* formed compatibilizers at the interface could reduce the interfacial tension and thus facilitate the drop deformation and breakup [27, 30]. From the morphology discrepancies of the blends compatibilized by a series of RL or RC polymers (Figure 10.7b–f), It could be estimated that the blend compatibilized by C-M-1-2-7(S-4800) had the lowest interfacial tension, because the drops exhibited the largest deformation. Recently, Zhang *et al.*

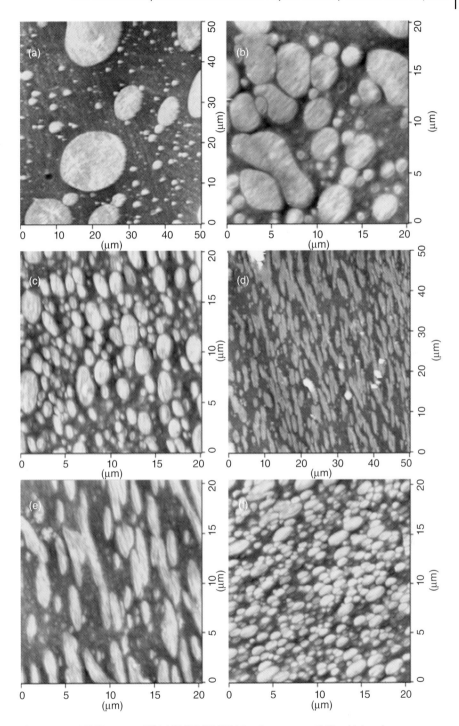

Figure 10.7 AFM images of PLLA/PVDF (50/50) blends uncompatibilized (a) and compatibilized by 3 wt% of (b) L-M-0-2-8; (c) C-M-1-2-7(S-2400); (d, e) C-M-1-2-7(S-4800); and (f) C-M-1-2-7(S-6300).

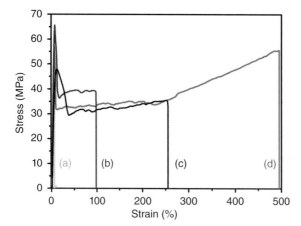

Figure 10.8 Representative stress–strain curves of (a) PLLA/PVDF (50/50), (b) PLLA/PVDF/L-M-0-2-8 (50/50/3), (c) PVDF, and (d) PLLA/PVDF/C-M-1-2-7(S-4800) (50/50/3).

found that the entanglement molecular weight between the dissimilar chains in the PMMA/PVDF blend (M_{e12}) was less than that of the blend component; it implied that the PMMA and PVDF chains were more likely to interact with each other due to the specific interaction between PMMA and PVDF. When the weight ratio of PVDF/PMMA was 90/10, the M_{e12} was 5700 g mol^{-1} and this value had a tendency to decrease with the decreasing of the weight ratio of PMMA [35]. It could be deduced that M_{e12} in the reactive compatibilized PLLA/PVDF system was slightly less than 5700 g mol^{-1}, because the weight ratio of the PVDF/PMMA-type compatibilizer was 100/6. Therefore, the reason that C-M-1-2-7(S-4800) exhibited the greatest ability to reduce the interfacial tension can be understood. Compared to C-M-1-2-7(S-2400), the molecular weight of the side chains in C-M-1-2-7(S-4800) was 4800 g mol^{-1}, which was more close to the entanglement weight between PMMA and PVDF ($M_{e12} = 5700$ g mol^{-1} when the PVDF/PMMA = 100/10).

The typical stress-strain curves and the tensile properties of PLLA/PVDF blends uncompatibilized and compatibilized by 3wt% of RL or RC polymers are shown in Figure 10.8. It was found that all the compatibilized blends exhibited necking, while the uncompatibilized blend fractured without yielding. An interesting difference between the blends compatibilized by RL and those by RC polymers was that the stress–strain curves showed a strain hardening when the strain was over 220% for the RC-compatibilized blends. Especially, the strain hardening was pronounced for the blends compatibilized with 3 wt% of C-M-1-2-7(S-4800) and the fracture stress was almost identical to the yield stress.

Although there existed some discrepancies in the compatibilization efficiencies of RC polymers with different molecular structures, all of them were significantly higher than that of RL polymers. TEM investigations on the microstructures of the PLLA/PVDF blends further showed that the interfacial area of the blends compatibilized by RC polymers was slightly thicker than that by RL polymers (Figure 10.9). A significant difference between the blends compatibilized by RL and RC was that a few white particles with a diameter of about 50 nm existed in the PVDF phase for RL polymers (Figure 10.9) [36]. During reactive blending, the epoxide groups of GMA reacted with the end groups of PLLA and the PLLA chains were thus grafted onto the backbone of PMMA. As shown in Figure 10.10,

10.3 Reactive Compatibilization of Immiscible Polymer Blends by Reactive Comb Polymers

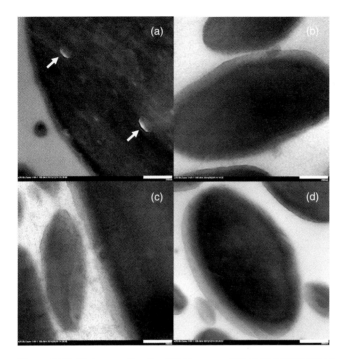

Figure 10.9 High-magnification TEM images of PLLA/PVDF (50/50) blends compatibilized by 3 wt% of (a) RL, (b) RC (C-M-1-2-7(S-2400)), (c) RC (C-M-1-2-7(S-4800)), and (d) RC C-M-1-2-7(S-6300). The scale bar was 200 nm.

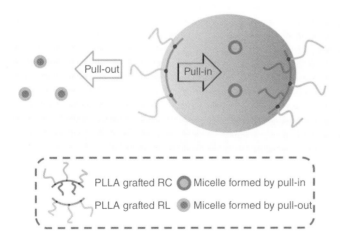

Figure 10.10 The micelles formed by pull-in have a core of PLLA (light) and a shell of PMMA (dark gray); those by pull-out have an inverse inversion phase.

for RL polymer, the PMMA backbone resided in the PVDF dispersed phase and the grafted PLLA chains extended into and entangled with the PLLA matrix. As the PLLA grafting reaction proceeded, the average molecular weight between two grafting sites in the PMMA backbone decreased. When this value was approaching the entanglement molecular weight between PMMA and PVDF ($M_{e12} = 5700$ g mol^{-1}), the interaction between the PVDF phase and the PMMA backbone weakened and the PLLA grafted RL polymers tended to be expelled from the interface and pulled into the PVDF phase (indicated by arrows in Figure 10.9) or pulled out to the PLLA matrix to form micelles. As shown in Figure 10.10, for PLLA grafted RC polymers, the molecular structure was totally different from that of RL polymers, because they were more symmetric and had two kinds of side chains. One of them was PMMA, which was derived from the MMA macromer, and the other was PLLA from the grafting reaction. The PMMA and PLLA side chains penetrated into the PVDF and PLLA phase, respectively. This in some extent could counterbalance the weakening of the interaction between the PMMA backbone and the PVDF phase, incurred by the grafting reaction. Thus, it was the appropriate number and length of the PMMA side chains that increased the stability of the RC polymers at the interface and efficiently lowered the interfacial tension and inhibited coalescence of the dispersed phase. At the same time, the PMMA side chains were also found to affect the crystallization behavior of the PVDF phase in the RC-compatibilized blends. Unlike the neat PVDF, which had well-developed spherulites, the PMMA side chain might induce the formation of some imperfect crystallites (i.e., needle-like), which were prone to orient toward or rotate along the stress field. The energy was consumed during this process and thus the mechanical properties of the blends were more or less improved after compatibilization.

10.3.2 PLLA/ABS Blends Compatibilized by Reactive Comb Polymers

It was reported that PMMA could entangle with SAN molecular chains at low temperatures. Therefore, the RC polymers could also be applicable in the PLLA/ABS system. The morphologies of the PLLA/ABS blends at various compositions compatibilized by RC (C-M-2-2-6(S-4800)) polymers (Table 10.1) are illustrated in Figure 10.11; the rubber particles appeared dark, the SAN phase appeared gray and the PLLA phase appeared white, due to their different staining rates. It was found that when the weight ratio of PLLA/ABS was larger than 50/50, the ABS phase dispersed homogeneously in the PLLA matrix and at 50/50, the morphology transformed from matrix dispersed to co-continuous morphology. With further decreasing the weight ratio of PLLA, phase inversion occurred and the PLLA phase dispersed in the ABS matrix. The co-continuous phase was particularly interesting, because both components could contribute to the properties of the blend [37, 38].

As shown in Scheme 10.2, the RC polymer applied as compatibilizers was composed of a PMMA backbone, two PMMA side chains, and a few epoxy groups that are distributed randomly along the backbone. During reactive blending, the carboxyl ends of PLLA reacted with the epoxy groups of RC, and PLLA grafted RC polymers (RC-*g*-PLLA) was *in situ* formed at the interface of PLLA and ABS. The

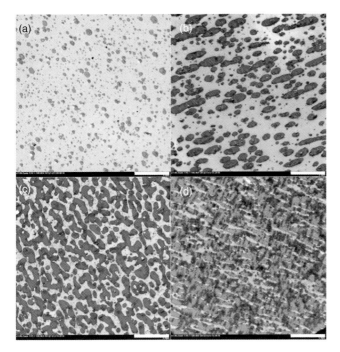

Figure 10.11 TEM images of PLLA/ABS blends compatibilized by RC (C-M-2-2-6(S-4800)) polymers at different compositions; (a) PLLA/ABS/RC = 90/10/3, (b) PLLA/ABS/RC = 70/30/3, (c) PLLA/ABS/RC = 50/50/3, and (d) PLLA/ABS/RC = 30/70/3. The scale bar was 5 μm.

Scheme 10.2 Reaction scheme between the carboxyl groups in PLLA with epoxy groups in RC polymer.

grafted PLLA chains protruded into and interacted with the PLLA phase and the PMMA side chains and backbones interacted with the SAN phase in ABS. Under external shear field, the RC-g-PLLA polymers anchored tightly at the PLLA/ABS interface and inhibited the dispersed phase from coalescence. Therefore, it could be found that in the RC-compatibilized PLLA/ABS system, the dispersed phase was more homogeneous than our previously reported system compatibilized by SAN–GMA [39]. At the same time, the interfacial tension was also reduced due to the emulsification effect of the RC-g-PLLA polymers at the interface.

The typical stress–strain curves of the neat blend components and the PLLA/ABS blends are illustrated in Figure 10.12a. Based on the tensile results, the ductility of the PLLA/ABS/RC blends was determined by the following equation: *Ductility = Failure strain/Yielding strain*. PLLA is a brittle polymer

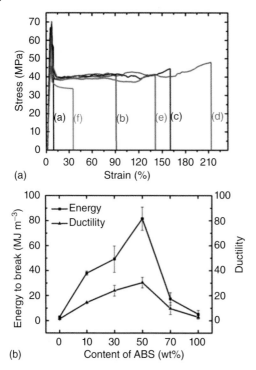

Figure 10.12 (A) Tensile stress–strain curves of neat components and compatibilized blends by RC (C-M-2-2-6(S-4800)) polymers; (a) PLLA, (b) PLLA/ABS/RC=90/10/3, (c) PLLA/ABS/RC = 70/30/3, (d) PLLA/ABS/RC = 50/50/3, (e) PLLA/ABS/RC = 30/70/3, and (f) ABS. (B) Effect of the ABS content on the energy to break and the ductility in the compatibilized blends.

and the specimen fractured suddenly when the deformation was just beyond the yield point. In our former reports, the PLLA/ABS blends compatibilized by SAN–GMA (styrene–acrylonitrile–glycidyl methacrylate) and catalyzed by ETPB (ethyltriphenylphosphonium bromide) displayed a maximum fracture strain of about 23.8%, when the weight ratio of PLLA/ABS/SAN–GMA/ETPB was 70/30/5/0.02; this value was only one-fifth of the applied ABS (D-100, high-impact grade, provided by Grand Pacific Petrochemical Corporation) [39]. At the same time, the yield stress decreased by 32% from 65.5 MPa (neat PLLA) to 44.6 MPa; the Young's modulus decreased by 33% from 2.024 GPa (neat PLLA) to 1.352 GPa. Interestingly, as shown in Table 10.2, when the RC (C-M-2-2-6(S-4800)) polymer was applied as compatibilizers at a weight ratio of 50/50/3 (PLLA/ABS/RC), the maximum fracture strain was 210%, almost seven times higher than that of the applied ABS (TR558A, LG Chemical). The energy to break and the ductility as a function of ABS content are plotted in Figure 10.12b. It is clear that all the compatibilized blends show significantly increased ductility and toughness than both neat PLLA and neat ABS, indicating the drastic synergistic effects of PLLA and ABS compatibilized by RC polymer. To our best knowledge, such behavior has not been reported for PLLA alloys so far. In addition, it was also found that the toughness and ductility of the alloys increased with the increasing of the ABS content and reached the maximum at 50 wt%, then decreased rapidly when the ABS was 70 wt%.

The morphologies of the tensile fracture surfaces were observed by SEM to investigate the toughening mechanism. As shown in Figure 10.13a,b, there was

Table 10.2 Mechanical properties of neat components and compatibilized blends by RC (C-M-2-2-6(S-4800)).

Sample	Young's modulus (GPa)	Yield stress (MPa)	Fracture stress (MPa)	Fracture strain (%)
PLLA	1.88	67.9	56.5	9.4
PLLA/ABS/RC (90/10/3)	1.70	70.9	41.7	91.0
PLLA/ABS/RC (70/30/3)	1.68	69.2	45.2	160.1
PLLA/ABS/RC (50/50/3)	1.67	62.9	48.1	213.3
PLLA/ABS/RC (30/70/3)	1.39	55.4	40.9	140.8
ABS	1.28	45.9	34.9	33.5

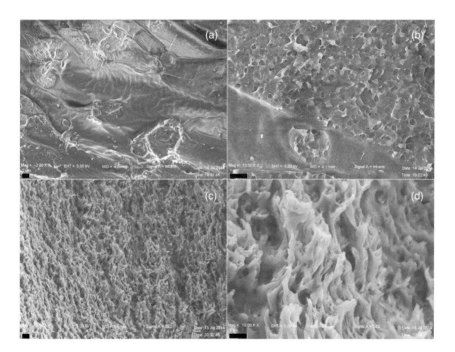

Figure 10.13 SEM images of the tensile fracture surfaces of (a, b) uncompatibilized PLLA/ABS = 50/50 and (c, d) compatibilized PLLA/ABS/RC(C-M-2-2-6(S-4800)) = 50/50/3 blend. The scale bar was 2 μm in (a, c) and 1 μm in (b, d).

an obvious difference between the PLLA and ABS phase in the uncompatibilized PLLA/ABS (50/50) blends; the PLLA phase displayed a smooth surface, characteristic of brittle fracture, while the ABS phase displayed a rough surface with many voids, which was due to the rubber phase being pulled out from the ABS matrix [40, 41]. This was because in the uncompatibilized blend, the interfacial adhesion was poor and the stress was hard to transfer across the interface. In the compatibilized PLLA/ABS/RC (C-M-2-2-6(S-4800)) (50/50/3) blends, the *in situ* formed PLLA grafted RC polymers at the interface, improved the interfacial strength, and decreased the interfacial tension efficiently. Therefore, the morphology discrimination between the PLLA and ABS phase was not clear anymore and the whole fracture surface was constituted by homogeneously and densely distributed, oriented fibrillars that were parallel to the stress direction. The formation of the fibrillar structure was a typical plastic deformation and much energy was dissipated during this process, which resulted in a significantly improved ductility [42, 43].

Oyama reported a supertough poly(lactic acid) material by blending PLLA with EGMA and found that the presence of the dispersed phase facilitated the crystallization of PLLA; it was further deduced that the crystallization of the PLLA matrix played a key role in toughening [44]. Bai *et al.* recently reported a PCL/PLLA blend, in which PCL played the role of an impact modifier and the crystallization of the PLLA matrix was tailored by a tiny amount of nucleating agent; they found that the toughness of the PLLA matrix increased linearly with the increasing of the PLLA crystalline content, while for the blends with amorphous PLLA matrix, the toughness was almost unchanged [45].

The thermal properties of the uncompatibilized and compatibilized PLLA/ABS blends with a series of compositions are summarized in Table 10.3. It was found that in the uncompatibilized blends, the ABS phase had a positive influence on the crystallization of PLLA and the maximum crystallinity (χ) was 14.8%, when the weight ratio of PLLA/ABS was 70/30, while in the compatibilized blends, the crystallinity of PLLA in the blends was always less than that of the neat PLLA. This leads to the assumption that some other factors exist, which contribute to the improved ductility of the PLLA/ABS blends. It was also found that the melting temperature (T_m) of PLLA decreased after reactive blending with ABS and this was because the grafted PLLA chains of RC-*g*-PLLA penetrated into the PLLA phase and interfered with the crystallization of the PLLA [22, 46].

Table 10.3 Thermal properties of neat components and compatibilized blends.

Material	$T_{m, PLLA}$ [a] (°C)	χ [a] (%)	$T_{g, PLLA}$ [b] (°C)	$T_{g, SAN}$ [b] (°C)
PLLA	169.6	7.5	73.6	—
PLLA/ABS (90/10)	167.7	2.9	—	—
PLLA/ABS/RC (90/10/3)	162.2	2.4	73.4	111.1
PLLA/ABS (70/30)	167.3	14.8	—	—
PLLA/ABS/RC (70/30/3)	165.0	4.6	73.5	111.9
PLLA/ABS (50/50)	167.0	10.7	—	—
PLLA/ABS/RC (50/50/3)	164.5	6.2	69.8	111.8
PLLA/ABS (30/70)	167.7	8.2	—	—
PLLA/ABS/RC (30/70/3)	162.0	6.0	70.2	115.3
ABS	—	—	—	115.6

a) Determined by DSC.
b) Determined by DMA.

As reported by some researchers, when a hard matrix was toughened by a soft rubber, the differentiation of the thermal contraction between the two phases would generate a negative pressure in the rubber phase, which resulted in an increase of the free volume and a corresponding decrease of the T_g of the rubber phase. The internal pressure also exerted a dilational stress field around the rubber phase and enhanced the local segmental motions in the matrix, which resulted in a depression of the β relaxation temperature ($T_β$) in the matrix and an improved ductility of the blend [47–51]. While in our previously reported PLLA/POM system, there existed strong interactions between the —C=O of PLLA and —CH$_2$ of POM. Therefore, in the PLLA/POM blend, the interfacial adhesion was strong and the negative pressure would significantly influence the contraction of both the POM and the PLLA phase and thus a double glass transition temperature (T_g) depression phenomenon was observed [52]. It has been demonstrated in our previous report that RC polymer was a stable compatibilizer at the interface of the immiscible blend during reactive blending, because the PMMA side chains introduced by the macromer was miscible with the SAN phase of ABS and the epoxide groups could react with the carboxyl ends of PLLA and the grafted PLLA chains protruded into and entangled with the PLLA phase. The *in situ* formed RC-*g*-PLLA polymers anchored stably at the interface of ABS and PLLA and these two phases were held tightly by the RC-*g*-PLLA polymers. The strong interface was further confirmed by TEM as shown in Figure 10.14. Compared to the uncompatibilized blends, the compatibilized blends displayed a black and thick interface [53].

As shown in Scheme 10.3, when the RC-compatibilized PLLA/ABS blend was cooled down from the molten state to the solid state, the internal pressure was so strong that the mobility of the local segmental chains, in not just one of the two phases, but in both, the matrix and the dispersed phase, was significantly improved, and therefore a double T_g depression phenomenon was also observed in the RC-compatibilized PLLA/ABS blend (Figure 10.15 and the corresponding results are summarized in Table 10.3). It could be deduced that the origin of the strong internal pressure in the RC-compatibilized PLLA/ABS blend was that first both PLLA and ABS had approximate and high Young's modulus, and second RC-*g*-PLLA polymers provided a strong interfacial adhesion. Some authors have indicated that the amount and the distribution of the free volumes were related to the mechanical and the rheological properties of polymers [54, 55]. As shown in Table 10.3, the T_g depressions of both the PLLA and ABS phase were most significant when the weight ratio of PLLA/ABS/RC (C-M-2-2-6(S-4800)) polymers was 50/50/3; it was because the force exerted on both the PLLA and ABS phase was proportional to the area of PLLA/ABS interface ($A_{interface}$) and it was obvious that the PLLA/ABS/RC (50/50/3) blend had the largest $A_{interface}$; therefore, the highest toughness was obtained at this weight ratio. The reason for the maximum fracture strain of the RC-compatibilized PLLA/ABS blend being more than seven times higher than the applied ABS (TR558A) is now clear. The increase of the free volume in both the PLLA and the ABS phase resulted in an increase of the mobility of the molecular chains, especially in the region adjacent to the interface and this was beneficial for cold drawing [56].

288 | *10 Reactive Comb Compatibilizers for Immiscible Polymer Blends*

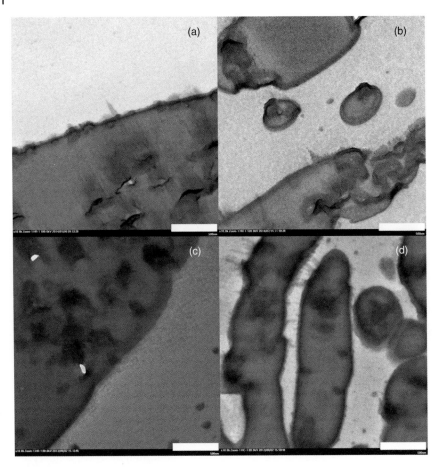

Figure 10.14 TEM images of the uncompatibilized blends and compatibilized PLLA/ABS by RC (C-M-2-2-6(S-4800)) polymers; (a) PLLA/ABS = 70/30, (b) PLLA/ABS/RC = 70/30/3, (c) PLLA/ABS = 50/50, and (d) PLLA/ABS/RC = 50/50/3. The scale bar was 500 nm.

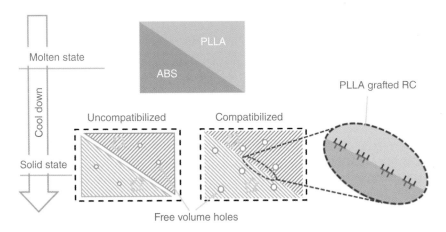

Scheme 10.3 Schematic of the generation of the internal pressure and the enlargement of the free volume in the compatibilized blend.

Figure 10.15 Tan δ-temperature curves of neat components and compatibilized blends by RC (C-M-2-2-6(S-4800)) polymers; (a) PLLA, (b) PLLA/ABS/RC = 90/10/3, (c) PLLA/ABS/RC = 70/30/3, (d) PLLA/ABS/RC = 50/50/3, (e) PLLA/ABS/RC = 30/70/3, and (f) ABS.

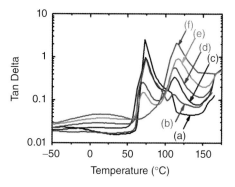

10.4 Immiscible Polymer Blends Compatiblized by Janus Nanomicelles

As mentioned earlier, when the compatibilizers were dragged from the interface under external shear condition to form micelles in one of the phases, they could not contribute to the compatibilization. Many studies have attempted to minimize the formation of micelles in reactive blending [57, 58]. Nevertheless, some questions may arise: Were the micelles forming in conventional reactive blending of potential benefit? Could the micelles enhance polymer miscibility if the micelles were located at the interface?

In fact, Janus nanoparticles (JNPs) with heterogeneous surface composition have already been used as emulsifying agents for immiscible blends. JNPs are a kind of colloidal particles, and due to the combination of their inherent character (Pickering effect) with amphipathy, are located exclusively at polymer–polymer interfaces [59–61]. Müller *et al.* achieved efficient compatibilization of immiscible polymer blends using pre-made JNPs and they found that JNPs with poly(butadiene) (PB) as the core and PMMA and PS chains as the Janus shell were effective compatibilizers for PMMA/PS and SAN/PPE blends [62–64]. The JNPs immigrated against high shear to anchor at the polymer–polymer interface, which was attributed to their high interfacial activity. Nevertheless, it was the complicated prefabrication and their large content (usually ≥8 wt%) used in this strategy that made the compatibilization by JNPs extremely expensive. Further, the harsh compounding conditions such as high shear, ultrasound, or high pressure that were needed to anchor the JNPs at the interface were also responsible for impeding the possible industrial application of this technique.

Inspired by the ability of the pre-made JNPs to compatibilize immiscible polymer blends and the formation of micelles in reactive blending, it was of great interest to understand whether the *in situ* formed JNMs (Janus nanomicelles) could behave as effective compatibilizers for immiscible polymer blends or not. Herein, a PS-type RC (C-St-1-2-7(S-6300)) polymer is chosen as compatibilizers in an immiscible polymeric system PLLA/PVDF. As shown in Table 10.1, it was made up of a PS backbone, PMMA side chains, and a few epoxy groups that are distributed along the backbone. Scheme 10.4 displays a representation of reaction-formed JNMs and their interfacial behaviors, which implied that in a

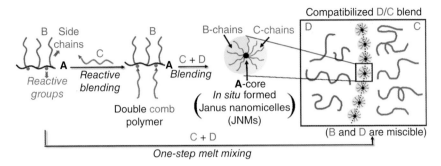

Scheme 10.4 Proposed mechanism of *in situ* formed JNMs as compatibilizers in immiscible polymer blends.

proper immiscible polymeric system, JNMs formed by reaction should be able to compatibilize immiscible polymer blends.

Figure 10.16 shows the phase morphologies of PLLA/PVDF blends with and without PS-type RC (C-St-1-2-7(S-6300)) polymers. The mixing time of melt blending was 10 min for all the samples in Figure 10.16. Both SEM images (Figure 10.16a) and TEM images (Figure 10.16d) indicated that PLLA and PVDF are totally immiscible, as evidenced by the large domain size and very weak interfacial adhesion. PVDF has higher density and viscosity than PLLA. Therefore, PVDF forms domains in the PLLA matrix even at 50/50 weight ratio.

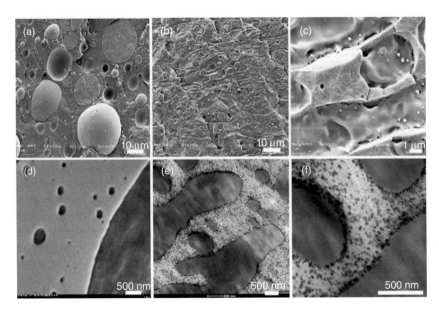

Figure 10.16 PLLA/PVDF blends uncompatibilized and compatibilized by JNMs. SEM images of (a) PVDF/PLLA (50/50), (b) and (c) PLLA/PVDF/C-St-1-2-7(S-6300) (50/50/3); TEM images of (d) PLLA/PVDF (50/50), (e) and (f) PLLA/PVDF/C-St-1-2-7(S-6300) (50/50/3). The blend samples were ultramicrotomed to a thickness of 80–90 nm and then stained by RuO_4 for 4 h in order to selectively stain the PVDF phase as well as the JNMs (the white phase was PLLA, the black phase was JNMs and the gray phase was PVDF).

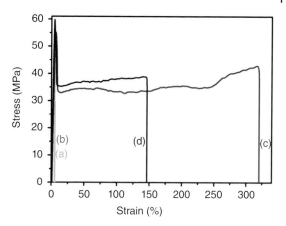

Figure 10.17 Tensile stress–strain curves of PLLA/PVDF (50/50) blends uncompatibilized (a) and compatibilized at a weight ratio of PLLA/PVDF/C-St-1-2-7 (S-6300) = 50/50/3 for (b) 0.1 min, (c) 10 min, and (d) 15 min.

Interestingly, incorporation of 3 wt% C-St-1-2-7(S-6300) dramatically changed the phase morphology. It could be observed that a co-continuous structure with a few smaller phases was formed (Figure 10.16b). The interface between the phases became blurred, indicating significant improvement of the interface adhesion (Figure 10.16c). It is worth noting that numerous nanomicelles (shown in black) were located at the interface as well as in the PLLA matrix (Figure 10.16e); the density of the micelles at the interface was much higher than that in the PLLA phase and even formed a dense layer of micelles at the interface of PLLA and PVDF. Such a "co-continuous phase encapsulated by nanomicelles" morphology formed by simple reactive blending was unprecedented. The nanomicelles within PLLA and those at the interface possessed different structures (Figure 10.16f). Compared with the nanomicelles in PLLA, the interfacial ones were larger with darker cores, which might be related to different self-assembly behavior.

The mechanical properties of PLLA, PVDF, and the blends are presented in Figure 10.17. It was observed that the ductility of the blend compatibilized by PS-type RC polymers for 10 min was remarkably enhanced. The fracture strain was about 320%, compared with 5% for the uncompatibilized blend and 8% for the blend formed by melt mixing for 0.5 min. The significant improvement of the mechanical properties indicated that reaction-formed JNMs could compatibilize the immiscible PLLA/PVDF blends effectively. With further increasing the melt mixing time to 15 min, the fracture strain decreased to about 140%, which was lower than that of the blend by melt mixing for 10 min. This was because part of the nanomicelles departed from the interface and moved into the PLLA phase, and therefore fewer JNMs were present at the interface to compatibilize the PLLA/PVDF system.

The results mentioned showed that PLLA/PVDF blends could be efficiently emulsified through the gradual formation of interfacial JNMs, as illustrated in Scheme 10.5. During reactive blending, the epoxide groups in PS-type RC (C-St-1-2-7(S-6300)) polymers reacted readily with the carboxyl ends of PLLA, to form a kind of double-grafted structure with both PLLA and PMMA as side chains. The kinetic process could be tentatively described as follows. The PS-type RC polymers were first located at the PLLA/PVDF interface as stacked domains before grafting reaction (State 1 in Scheme 10.5). Therefore, the reaction between

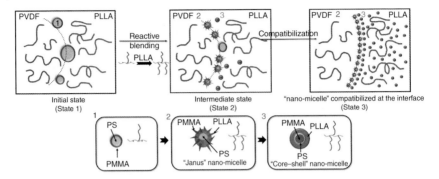

Scheme 10.5 Schematic view of the *in situ* formation of JNMs and the morphology development of binary blends from "droplet stacks domains" to "co-continuous phase encapsulated by nanomicelles" via reactive blending.

PLLA and the PS-type RC could easily happen at the interface. Then the stacked domains broke up gradually to form small JNMs with both PMMA and PLLA side chains, with the help of shearing and grafting reaction. These JNMs had a spherical structure with one hemispherical part constituted by PMMA chains and the other by PLLA chains. The PMMA chains were miscible with PVDF because of the well-known specific interactions between CF_2 and carboxyl groups, while the grafted PLLA chains of the micelles were readily entangled with themselves in the PLLA phase. Therefore, the JNMs acted as effective compatibilizers at the PLLA/PVDF interface (State 2 to State 3 in Scheme 10.5). The formation of the co-continuous domains encapsulated by JNMs could be ascribed to the decrease in Laplace pressure caused by the lowering of interfacial tension by JNMs.

It was also found that more micelles appeared in the PLLA phase with the increasing of the mixing time and this was due to the micelles at the interface being pulled out into the PLLA phase under shear condition as more PLLA chains were grafted onto the micelles. In addition, the structures of the micelles in the PLLA phase were slightly different from that of the micelles at the interface. It could be deduced that the micelles in the PLLA phase were more likely to have a PLLA shell to mix with the bulk PLLA phase, a PMMA intermediate shell, and a PS core, compared to the JNMs at the interface which were constituted by PMMA and PLLA chains as the two hemispheres and a PS backbone as the core. This speculation was consistent with the high-magnification TEM image in Figure 10.16f, which showed distinctive structures between nanomicelles in the PLLA phase and those at the interface. Obviously, the JNMs at the interface contributed more to the compatibilization of PLLA/PVDF blends and this should be the reason why the blend prepared by melt mixing for 15 min had deteriorated mechanical properties, compared to that prepared for 10 min; the lower melt mixing timing was adopted in cases where there was lower interfacial adhesion between the PLLA and PVDF phase. It was worth noting that compared with the core–shell structured nanomicelles in the PLLA phase which displayed a homogeneous distribution, the JNMs at the interface were crowded much closer with each other. Müller *et al.* compared the desorption energies of pre-made nanoparticles with Janus and homogeneous surface structures from the interface of an immiscible polymer blend. They found that the pre-made nanoparticles with JNPs were 20 times less likely to immigrate from the interface than those with a homogeneous surface structure [62–64]. In our case, the PLLA was gradually grafted onto the micelles, and therefore the micelles immigrated gradually from the interface into the PLLA phase. Such immigration was attributed to both dynamic shear and the increased interaction between the micelles and the PLLA phase.

10.5 Conclusions and Further Remarks

A series of PMMA-type RC polymers with different molecular structures were reported here as compatibilizers in some typical immiscible systems. It was found that compared to their linear counterparts, RC polymers exhibited a significant enhancement in their compatibilization efficiency. The superiorities of RC polymers are ascribed to the existence of side chains, which in some extent could counterbalance the weakening of the interaction between the PMMA backbone and the PVDF phase, as the average molecular weight between two grafting sites decreased with grafting reaction. On the other hand, it was also found that the parameters of the molecular structures (e.g., length of side chains, content of reactive groups) of RC polymers influenced their compatibilization efficiencies more or less. Furthermore, PS-type RC polymers were also applied as compatibilizers in immiscible blends and the reaction-formed JNMs that were constituted by a PS core, hemispherical PMMA, and PLLA shells efficiently emulsified the immiscible systems. This novel approach revealed new opportunities to exploit interfacial emulsification using nanomicelles, which was of great significance for engineering plastic alloys. Further investigation should be carried out to achieve JNMs located exclusively at the interface of immiscible polymer blends to be used as compatibilizers, and this is possible through careful design of the molecular structure.

References

1 Utracki, L.A. (1998) *Commercial Polymer Blends*, Chapman and Hall, London.
2 Paul, D.R. and Bucknal, C.B. (2000) *Polymer Blends*, 2nd edn, vols 1 and 2, Wiley, New York.
3 Jeon, H.K., Macosko, C.W., Moon, B., Hoye, T.R., and Yin, Z.H. (2004) Coupling reactions of end- vs mid-functional polymers. *Macromolecules*, **37**, 2563.
4 Retsos, H., Anastasiadis, S.H., Pispa, S., Mays, J.W., and Hadjichristidis, N. (2004) Interfacial tension in binary polymer blends in the presence of block copolymers. 2. Effects of additive architecture and composition. *Macromolecules*, **37**, 524.
5 Pernot, H., Baumert, M., Court, F., and Leibler, L. (2002) Design and properties of co-continuous nanostructured polymers by reactive blending. *Nat. Mater.*, **1**, 54.

6 Shi, H.C., Shi, D.A., Wang, X.Y., Yin, L.G., Yin, J.H., and Mai, Y.W. (2010) A facile route for preparing stable co-continuous morphology of LLDPE/PA6 blends with low PA6 content. *Polymer*, **51**, 4958.

7 Sailer, C. and Handge, U.A. (2007) Melt viscosity, elasticity, and morphology of reactively compatibilized polyamide 6/styrene–acrylonitrile blends in shear and elongation. *Macromolecules*, **40**, 2019.

8 Dedecker, K. and Groeninckx, G. (1999) Interfacial graft copolymer formation during reactive melt blending of polyamide 6 and styrene-maleic anhydride copolymers. *Macromolecules*, **32**, 2472.

9 Díaz, M.F., Barbosa, S.E., and Capiati, N.J. (2007) Reactive compatibilization of PE/PS blends. Effect of copolymer chain length on interfacial adhesion and mechanical behavior. *Polymer*, **48**, 1058.

10 Zhang, C.-L., Feng, L.-F., Zhao, J., Huang, H., Hoppe, S., and Hu, G.-H. (2008) Efficiency of graft copolymers at stabilizing co-continuous polymer blends during quiescent annealing. *Polymer*, **49**, 3462.

11 Xu, Y.W., Thurber, C.M., Lodge, T.P., and Hillmyer, M.A. (2012) Synthesis and utility of model polyethylene-*graft*-poly(methyl methacrylate) copolymers as compatibilizers in polyethylene/poly(methyl methacrylate) blends. *Macromolecules*, **45**, 9604.

12 Gersappe, D., Harm, P.K., Irvine, D., and Balazs, A.C. (1994) Contrasting the compatibilizing activity of comb and linear copolymers. *Macromolecules*, **27**, 720.

13 Dong, W.Y., Wang, H.T., He, M.F., Ren, F.L., Wu, T., Zheng, Q.R., and Li, Y.J. (2015) Synthesis of reactive comb polymers and their applications as a highly efficient compatibilizer in immiscible polymer blends. *Ind. Eng. Chem. Res.*, **54**, 2081.

14 Dong, W.Y., He, M.F., Wang, H.T., Ren, F.L., Zhang, J.Q., Zhao, X.W., and Li, Y.J. (2015) PLLA/ABS blends compatibilized by reactive comb polymers: double T_g depression and significantly improved toughness. *ACS Sustainable Chem. Eng.*, **3**, 2542.

15 Wang, H.T., Dong, W.Y., and Li, Y.J. (2015) Compatibilization of Immiscible polymer blends using in situformed Janus nanomicelles by reactive blending. *ACS Macro Lett.*, **4**, 1398.

16 Dong, W.Y., Wang, H.T., Ren, F.L., Zhang, J.Q., He, M.F., Wu, T., and Li, Y.J. (2016) Dramatic improvement in toughness of PLLA/PVDF blends: the effect of compatibilizer architectures. *ACS Sustainable Chem. Eng.*, **4**, 4480.

17 Moussaif, N. and Jérôme, R. (1999) Compatibilization of immiscible polymer blends (PC/PVDF) by the addition of a third polymer (PMMA): Analysis of phase morphology and mechanical properties. *Polymer*, **40**, 3919.

18 Macosko, C.W., Jeon, H.K., and Hoye, T.R. (2005) Reactions at polymer-polymer interfaces for blend compatibilization. *Prog. Polym. Sci.*, **30**, 939.

19 Dong, W.Y., Jiang, F.H., Zhao, L.P., You, J.C., Cao, X.J., and Li, Y. (2012) PLLA microalloys versus PLLA nanoalloys: preparation, morphologies, and properties. *J. ACS Appl. Mater. Interfaces*, **4**, 3667.

20 Kudva, R.A., Keskkula, H., and Paul, D.R. (1998) Compatibilization of nylon 6/ABS blends using glycidyl methacrylate/methyl methacrylate copolymers. *Polymer*, **39**, 2447.

21 Hale, W., Keskkula, H., and Paul, D.R. (1999) Compatibilization of PBT/ABS blends by methyl methacrylate-glycidyl methacrylate-ethyl acrylate terpolymers. *Polymer*, **40**, 365.

22 Nishi, T. and Wang, T.T. (1975) Melting point depression and kinetic effects of cooling on crystallization in poly(vinylidene fluoride)–poly(methyl methacrylate) mixtures. *Macromolecules*, **8**, 909.

23 Polizu, S., Favis, B.D., and Vu-Khanh, T. (1999) Morphology-interface-property relationships in polystyrene/ethylene-propylene rubber blends. 2. Influence of areal density and interfacial saturation of diblock and triblock copolymer interfacial modifiers. *Macromolecules*, **32**, 3448.

24 Lepers, J.C. and Favis, B.D. (1999) Interfacial tension reduction and coalescence suppression in compatibilized polymer blends. *AIChE J.*, **45**, 887.

25 Sundararajs, U. and Macosko, C.W. (1996) Drop breakup and coalescence in polymer blends: the effects of concentration and compatibilization. *Macromolecules*, **28**, 2647.

26 Charoensirisomboon, P., Inoue, T., and Weber, M. (2000) Pull-out of copolymer in situ-formed during reactive blending: effect of the copolymer architecture. *Polymer*, **41**, 6907.

27 Lin, B., Mighri, F., Huneault, M.A., and Sundararaj, U. (2005) Effect of pre-made compatibilizer and reactive polymers on polystyrene drop deformation and breakup in simple shear. *Macromolecules*, **38**, 5609.

28 Xu, X., Yan, X., Zhu, T., Zhang, C., and Sheng, J. (2007) Phase morphology development of polypropylene/ethylene-octene copolymer blends: effects of blend composition and processing conditions. *Polym. Bull.*, **58**, 465.

29 Yin, B., Zhao, Y., Yu, R.Z., An, H.N., and Yang, M.B. (2007) Morphology development of PC/PE blends during compounding in a twin-screw extruder. *Polym. Eng. Sci.*, **47**, 14.

30 Macosko, C.W. (2000) Morphology development and control in immiscible polymer blends. *Macromol. Symp.*, **149**, 171.

31 Li, J.M., Ma, P.L., and Favis, B.D. (2002) The role of the blend interface type on morphology in cocontinuous polymer blends. *Macromolecules*, **35**, 2005.

32 Macosko, C.W., Guégan, P., Khandpur, A.K., Nakayama, A., Marechal, P., and Inoue, T. (1996) Compatibilizers for melt blending: Premade block copolymers. *Macromolecules*, **29**, 5590.

33 Harrats, C., Thomas, S., and Groeninckx, G. (2006) *Micro- and Nanostructured Multiphase Polymer Blend Systems: Phase Morphology and Interfaces*, CRC Press, Taylor & Francis.

34 Burch, H.E. and Scott, C.E. (2001) Effect of viscosity ratio on structure evolution in miscible polymer blends. *Polymer*, **42**, 7313.

35 Zhang, Y.X., Zuo, M., Song, Y.H., Yan, X.P., and Zheng, Q. (2015) Dynamic rheology and dielectric relaxation of poly(vinylidene fluoride)/poly(methyl methacrylate) blends. *Compos. Sci. Technol.*, **106**, 39.

36 Charoensirisomboon, P., Chiba, T., Inoue, T., and Weber, M. (2000) In situ formed copolymers as emulsifier and phase-inversion-aid in reactive polysulfone/polyamide blends. *Polymer*, **41**, 5977.

37 Steinmann, S., Gronski, W., and Friedrich, C. (2001) Cocontinuous polymer blends: influence of viscosity and elasticity ratios of the constituent polymers on phase inversion. *Polymer*, **42**, 6619.

38 Lee, J.K. and Han, C.D. (1999) Evolution of polymer blend morphology during compounding in an internal mixer. *Polymer*, **40**, 6277.

39 Li, Y.J. and Shimizu, H. (2009) Improvement in toughness of poly(L-lactide) (PLLA) through reactive blending with acrylonitrile–butadiene–styrene copolymer (ABS): morphology and properties. *Eur. Polym. J.*, **45**, 738.

40 Bellinger, M.A., Sauer, J.A., and Hara, M. (1994) Tensile fracture properties of rigid–rigid blends made of sulfonated polystyrene ionomer and polystyrene. *Macromolecules*, **27**, 6147–6155.

41 Chen, C.C. and Sauer, J.A. (1990) Yield and fracture mechanisms in ABS. *J. Appl. Polym. Sci.*, **40**, 503.

42 Angnanon, S., Prasassarakich, P., and Hinchiranan, N. (2011) Styrene/acrylonitrile graft natural rubber as compatibilizer in rubber blends. *Polym. Plast. Technol. Eng.*, **50**, 1170.

43 Feng, X.L., Zhang, S., Zhu, S., Han, K.Q., Jiao, M.L., Song, J., Ma, Y., and Yu, M.H. (2013) Study on biocompatible PLLA–PEG blends with high toughness and strength via pressure-induced-flow processing. *RSC Adv.*, **3**, 11738.

44 Oyama, H.T. (2009) Super-tough poly(lactic acid) materials: reactive blending with ethylene copolymer. *Polymer*, **50**, 747.

45 Bai, H.W., Xiu, H., Gao, J., Deng, H., Zhang, Q., Yang, M.B., and Fu, Q. (2012) Tailoring impact toughness of poly(L-lactide)/poly(ε-caprolactone) (PLLA/PCL) blends by controlling crystallization of PLLA matrix. *ACS Appl. Mater. Interfaces*, **4**, 897.

46 Chiou, K.C. and Chang, F.C. (2000) Reactive compatibilization of polyamide-6 (PA 6)/polybutylene terephthalate (PBT) blends by a multifunctional epoxy resin. *J. Polym. Sci. Pol. Phys.*, **38**, 23.

47 Hashima, K., Nishitsuji, S., and Inoue, T. (2010) Structure-properties of super-tough PLA alloy with excellent heat resistance. *Polymer*, **51**, 3934.

48 Inoue, T. and Kobayashi, S. (2012) A super impact-absorbing polymer alloy by reactive blending of nylon with poly(ethylene-*co*-glycidyl methacrylate). *Recent Res. Devel. Polym. Sci.*, **11**, 1.

49 Hashima, K., Usui, K., Fu, L.X., Inoue, T., Fujimoto, K., Segawa, K., Abe, T., and Kimura, H. (2008) Super-ductile PBT alloy with excellent heat resistance. *Polym. Eng. Sci.*, **48**, 1207.

50 Hertler, W.R., Sogah, D.Y., Webster, O.W., and Trost, B.M. (1984) Depression of glass transition temperature in aramid-polybutadiene multiblock copolymers. *Macromolecules*, **17**, 1417.

51 Su, Z.Z., Li, Q.Y., Liu, Y.J., Hu, G.H., and Wu, C.F. (2009) Compatibility and phase structure of binary blends of poly(lactic acid) and glycidyl methacrylate grafted poly(ethylene octane). *Eur. Polym. J.*, **45**, 2428.

52 Qiu, J.S., Xing, C.Y., Cao, X.J., Wang, H.T., Wang, L., Zhao, L.P., and Li, Y.J. (2013) Miscibility and double glass transition temperature depression of poly(L-lactic acid) (PLLA)/poly(oxymethylene) (POM) blends. *Macromolecules*, **46**, 5806.

53 Lyu, S.P., Jones, T.D., Bates, F.S., and Macosko, C.W. (2002) Role of block copolymers on suppression of droplet coalescence. *Macromolecules*, **35**, 7845.

54 Onogi, S., Masuda, T., and Kitagawa, K. (1970) Rheological properties of anionic polystyrenes. I. Dynamic viscoelasticity of narrow-distribution polystyrenes. *Macromolecules*, **3**, 109.

55 Dong, H. and Jacob, K.I. (2003) Effect of molecular orientation on polymer free volume distribution: an atomistic approach. *Macromolecules*, **36**, 8881.

56 Newman, S. and Strella, S. (1965) Stress–strain behavior of rubber-reinforced glassy polymers. *J. Appl. Polym. Sci.*, **9**, 2297.

57 Pan, L.H., Chiba, T., and Inoue, T. (2001) Reactive blending of polyamide with polyethylene: pull-out of in situ-formed graft copolymer. *Polymer*, **42**, 8825.

58 Pan, L.H., Inoue, T., Hayami, H., and Nishikawa, S. (2002) Reactive blending of polyamide with polyethylene: pull-out of in situ-formed graft copolymers and its application for high-temperature materials. *Polymer*, **43**, 337.

59 Yan, L.T., Popp, N., Ghosh, S.K., and Böker, A. (2010) Self-assembly of Janus nanoparticles in diblock copolymers. *ACS Nano*, **4**, 913.

60 Walther, A. and Müller, A.H.E. (2013) Janus particles: synthesis, self-assembly, physical properties, and applications. *Chem. Rev.*, **113**, 5194.

61 Lattuada, M. and Hatton, T.A. (2011) Janus particles: synthesis, self-assembly, physical properties, and applications. *Nano Today*, **6**, 286.

62 Walther, A., Matussek, K., and Müller, A.H.E. (2008) Engineering nanostructured polymer blends with controlled nanoparticle location using Janus particles. *ACS Nano*, **2**, 1167.

63 Bahrami, R., Löbling, T.I., Gröschel, A.H., Schmalz, H., Müller, A.H.E., and Altstädt, V. (2014) The impact of Janus nanoparticles on the compatibilization of immiscible polymer blends under technologically relevant conditions. *ACS Nano*, **8**, 10048.

64 Bryson, K.C., Löbling, T.I., Müller, A.H.E., Russell, T.P., and Hayward, R.C. (2015) Using Janus nanoparticles to trap polymer blend morphologies during solvent-evaporation-induced demixing. *Macromolecules*, **48**, 4220.

11

Reactive Compounding of Highly Filled Flame Retardant Wire and Cable Compounds

Mario Neuenhaus[1] and Andreas Niklaus[2]

[1] *Martinswerk GmbH, Kölner Strasse 110, 50127 Bergheim, Germany*
[2] *Buss AG, Hohenrainstrasse 10, 4133 Pratteln, Switzerland*

11.1 Introduction

"Low Smoke Zero Halogen" (LSOH) or "Halogen Free Flame Retardant" (HFFR) compounds are becoming more and more important in the cable industry. In the past 30 years the consumption of HFFR compounds in energy cables has increased to about 170 000 mt per year in Europe and to 394 000 mt per year worldwide[1] and a growth rate of more than 10% in the coming years has been forecasted for these compounds (Figure 11.1).

Most of the HFFR compounds are used as thermoplastic material in jacketing applications. However, due to continuously increasing stringent requirements in both mechanical and thermal properties, improved solutions were required to be developed. One of the most widely used approaches is the crosslinking of polyolefin compounds, which is the focus of this chapter [23]. In principle, there are four ways of crosslinking polyolefin polymers [2]:

- Peroxide crosslinking, often referred to as *PEX-a*
- Silane crosslinking, often referred to as *PEX-b*
- Electron-beam crosslinking, often referred to as *PEX-c*
- Crosslinking with azo compounds, often referred to as *PEX-d*.

Crosslinking with CV is difficult for HFFR compounds, because the processing conditions are limited by the decomposition temperature of the organic peroxide, which is typically 150 °C. High loadings of mineral flame retardants in olefinic compounds typically result in higher processing temperatures, which may result in undesired pre-crosslinking. Crosslinking with electron-beam (PEX-c) needs a separate crosslinking step and only small wire diameters with insulation thickness of maximum 3 mm can be crosslinked at speed [3]. PEX-d is produced by the Lubonyl method, where an azo compound is added to the PE base. The crosslinking via the nitrogen bridges is done in a hot salt bath after the extrusion. This method is almost extinct now.

The most frequently used method for increasing the temperature when using an HFFR compound is by silane crosslinking (PEX-b). After producing a Sioplas

Reactive Extrusion: Principles and Applications, First Edition. Edited by Günter Beyer and Christian Hopmann.
© 2018 Wiley-VCH Verlag GmbH & Co. KGaA. Published 2018 by Wiley-VCH Verlag GmbH & Co. KGaA.

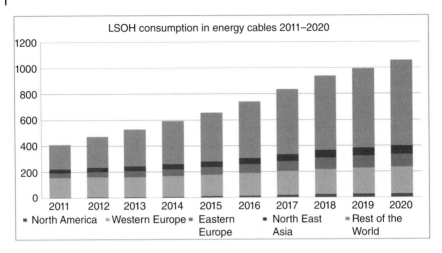

Figure 11.1 Wire and cable market outlook 2015 Q4 [1].

graft compound (see Section 11.2.2) this grafted material is filled with a flame retardant. Although, traditionally, these processing tasks are performed in two individual steps there are advantages of employing an integrated one-step process, which will also be discussed in this chapter.

The research conducted on silane crosslinkable wire and cable compounds described in this chapter is based on a collaboration between the companies Martinswerk GmbH, Buss AG, Evonik Industries AG, and ExxonMobil Chemical[4]. This chapter gives an overview about the basics of reactive compounding, the difference between polar and non-polar systems as well as the compounding performance of ATH with different specific surface areas. The FR performance was measured according to ISO 5660 using a cone calorimeter. Some selected compounds have been extruded as cables and were tested according to the new European standard for construction cables EN 50399.

11.2 Formulations and Ingredients

11.2.1 Typical Formulation and Variations for the Evaluation

Blends of PE (polyethylene) and EVA (ethyl vinyl acetate) copolymers are widely used for highly filled wire and cable compounds. Due to the polarity of EVA, such functional polymers are the preferred choice for crosslinked systems. Additionally, the higher the branching level of the EVA polymer chains, the higher the impact on flexibility and the flame retardant performance.

In addition to this typical recipe for a setup, a non-polar compound based on PE/octene block copolymer (plastomer) was tested for comparison. A blend of 75 phr EVA and 25 phr LLDPE was chosen as the polar base resin. The non-polar resin contained two ethylene–octene copolymers with different melt index (see also Figure 11.2). The comparison clearly shows the impact of the polar groups on crosslinking behavior, the mechanical performance, andon the flame retardancy.

Ingredients			Loading in phr	
Polar system	EVA (28 wt% VA)	MFI 7	75	
	LLDPE	MFI 1	25	
Non-polar system	Ethylene-octene copolymer	MFI 1		65
	Ethylene octene copolymer	MFI 30		35
Flame retardant	Aluminum trihydrate (ATH)		160	175
Coupling agent	Blend of 37.5% Silfin13 and 62.5% Silfin 25 [5]		About 1.5	About 1.7
Antioxidants			3	3

Figure 11.2 Formulations tested (polar and non-polar polymer base).

By varying the ratio of the two different plastomers, it is possible to adjust the mechanical properties (i.e., tensile strength, elongation), the hot-set results that indicate the degree of crosslinking and the Melt Flow Index (measured according to ISO 1133) of the compound. The mechanical properties can also be influenced by the amount of silane added. Silane addition has been varied by changing addition levels between 10% and 20% to note the effect on the tensile testing and the crosslinking degree measured by hot-set values. Due to the lower reactivity of the ethylene–octene copolymer, a slightly increased amount of the silane peroxide mixture is required.

As mentioned in [24], pure polyolefin requires high loading to achieve a certain flame retardancy. As soon as the functional side groups are attached to the main molecular chain the required loading to achieve similar flame retardancy is lowered. This is the reason why the polar formulation requires for a lower loading than the non-polar system.

11.2.2 Principle of Silane Crosslinking by Reactive Extrusion

The crosslinking of PE and its copolymers creates a three dimensional network that leads to a significant improvement of the material properties. In wire and cable applications, it improves not only the mechanical properties but also the thermal performance, aging properties, and chemical resistance.

Usually, thermally degradable organic peroxides are added to the process to generate free radicals on the polymer chain by abstracting H atoms. Subsequently, vinyl silanes will occupy these radical sites by a chemical reaction with the C=C groups. The final crosslinking step takes place later on the extruded cable product by a hydrolysis and condensation reaction in water, which is initiated by a catalyst [6, 7] (Figure 11.3).

11.2.3 Production of Aluminum Trihydroxide (ATH)

Huge quantities of ATH have been used in polyolefin compounds for many years. It is well known as a reliable, environmentally friendly, and non-toxic flame retardant.

Synthetic aluminum trihydrate is normally produced with the so-called Bayer process. This process uses bauxite as raw material that is leached in caustic liqueur. Bauxite contains approximately 50–60% $Al(OH)_3$ and $AlO(OH)$, and

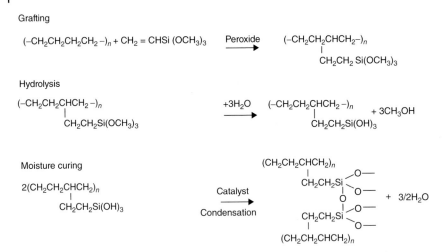

Figure 11.3 Principles of silane grafting.

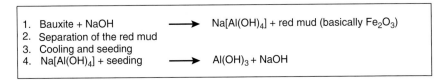

Figure 11.4 Simplified description of the Bayer process.

the rest is mainly Fe_2O_3. To produce aluminum trihydrate, finely ground bauxite is leached in hot NaOH to form $Na[Al(OH)_4]$. The insoluble Fe_2O_3 and other residuals are removed from the liqueur by several filtration processes. The remaining clear and super saturated solution of $Na[Al(OH)_4]$ is diluted with water and cooled down to a certain temperature. Normally, higher temperatures result in coarser particles. To accelerate the crystallization process and also to modify the final product special seeding $Al(OH)_3$ is added. After a residence time of several days the solid $Al(OH)_3$ is separated from the liqueur by filtration and washing steps (Figure 11.4).

Nowadays, producers specializing in the production of fine precipitated $Al(OH)_3$ no longer use bauxite as raw material but use a semi-finished hydroxide. This hydroxide is still leached in NaOH and crystalized according to the Bayer process but does not require the separation of the red mud.

11.2.4 Mode of Action of Aluminum Trihydroxide

The flame retardant properties of aluminum hydroxide are based on its endothermic decomposition at temperatures above approximately 200 °C according to the equation shown in Figure 11.5.

The energy consumed by the endothermic decomposition cools down the compound and slows down the pyrolysis of the polymer(s). The generated water vapor reduces the concentration of burnable gases. The Al_2O_3 builds a protective layer

Figure 11.5 Endothermic decomposition of ATH.

on the surface that, on one hand, keeps the oxygen away and, on the other hand, reduces the release of decomposition products. Furthermore, the relatively high necessary loadings of greater than 50 wt% lead to a dilution of the polymer in the final product, which obviously reduces the amount of burnable material.

11.2.5 Selection of Suitable ATH Grades

In general, one distinguishes between standard coarse and fine precipitated grades. The coarser standard grades are characterized by their median particle size d_{50}. The d_{50} indicates that 50 vol.% of the particles are smaller than the indicated value.

Typical d_{50} values of the coarsest available grades are in the range of 40–100 μm. The use of these grades in wire and cable applications is very limited. However, they are the starting point for the production of ground materials and in some cases are the raw materials for an adopted Bayer process (see Section 11.2.3).

Commercially available ground ATH grades typically have d_{50} values between 5 and 25 μm. They are produced by de-agglomeration of the coarse particles mentioned earlier. Applications for these grades in the wire and cable industry typically are limited to bedding compounds where they can replace $CaCO_3$. The use of ATH in bedding compounds improves the flame retardant performance of the finished cables.

To meet the highest requirements with respect to flame retardant performance as well as to mechanical properties with aluminum trihydrate, fine precipitated grades are preferred (d_{50} values below 2.5 μm). These grades allow the highest possible loadings to achieve the necessary flame retardant standards while maintaining the processing window of the specific polymers. Main applications for this kind of aluminum trihydrate are sheathing and insulation compounds.

Figure 11.6 gives an overview of distinctive properties of commercially available fine precipitated ATH grades.

The use of ATH in reactive compounding is classically limited to its thermal stability. The thermal stability of a product can be measured by TGA, which is typically performed at a heating rate of 1 K min^{-1}, while the thermal stability is expressed as the temperature at 2% weight loss. With standard fine precipitated ATH grades (e.g., Martinal® LE grades), the thermal stability is normally in the range of 180–200 °C, which limits their use in compounding. Usually, the decomposition temperature of the ATH decreases with increasing specific surface area.

Martinal® OL-104 LEO and OL-107 LEO mark the latest development in fine precipitated aluminum hydroxide. Besides optimized compounding performance, these LEO grades show extraordinarily high decomposition temperatures. For the OL-104 LEO, the temperature at 2% weight loss is approximately 5 °C higher than with a standard 4 m^2 g^{-1} material; OL-107 LEO

		Martinal® OL-104 LEO	Martinal® OL-107 LEO	Martinal® OL-111 LE
Al(OH)$_3$ content	%	99.4	99.4	99.4
Loss on Ignition (1200 °C)	%	34.5	34.5	34.5
Moisture (105 °C)	%	0.35	0.4	0.7
Median particle size d_{50}	μm	1.7–2.1	1.6–1.9	0.9–1.4
Specific surface area (BET)	m^2	3–5	6–8	10–12
Electrical conductivity (10% in H$_2$O)	μS cm^{-1}	60	70	200
Bulk density bag	kg m^{-3}	450	400	270
Bulk density big bag	kg m^{-3}	550	500	350
Thermal stability (2wt%. loss @ 1K min^{-1})	°C	225	220	210

Figure 11.6 Technical data of different fine precipitated ATH grades.

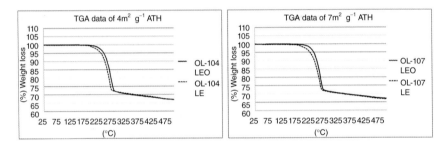

Figure 11.7 TGA at 1 K min^{-1} of fine precipitated Martinal® grades.

gains a 15–20 °C higher decomposition temperature than other ATH grades of the same specific surface area. Figure 11.7 compares the weight loss over the temperature range (TGA @ 1 K min^{-1}) of Martinal® LEO grades with the corresponding predecessor grades.

In a compounding process or during extrusion, the higher thermal stability enables end users to run the process at higher melt temperatures. The higher melt temperature leads to lower viscosity of the molten compound and thus to lower frictional heat. Finally, the combination of the before mentioned properties result in potentially higher throughput during compounding or improved line speed in an extrusion process [24].

The absolute temperature and the residence time in the compounding line have a significant impact on the thermal stability. Figure 11.8 illustrates the weight loss over time of two different fine precipitated ATH grades at different temperatures, which obviously increases with increasing temperature. Compared to OL-104 LE at the same temperature, the weight loss with OL-104 LEO is lower and slightly delayed. Since the total weight loss is still the same it does not have an impact on the flame retardant performance but has an advantage in processing.

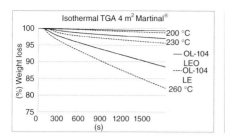 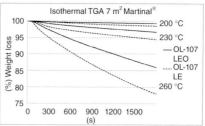

Figure 11.8 Isothermal TGA of fine precipitated Martinal® grades.

Martinal® OL-104 LEO and OL-107 LEO are optimized fine precipitated ATH grades with enhanced compounding performance, improved electrical properties, improved bulk density after conveying, improved thermal stability, and lower compound viscosity [8]. Their high thermal stability in combination with the improved viscosity enables compounders to run continuous compounding lines up to the limits. Even at processing temperatures where standard ATH grades already release a significant amount of water, Martinal® can still be used.

End users can adjust the mechanical properties of wire and cable compounds significantly by choosing ATH grades with an alternative specific surface area. In general, the tensile strength increases by enlarging the specific surface area. Additionally, smaller particle size ATH grades result in improved FR performance (highest in BET value). However, such formulation always needs to be optimized. At a certain point the viscosity of the compounds will start to increase, the processing of the formulation will become more difficult, and the elongation at break of the final compound may decrease. Figure 11.9 shows the impact of different

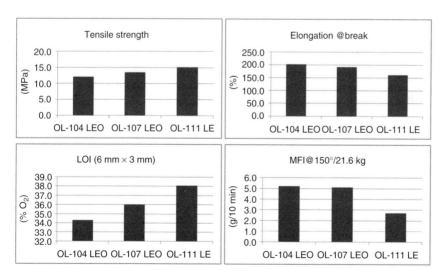

Figure 11.9 Impact of specific surface area on compound properties.

specific surface areas on the compound properties in a standard EVA lab formulation (60 wt% of Martinal in Escorene UL00119).

11.3 Processing

11.3.1 Compounding Line

The reciprocating single-screw machine, the Buss Kneader, where the screw simultaneously executes an axial oscillating motion with every rotation, has been found to be the most preferable machine system to perform HFFR compounding among others [9]. The combination of rotation and axial motion of the screw produces extensional flow with highly dispersive mixing action between the kneading flights and kneading bolts (Figure 11.10). The efficiency of the distributive mixing caused by the combination of the radial and longitudinal actions at moderate shear rates is testimony to this. The striation thickness is a scientific mathematical figure that represents the mixing efficiency of a system. With the Buss Kneader, 12 pairs of kneading flights and kneading teeth follow each other (both radially and axially) within a processing length of 1 L/D. For a processing length of 4 L/D this results in an astronomical striation thickness of 2^{48} or 2.8 times 10^{14} [10].

A further special feature of the Buss Kneader enables exact scale-up scenarios: the shear gap s_z (Figure 11.10) in the Buss Kneader is proportional to machine size, that is, outer shaft diameter D. This makes the Buss Kneader unique as the shear gradient is independent of machine size. It depends only on the screw speed, which decisively simplifies scale-up and process transferability to various machine sizes.

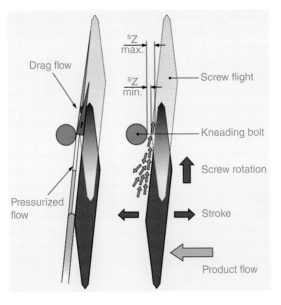

Figure 11.10 Mixing/shearing mechanism on Kneaders [9, 10].

In the earlier Buss Kneader (i.e., MKS type), each screw flight was divided into three kneading segments. Technical and procedural characteristics have been further refined and optimized in recent years to comply better with specific user requirements. This involved optimizing the geometric relationships of the kneading zone (outer/inner diameter and outer diameter/stroke ratios) to further improve the input material feed and the required torque transmission. With patented methodology [13, 14], screw elements were developed and used most successfully for systematically improving the material transport efficiency and dispersing action.

The latest high-performance development features four segments per revolution (MX type). This expands the design possibilities in the processing section remarkably [11, 12]. As a result, the increased free volume gives a much larger processing chamber. The transition to four-flight screw technology, however, is the major reason for performance and quality improvement. The ratio of stroke to outside diameter is increased, permitting a high screw pitch. In addition, there are greater opportunities with regard to design and optimization than with the three-flight screw. Flights with longer flanks improve the conveying characteristics, and the flight geometry can be used to influence the mixing action in specific ways in terms of both distributive and dispersive mixing [10]. An open view of the mixing section is shown in Figure 11.11.

In reactive extrusion, in general, and in silane grafting, in particular, the control of activation, residence time and heat history of the process are essential. A design feature of the Kneader offers an ideal solution for these requirements. Every kneading bolt mounted in the barrel can be used as the injection point for liquid ingredients (Figure 11.12). Thus, the reactive additive can be injected at the

Figure 11.11 Details of the screw geometry on latest generation of Kneaders [13, 14]. (Reproduced with permission of Buss AG.)

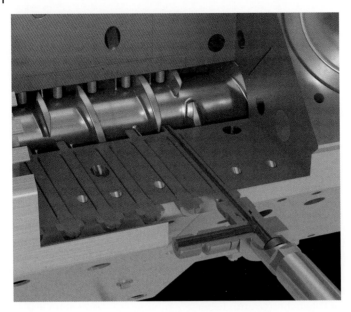

Figure 11.12 Injection nozzle for liquid components [15]. (Reproduced with permission of Buss AG.)

optimum position. The special design of the bolt allows the injection of the liquid *into* the polymer melt. The mixing starts immediately as compared to alternative systems and assures safe operation.

11.3.2 Compounding Process for Cross Linkable HFFR Products

The most frequently used method for increasing the thermal stability of an HFFR compound is by silane crosslinking (PEX-b) as mentioned in Section 11.2.1. The grafting of the Sioplas compound and the filling with ATH flame retardant are the basic requirements that need to be met. Traditionally, these processing tasks are performed in two individual steps.

11.3.2.1 Two-Step Compounding Process

The so-called two-step process is conventionally in use for the production of silane-grafted HFFR compounds. As the first processing step the silanes are grafted with the polymers. At temperatures of 180–200 °C, peroxides abstract H atoms from the polymer chains, creating free radicals. Organo functional silanes can link up at these activated sites with their vinyl groups. These radicals react with the vinyl silane forming covalent bonds. During the first processing step, it is important to achieve a melt temperature of greater than 180 °C such that the grafting reactions starts.

During the second processing step, the ATH and other additives are mixed with the grafted polymers. New European flame retardant standards require high loadings of ATH. Due to the relatively low bulk density, a large volume has to be fed into the compounding lines. Therefore, the feed of the ATH is split into two feeders. With relatively short processing length, for example,

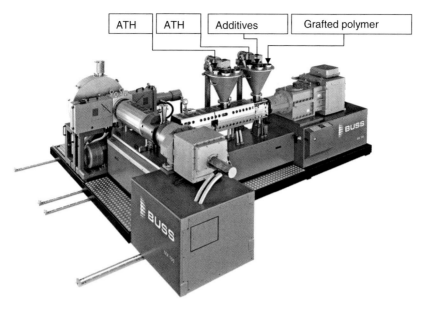

Figure 11.13 Buss Kneader with 15 L/D configuration for the second step of the two-step process [15]. (Reproduced with permission of Buss AG.)

with an $L/D = 11-15$, approximately 60% of the total filler amount is added together with the polymer into the first hopper. The remaining ATH enters the line via the second hopper. Incorporated air is released backward either via special openings in the kneaders housing or via the feeding hoppers. Further downstream, a vacuum degassing removes the remaining air and volatiles. See also the Figures 11.13 and 11.14.

After compounding, the material is still thermoplastic and can be processed as such. Typically, the final shaping is an extrusion process. To accelerate the crosslinking after extrusion, special catalysts are added during the extrusion. On the industrial scale, master batches of di-butyl-tin-laurate (DBTL) with different carriers are very common.

11.3.2.2 One-Step Compounding Process

The one-step compounding process requires optimized processing conditions. Compounding lines need to have a 50% longer processing length than traditional lines. It has to be set up with three feeding ports for solids and a liquid injection point slightly downstream of port 1 (see Figure 11.15).

Only the polymers are added to the first feeding port. The silane peroxide mixture is added further downstream directly into the melt. Again it is very important to achieve the necessary melt temperature of greater than 180 °C so that the peroxide can decompose and initiate the graft reaction. As the polymers make about 50 vol.% of the total formulation the residence time in the first 1/3 of the compounding line is 100% longer than further downstream. This is normally sufficient for the grafting process.

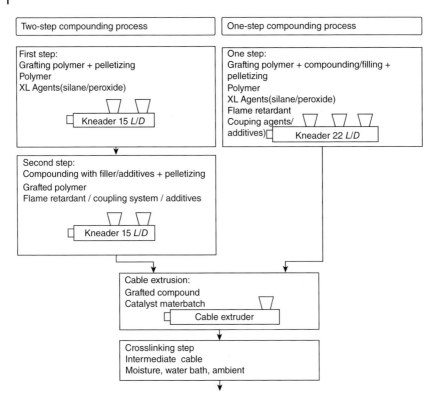

Figure 11.14 Flowchart of two-step process versus one-step process.

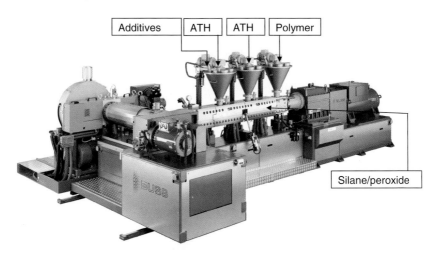

Figure 11.15 Buss Kneader with 22 *L/D* configuration for the one-step process [15]. (Reproduced with permission of Buss AG.)

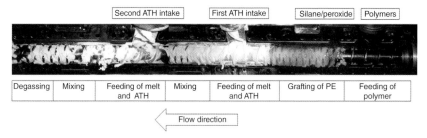

Figure 11.16 Observed (intermediate product conditions after a forced emergency stop along the processing zones (flow direction from right to left) [4, 15]. (Reproduced with permission of Buss AG.)

As with the two-step process, the relatively high volume of ATH has to be split into two parts. Approximately 50% of the total ATH is added to the second feeding port. The remaining ATH is added at the final feeder together with other additives such as stabilizers and potential lubricants. Incorporated air is released backward via special openings in the vicinity of the feeding ports 2 and 3. A degassing port with or without vacuum removes the remaining volatiles generated in the grafting process. Figure 11.15 illustrates the one-step compounding process.

Figure 11.16 gives an interior view of the compounding process. The Kneader was stopped by the emergency stop button when running at capacity and opened immediately. On the right-hand side of Figure 11.16, the feeding port of the polymers is located. The first section of the Kneader screw is equipped with conveying elements to make sure that the granules are fed into the Kneader and do not block the feeding zone. As soon as the polymer is molten the silane/peroxide mixture is added directly into the melt via a kneading pin with an injection needle. Grafting reactions (see Section 11.2.3) take place with a length of approximately 3 L/D. which is described in the following section. Further downstream, the first portion of the ATH is added. Conveying elements make sure that the ATH is fed away from the intake as soon as possible. A backward degassing removes most of the incorporated air and improves the in-take of the ATH. Downstream kneading elements help to disperse the ATH in the polymer matrix. At port 3, the second portion of the ATH is added together with other additives. It is essential for the process to add lubricants and any stabilizer additives as late as possible into the process, so that they do not hinder the grafting reactions.

One of the big advantages of the Kneader is that it allows precise temperature control and measurement. Thermocouples are installed in hollow kneading pins that are immersed directly into the melt (Figure 11.17). This ensures that the operators achieve the necessary temperature of 180 °C for the grafting process. It also helps to take the necessary alternative steps if the melt temperature exceeds the decomposition temperature of the ATH.

The graphs in Figure 11.18 show the temperature of the melt in the processing unit of the Kneader with Martinal® OL-104 LEO and OL-107 LEO at the different pin positions. Even at 230 °C no decomposition of the ATH was visible; the granules are of high quality and show that no gas bubbles are formed by water

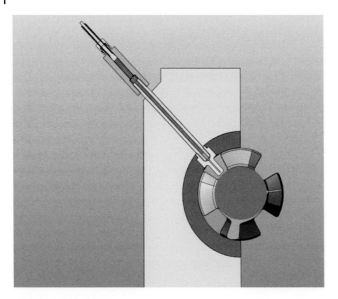

Figure 11.17 Thermocouple installed in a hollow kneading tooth at selected position along the process length [4]. (Reproduced with permission of Buss AG.)

vapor. This value is 5–10 °C higher than that described in the technical specification (Figure 11.6). The very short residence time and narrow residence time distribution might be responsible for this finding. Nevertheless, this allows a very broad processing window.

At pin position 1, the polymers are added into the Kneader. Right after pin position 2 the silane/peroxide mix is injected directly into the melt. Here the volumetric flow rate is 50% of the final volume only resulting in longer residence time. The melt temperature of greater than 180 °C is sufficient for the grafting process described in Section 11.1.3. Just before pin position 4 approximately 50% of the total ATH is added. The frictional heat generated during the compounding process is responsible for the temperature increase at pin position 5. The second portion of the ATH is added between pins 5 and 6. The relatively low temperature of the ATH that is added to the process is responsible for the temperature drop at the feeding ports. A higher compounding throughput generates more frictional heat so that the melt temperature increases. Due to the remarkably high thermal stability of the Martinal® LEO grades used in this study, no decomposition has been observed even at a melt temperature of approximately 230 °C. This temperature is a little higher than mentioned in Section 11.2.5. The reason is that TGA is measured at a standard pressure of 1013 mbar; however, the pressure in the kneader is slightly increased. A higher pressure typically shifts the decomposition temperature to higher temperatures.

One would expect the necessity of an adjusted screw configuration, a different temperature setting in the compounding zones, or even a lower throughput with an increase of the specific surface area of the ATH filler. However, this was not

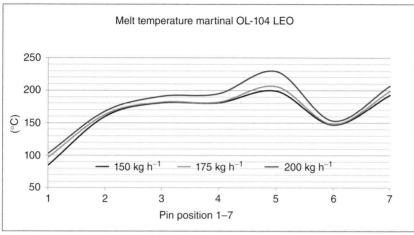
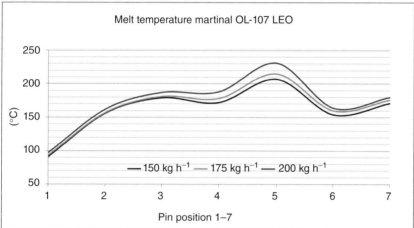

Figure 11.18 Example of the melt temperature at different points in a Kneader at various throughputs.

the case with LEO ATH grades. When utilizing Martinal® with higher specific surface area the melt temperature only increases up to about 15 °C.

11.3.2.3 Advantages and Disadvantages of the Two Process Concepts (Two-Step vs One-Step)

Advantages of the one-step process over the two-step include the remarkably lower energy consumption due to the absence of a second cooling/melting process. Detailed figures on the possible cost savings have been published in [4]. The reduced heat history and thermal stress of the polymer could lower the demand for stabilizers. The simplified internal logistics because of the absence of intermediates allows for lean manufacturing concepts and a smaller plant footprint.

A possible disadvantage is the need for a longer compounding line (22 vs 15 L/D). This issue might be compensated by the greater flexibility and wider usability of plants like this.

Regardless of the compounding technology, a catalyst batch has to be added during the extrusion on wire. To accelerate crosslinking, finished cables are typically extruded into a steam atmosphere or a water bath (Figure 11.14).

11.4 Evaluation and Results on the Compound

The following evaluations have been performed to confirm compliance with product requirements.

11.4.1 Crosslinking Density

The hot-set test is a well-known method to measure crosslinking density. According to EN 60811-2-1 test specimens ballasted with 0.2 MPa and placed in an oven at 200 °C for 15 min. The length of the test specimens is measured prior to the heat treatment, immediately after the treatment, and when samples have been conditioned at room temperature for a certain time. The change in length measured immediately after the heat treatment must not exceed 100% and the permanent length change measured after conditioning must not exceed 15%. In general, a lower change in length indicates a higher crosslinking density.

The silane dosage amount has been adapted stepwise depending on the specific compound properties. Besides the compound viscosity and mechanical properties the hot-set was used to provide the key value to adjust the Silane amount. All specimens were cut from extruded tapes. A catalyst batch made of 5% DBTL in Escorene Ultra UL00728 was added to the compounds at a fixed ratio prior to extrusion. The pressure at the die of the extruder was another indicator of the crosslinking density. The increasing pressure indicates an increased viscosity due to partial crosslinking. Figure 11.19 illustrates the response of the compound properties on decreasing levels of silane/peroxide (100%, 90%, and 80%). A 100% silane mixture equals 1.6 phr and resulted in very good hot-set values' however,

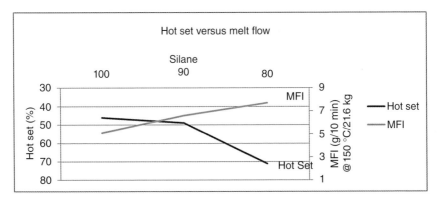

Figure 11.19 Impact of silane addition level on hot-set and melt flow of a polar compound.

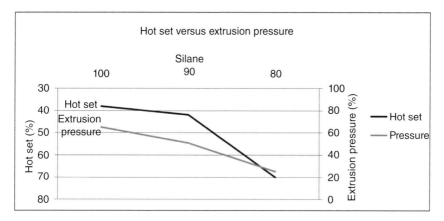

Figure 11.20 Impact of silane addition level on hot-set and extrusion pressure of a polar compound.

the viscosity was on the high side so that extrusion pressure has been close to the extruder's limit. At 80% (1.28 phr) silane mixture, exactly the opposite trend can be observed; that is, improved viscosity behavior, althouh at hot-set values close to the upper allowed limit. At 90% silane mixture (1.44 phr), these properties are balanced, that is, hot-set was almost at the minimum and extrusion pressure was acceptable.

Since extrusion pressure is a function of the compound viscosity, the melt flow index increases with reductionin the addition of silane peroxide as well. Again, 1.44 phr of the above mentioned catalyst batch turned out to be the best compromise (Figure 11.20). This ratio has been chosen for all following compounding trials.

11.4.2 Mechanical Properties

The choice and amount of coupling agent are not the only factors that impact the mechanical performance of mineral filled polyolefin compounds (see Section 11.2.5). Increasing the specific surface area of the mineral flame retardant from 4 to $7 m^2 g^{-1}$ (e.g., from Martinal OL-104 LEO to Martinal OL-107 LEO) increases the tensile strength by approximately 7%. A similar boost is achieved by increasing the loading of a $4 m^2 g^{-1}$ material from 160 to 187.5 phr due to the reinforcing properties of ATH (Figure 11.21).

At 160 phr, loading both ATH grades achieves an elongation at break well within the requirements of DIN VDE 0276-622 and DIN VDE 0276-604 HM4 (>125%). However, typically wire and cable manufacturers require for an elongation at break of at least 150%. Increasing the loading results in a drop of 22%; The resulting value is already borderline for some applications (Figure 11.22).

11.4.3 Aging Performance

An important application of silane crosslinked wire and cable compounds is in building cables. Once installed these cables are very difficult to replace, and therefore the aging performance plays an important role. Improving the long-term

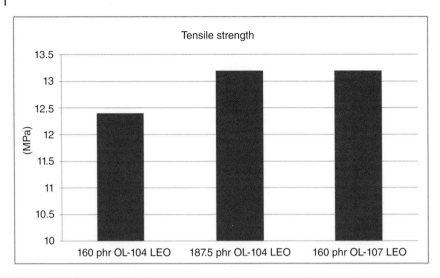

Figure 11.21 Impact of specific surface area and loading on tensile strength.

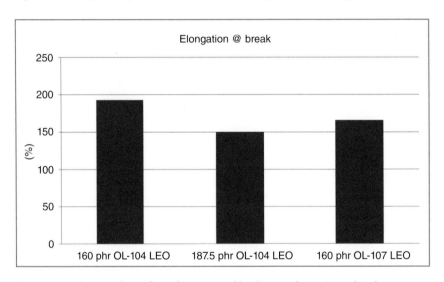

Figure 11.22 Impact of specific surface area and loading on elongation at break.

aging properties of any polyolefin compound typically requires for the addition of certain stabilizers such as hindered phenols (HALS) [16, 17].

Improved long-term aging performance was not targeted by any compound in this study. However, with some selected materials accelerated aging tests have been performed to find out if there is an impact by the ATH itself, the loading, or the specific surface area.

Mechanical properties were measured after exposing specimens for 7 days in air at 135 °C. Typical wire and cable standards allow for changes in the mechanical properties of less than or equal to 30%. As illustrated in Figure 11.23 neither the loading nor the specific surface area had a negative impact on the aging

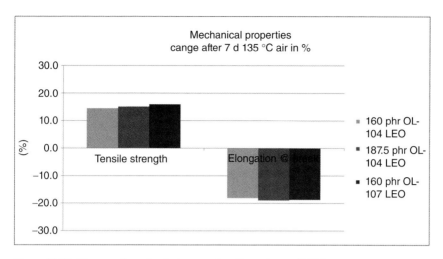

Figure 11.23 Change of mechanical properties after aging at 135 °C.

performance considering 160 phr ATH as the standard. All compounds met the typical requirements on heat aging with low addition levels of antioxidants (0,75 phr HALS) that have been added as processing stabilizers only.

11.4.4 Fire Performance on Laboratory Scale

The Cone Calorimeter according to ISO 5660 has been established as a reliable and robust method for the measurement of the flame retardant properties of plastic compounds. Besides this, it measures the heat release rate (HRR) under specified conditions. Oxygen depletion calorimetry has been identified as the most suitable method to calculate the HRR [18]. The data measured in a cone calorimeter is useful for selecting materials with promising flame retardant performance for the more sophisticated full scale tests like the Fire Performance of Electrical Cables (FiPEC) test according to EN 50399, which is required by the construction products regulation (CPR). In the cone calorimeter specimens of 100 mm × 100 mm × 3 mm are horizontally exposed to the conical heater (Figure 11.24). Its radiation heat is responsible for the degradation and decomposition of the combustible material. The combustion gases are ignited by an electrical sparkler. Typical heat flux for measuring HFFR compounds is 35 kW m^{-2}.

Thermoplastic compounds start to melt followed by their decomposition as soon as they are exposed to the heat; crosslinked compounds do not melt and typically decompose at a later stage or higher temperature. Finally, the burnable decomposition products are ignited by an electrical sparkler. The time between the start of the test and the ignition of the specimen, the time to ignition (TTI) and the HRR already give good indications about the flame retardant performance of a compound. Smoke generated by the combustion process is measured by a light source installed in the ventilation system. Legislators around the world tend to require for lower smoke density. In case of a fire, less smoke makes it easier to identify emergency exits and to escape.

Figure 11.24 Burning chamber of a cone calorimeter [18].

The graphs in Figures 11.25 and 11.26 compare the flame retardant performance of crosslinked compounds described in Figure 11.2. By increasing the loading of Martinal® OL-104 LEO from 160 to 187.5 phr the TTI is delayed significantly. The higher loading also reduces the peak heat release rate (PHRR) by approximately 10%.

At 160 phr loading compounds with Martinal® OL-104 LEO and OL-107 LEO ignite after a comparable period of time (TTI). However, the higher specific surface area is responsible for a lower slope of the HRR curve and the delay of the maximum HRR. This indicates that the specimen burns less intensively and leads to lower flame spread and improved flame retardancy.

11.4.5 Results of the Non-Polar Compounds

All compounds with a non-polar polymeric base contained 175 phr of Martinal® OL-104 LEO according to Figure 11.2.

As mentioned in Section 11.2.1, non-polar polymers tend to have lower reactivity and thus require an increased amount of silane to achieve the necessary crosslinking density. The impact of decreasing the silane content on the viscosity and the hot-set of the compound (MFI) is less pronounced than with the polar EVA-based materials. However, the tendency is similar, that is, less coupling agent results in higher melt flow and higher (worse) hot-set values. Figure 11.27 illustrates the response of the compound when the silane addition was reduced stepwise from 1.7 to 1.36 phr (formulations according to Figure 11.2).

11.4 Evaluation and Results on the Compound

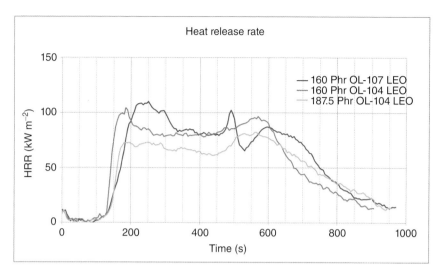

Figure 11.25 Heat release rate (HRR) over the time of cross-linked compounds according to Figure 11.2.

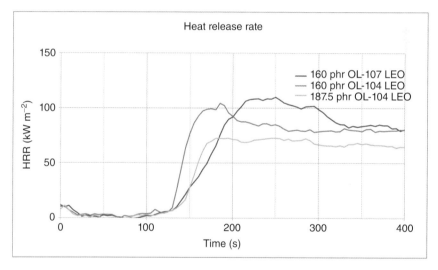

Figure 11.26 Heat release rate (HRR) measurement according to Figure 11.2.

Increasing the addition of the silane mixture above a certain limit results in lower tensile strength and elongation at break. With 80% silane the mechanical properties were fine but the hot-set (Figure 11.27) was getting close to the limit of 100%. As with the polar system, 90% silane mixture (1.53 phr) turned out to be the optimum with balanced mechanical properties. This compound achieved the highest (best) tensile strength at sufficient elongation at break (see Figure 11.28).

Despite the higher loading with ATH the non-polar system results in a slightly higher PHRR than comparable polar compounds based on EVA. These compounds also show a more pronounced second peak after aproximately 15 min.

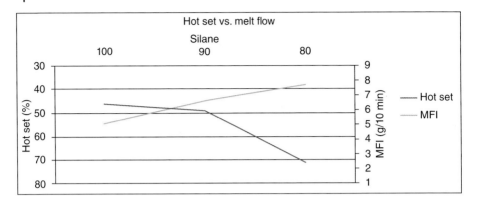

Figure 11.27 Impact of silane addition level on hot-set and melt flow of a non-polar compound.

Silane mixture according to Figure 2	Tensile strenth (MPa)	Elongation @ break (%)
1.7	13.8	256
1.53	14.3	223
1.36	12.5	226

Figure 11.28 Mechanical properties of crosslinked non-polar compounds.

Figure 11.29 Char of the polar compound.

This indicates the formation of cracks in the generated crust and thus a release of combusted material from the inside of the test specimen (Figures 11.29 and 11.30). Figure 11.31 compares the FR performance of a polar and a non-polar system according to Figure 11.2. Even at slightly increased loading the non-polar formulation does not meet the performance of the polar compound.

Figure 11.30 Char of the non-polar compound.

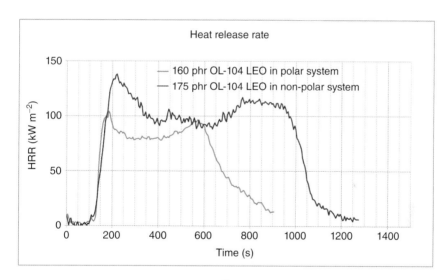

Figure 11.31 Cone performance of Martinal® OL-104 LEO filled cross-linked compounds in polar and non-polar polymer systems.

The non-polar compound shows a higher PHRR and an overall higher energy release.

As described in Section 11.4.4 lower PHRR indicates improved flame retardant properties. Since the polar EVA compounds show better FR performance than the non-polar compounds they have been chosen for cable extrusion and further burning tests only.

11.5 Cable Trials

The typical construction of building and construction cables consist of insulated twisted copper wires, a highly filled bedding compound and a sheathing. Bedding compounds are designed to fill the gaps between the single wires and are usually filled with high amounts (up to 80%) of cheap fillers like $CaCO_3$. Bedding compounds based on ATH can be used to boost the flame retardant properties of the finished cable.

To measure the fire performance of the sheathing compounds, a relatively simple cable design was chosen. Single copper wires of 1.5 mm^2 were insulated with non-flame retarded crosslinked PE. In a second extrusion step three of these wires were covered with the sheathing materials described in Figure 11.2. The final cable construction did not utilize a bedding compound.

The final extrusion step was done on a 20 L/D single-screw extruder equipped with a standard crosshead (Figure 11.32). The barrel temperature was set to 190–210 °C (from feeding port toward the die). At the highest possible line speed the maximum melt temperature of 230 °C was observed to have no negative impact on the cable's properties or surface. 2.5% of a masterbatch made of 5% di-octyl-tin-laurate (DOTL) and 95% Escorene Ultra UL00728 was added to the compounds prior to extrusion. Finally, the cables were crosslinked at 100 °C in a steam chamber for 4 h (Figure 11.33).

11.5.1 Fire Performance of Electrical Cables According to EN 50399

Any construction product sold in the European Union has to be CE marked. As part of the CE approval the flame retardant properties have to be measured. Wire and cable materials are tested according to EB 50399.

Figure 11.32 Standard crosshead [20].

Figure 11.33 Final cable construction [21]. (Reproduced with permission of Buss AG.)

Cable diameter	Mounting
≥ 20 mm	20 mm gap between the cables
5–20 mm	1 diameter gap between the cables
≤ 5 mm	Bundles of approximately 10 mm, not twisted with a gap of 10 mm

Figure 11.34 Mounting of the cables in the burning chamber.

For the test according to EN 50399, cables are fixed with a certain gap and steel wires on each rung of a steel ladder. The gap between the cables depends on their outer diameter and is listed in Figure 11.34. Figure 11.35 shows the setup with the cables. Prior to the test, airflow of 8000 l min^{-1} is applied. The gas burner installed on the lower end of the chamber has a power output of 20.5 kW for classes B2, C, and D. Class B1 requires a 30 kW flame and a fibre re-enforced concrete board installed behind the cables (Figure 11.34). The board reflects the heat and makes the test more severe. Due to the gap between the single cables almost the entire surface is exposed to the attacking flame of the burner.

After conditioning the chamber for 5 min the burner is ignited and the cables are exposed to the flame for 20 min (end of the test). Sheathing material decomposes and starts burning. Besides HRR the smoke density is indicated via the reduced transmission of a calibrated light source. Conductivity of the smoke can be measured to rate the acidity of the smoke but was not part of this investigation. Finally, burning or non-burning droplets are rated visually.

Besides the burning characteristics the damaged length of the cables is measured and rated. Figures 11.36 and 11.37 give an overview about the requirements of the different Euroclasses.

11.5.2 Burning Test on Experimental Cables According to EN 50399

The burning test compared cross-linked sheathing compounds with different loading as described in Figure 11.2: 160 phr and 187.5 phr. The test was performed according to EN 50399 as described earlier.

Figure 11.35 Cables mounted in the burning chamber according to EN 50399 [21]. (Reproduced with permission of Buss AG.)

Class	Method	Criteria	Additional criteria
Aca	EN ISO 1716 (Bomb calorimeter)	PCS ≤ 2.0 MJ kg^{-1}	
B1ca	EN 50399 (30 kW ignition source)	FS ≤ 1.75 m and THR1200s ≤ 10 MJ and PHRR ≤ 20 kW and FIGRA ≤ 120 W s^{-1}	s1 (s1a or s1b), s2 or s3 d0, d1, or d3 a1, a2, a3, or no declaration
	EN 60332-1-2	H ≤ 425 mm	
B2ca	EN 50399 (20.5 kW ignition source)	FS ≤ 1.5 m and THR1200 s ≤ 15 MJ and PHRR ≤ 30 kW and FIGRA ≤ 120 W s^{-1}	s1 (s1a or s1b), s2 or s3 d0, d1, or d3 a1, a2, a3, or no declaration
	EN 60332-1-2	H ≤ 425 mm	
Cca	EN 50399 (20.5 kW ignition source)	FS ≤ 2,0 m and THR1200s ≤ 30 MJ and PHRR ≤ 60 kW and FIGRA ≤ 300 W s^{-1}	s1 (s1a or s1b), s2 or s3 d0, d1, or d3 a1, a2, a3, or no declaration
	EN 60332-1-2	H ≤ 425 mm	
Dca	EN 50399 (20.5 kW ignition source)	THR1200s ≤ 70 MJ and PHRR ≤ 400 kW and FIGRA ≤ 1300 W s^{-1}	s1 (s1a or s1b), s2 or s3 d0, d1, or d3 a1, a2, a3, or no declaration
	EN 60332-1-2	H ≤ 425 mm	
Eca	EN 60332-1-2	H ≤ 425 mm	
Fca	No performance determined		

Figure 11.36 Classification of cables according to the construction products regulation.

Smoke	s1 = TSP ≤ 50 m² and PSPR ≤ 0.25 m² s⁻¹	
	s1a = s1 and transmittance* ≥ 80%	
	s1 b = s1 and transmittance* ≥ 60% < 80%	
	s2 = TSP 1200 ≤ 400 and PSPR ≤ 1.5 m² s⁻¹	
	s3 = not s1 or s2	
Droplets	d0 = no flaming droplets within 20 min	
	d1 = no flaming droplets persisting longer than 10 s within 20 min	
	d2 = not d0 or d1	
Acidity	a1 = conductivity < 2,5 μS mm⁻¹ and pH > 4.3	
	a2 = conductivity <10 μm mm⁻¹ and pH > 4.3	
	a3 = not a1 or a2	
	No declaration = no performance determined	

Figure 11.37 Additional criteria for the classification of cables.

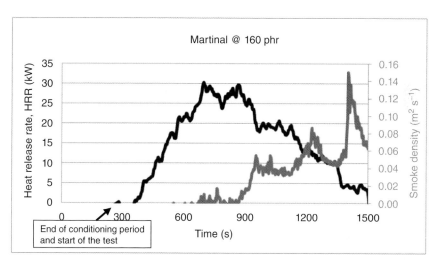

Figure 11.38 Burning test according to EN 50399; sheathing material flame retarded with 160 phr Martinal® OL-104 LEO (formulation according to Figure 11.2).

After a conditioning phase of 5 min the burner is mounted into the chamber. At 160 phr loading the cables start burning 1 min after the flame is applied. Five minutes later the maximum height of the flames is reached which is also the point in time of the PHRR. For the next 5 min the cables burn at an almost constant rate before the heat release drops. Just 3 min before the test ends the open flames disappear. However, some heat release can still be measured (Figure 11.38). This is due to some limited smoldering of the remaining sheathing material. As the sheathing was cross-linked, no dripping was observed but burning of some particles slackened. The pieces did not burn longer than 10 s so the cables were classified as non-dripping (class d1).

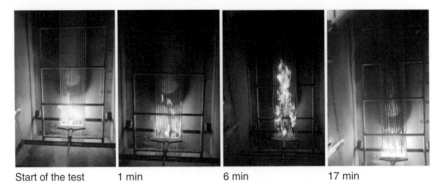

Start of the test   1 min   6 min   17 min

Figure 11.39 Burning test according to EN 50399; sheathing material flame retarded with 160 phr Martinal® OL-104 LEO (formulation according to Figure 11.2) [21]. (Reproduced with permission of Buss AG.)

	160 phr Martinal	187.5 phr Martinal	Requirements	
Flame spread (mm)	1640	1040	Max. 1500 mm	class B
			Max 2000 mm	class C
THR 1200 s (MJ)	18.5	10.2	Max 15 MJ	class B
			Max 30 MJ	class C
PHRR (kW)	30.3	17.7	Max 30 kW	class B
			Max 60 kW	class C
FIGRA (W/s)	77.9	50.3	Max 120 W s^{-1}	class B
			Max 300 W s^{-1}	class C

Figure 11.40 Comparison of the key properties measured according to EN 50399.

Over the 20 min test almost no smoke was detected by the naked eye and the cables in the chamber were visible all the time during the test (Figures 11.38 and 11.39).

After the test, the flame spread is rated by measuring the length of the sheathing damaged by the flames. At 160 phr loading, the flame spread was measured at 1640 mm which is far below the limit required for Euroclass C but slightly above the limit of Euroclass B2. The total HRR over 20 min and the PHRR also exceed the limits of Euroclass B2 but are well within the requirements of Euroclass C. The fire growth rate of 77.9 W s^{-1} is within the limits of class B2. Due to the fact that some values exceeded the limits of Euroclass B2 the cable has been rated Euroclass C_{ca}s1d1 (Figure 11.40).

Increasing the loading of ATH to 187.5 phr delayed the TTI by 2 min. The PHRR is significantly lower (40%) and was observed at a later stage of the test. Finally, the burning stops 15 min after the start of the test while the impact on smoke is again negligible. Similar to the first test at lower filler loading, some particles

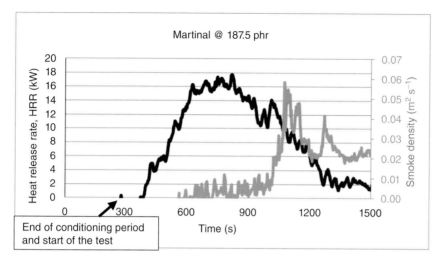

Figure 11.41 Burning test according to EN 50399; sheathing material flame retarded with 187.5 phr Martinal® OL-104 LEO (formulation according to Figure 11.2).

Figure 11.42 Burning test according to EN 50399; sheathing material flame retarded with 187.5 phr Martinal® OL-104 LEO [21]. (Reproduced with permission of Buss AG.)

slackened. With higher loading these particles were bigger and burned longer than 10 s (see Figures 11.41 and 11.42)

The damaged lengths of the cables were measured at 1040 mm and was well within the limits of Euroclass B2 (Figures 11.41 and 11.42). The total HRR over 20 min and the PHRR were approximately 40% lower compared to 160 phr loading; the FIGRA was 35% lower.

As expected the fire growth rate improved with higher loading. The cable would be rated Euroclass B_{ca}s1d2 (Figure 11.43).

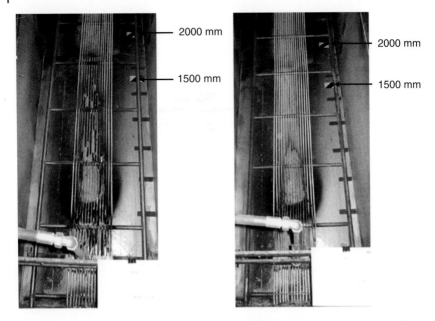

Figure 11.43 Flame spread at different loading [21]. (Reproduced with permission of Buss AG.)

11.6 Conclusions

The increased thermal stability of the Martinal® LEO grades, an improved compound viscosity in combination with the very short residence time, and narrow residence time distribution in the Ko-kneader enable end users to run the compounding process at its limit. It also helps to safely achieve the necessary temperature for the grafting reactions.

The reactive compounding in only one step requires for a 50% longer than usual compounding line. This is necessary to separate the chemical reaction of the silane and the polymers from the filling step. Due to the absence of a second melting and cooling steps the one-step process has a lower impact on the heat history of the polymers and helps to reduce energy costs.

The mechanical performance as well as the flame retardant properties can be adjusted by selecting ATH grades with different specific surface area. The specific surface area does not have any impact on the crosslinking density. Cone calorimetry clearly showed improved FR performance for compounds based on polar polymers. The higher specific surface area of Martinal OL-107 LEO versus OL-104 LEO slightly improves the flame retardant properties of the compound in the cone calorimeter and is a potential solution to meet Euroclass B2 at the traditional loading of 60 wt%.

For achieving Euroclass B2 with the cable design tested in this investigation the traditional loading of 60 wt% of Martinal OL-104 LEO was not sufficient. Increasing the loading to 65 wt% helps to clearly meet the requirements of Euroclass B2.

References

1 CRU wire & cable quarterly industry and market outlook, October 2015.
2 Saechtling Kunststoff Taschenbuch, 10/2013, ISBN: 978-3-446-43729-6, p. 443 ff.
3 Woods Robert J., Alexei K. Pikaev *Applied Radiation Chemistry: Radiation Processing*, p. 368.
4 Niklaus, A. (2012), *High Performance Cable Compounds – New Collaborative Results, Cables*.
5 Techncial data Evonic Industries.
6 Stefan Ultsch & Hans-Gerhard Fritz (1999) Crosslinking of LLDPE and VLDPE via graft polymerized vinyltrimethoxysilane. *Plast. Rubber Process. Appl.*, **13**, 81–91.
7 Jalil Morshedian and Pegah Mohammad Hoseinpour, *Polyethylene Crosslinking by Two-Step Silane Method: A Review* Iran Polymer and Petrochemical Institute Tehran.
8 Herbiet R. (2008) New ATH grades and the advantages for cable producers, AMI cables.
9 Rauwendaal, C. (1994) *Mixing and Compounding of Polymers*, Hanser, New York, pp. 735–760.
10 Jakopin, S. and Franz, P. (1983) *Flow Behaviour in Continuous Kneader and Its Effect on Mixing*, AICHE Diamond Jubilee, Washington, DC.
11 Gruetter, H., Trachsel, R., and Siegenthaler, H.-U. (2007) New generation of buss kneader for cable compounds. *Kunstst. Int.*, **9**, 207.
12 Siegenthaler, H.-U., Trachsel, R., and Nägele, S. (2012) 10 Jahre Vierflügel-Technologie. Teil 1: Durchsatz und Compoundqualität gesteigert, Produktionskosten gesenkt, Extrusion 6/2012, pp. 26–29; Teil 2: Innovative Kneterbauweise mit großem Zukunftspotential, Extrusion 7/2012, pp. 50–55.
13 Patent No. EP 1 185 958 A1 *Mixing and Kneading Machine*. Applicant: Buss Ltd, Pratteln, Switzerland. Inventor: H.-U. Siegenthaler.
14 Patent No. EP 2 018 946 A3: *Mixing and Kneading Machine and Method of Implementing Continual Compounding*. Applicant: Buss Ltd, Pratteln, Switzerland. Inventor: Heini Grütter, H.-U. Siegenthaler.
15 Picture courtesy of Buss AG.
16 Jacoby, P. The effect of hindered phenol stabilizers on Oxygen Induction Time (OIT) measurements and the use of OIT Measurements to predict long term thermal stability.
17 Erik Kleina, Vladimír Lukešb, Zuzana Cibulkováa, On the energetics of phenol antioxidants activity, *Pertoleum Coal*, (www.vurup.sk/pc).
18 FTT Fire Testing Technology, http://www.fire-testing.com/cone-calorimeter-dual.
19 http://www.fire-testing.com/cone-calorimeter-dual, Vytenis Babrauskas, Ten Years of Heat Release Research with the Cone Calorimeter March 2013.
20 Picture courtesy of KBE Elektrotechnik GmbH.

21 Picture courtesy of Martinswerk GmbH.
22 DIN EN 50399 (VDE 0482-399) Deutsches Institut für Normung e.V. und VDE Verband der Elektrotechnik Elektronik Informationstechnik e.V., Beuth Verlag GmbH.
23 S M Tamboli, S T Mhaske, D D Kale Cross-linked polyethylen.
24 René Herbiet, Martinal LEO: New aluminium hydroxides with improved performance, RFP 4/2008 – Volume 3.

12

Thermoplastic Vulcanizates (TPVs) by the Dynamic Vulcanization of Miscible or Highly Compatible Plastic/Rubber Blends

Yongjin Li and Yanchun Tang

Department of Polymer Science, College of Material, Chemistry and Chemical Engineering Hangzhou Normal University, No. 16 Xuelin Road, Hangzhou 310036, P.R. China

12.1 Introduction

A thermoplastic elastomer (TPE) is defined as a polymeric material with the elastomeric characteristics of vulcanized rubber and the processing properties of thermoplastic polymer [1]. TPEs are two-phase materials constituting elastomers that possess elastic and thermoplastic properties, which enable rubbery materials to be processed. TPEs have been widely used in the automotive sector, and in the manufacture of medical devices, mobile electronics, and household appliances and they are ideally suited for high-throughput thermoplastic techniques, such as extrusion, blow molding, injection molding, vacuum forming, and calendaring [2, 3].

Generally, the family of TPEs is categorized into two classes according to its chemistry and morphology: intrinsic copolymer TPEs and reactive blending TPEs [4]. The first group constitutes copolymers containing soft (low T_g) and thermoplastic segments. The soft blocks form a continuous crosslinking matrix that exhibits elastic behavior. The thermoplastic blocks turn to dispersed rigid nanoscale domains that form a three-dimensional network of physical crosslinks. The physical crosslinks do not exist permanently; they disappear at high temperature, providing the intrinsic copolymer TPEs thermal reversibility and processability (Figure 12.1) [5].

In the second group, the reactive blending TPEs, the thermoplastic vulcanizates (TPVs), play an important role because they form the *second largest class of TPEs after the styrenic-based block copolymers*. They are produced via dynamic vulcanization of blends that consist of a thermoplastic polymer and a relatively large volume fraction of elastomer ($V = 0.4$–0.9) [3]. The dynamic vulcanization procedure is carried out above the melting temperature of the thermoplastic component under high shear. At a sufficiently high temperature, the elastomeric component mixing with molten thermoplastics can be selectively vulcanized by the addition of curing agents. The viscosity of the elastomeric phase increases abruptly and tends to be encapsulated by the less viscous thermoplastics. Meanwhile, the applied high shear field prohibits the crosslinked rubber phase

Reactive Extrusion: Principles and Applications, First Edition. Edited by Günter Beyer and Christian Hopmann.
© 2018 Wiley-VCH Verlag GmbH & Co. KGaA. Published 2018 by Wiley-VCH Verlag GmbH & Co. KGaA.

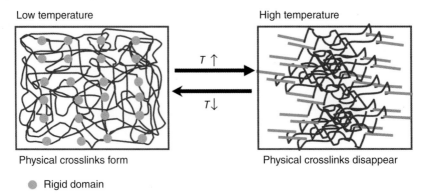

Figure 12.1 Schematic of thermal reversibility of copolymer TPEs.

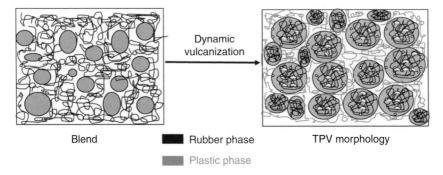

Figure 12.2 Schematic illustration of the transformation from blend to TPV in morphology.

coalescence of the thermoplastic polymer and the elastomer and breaks the TPEs down into microscale particles. Therefore, the phase inversion from a plastic phase dispersed structure into a typical TPV structure takes place, which is characterized by the presence of finely dispersed crosslinked rubber particles distributed in a continuous thermoplastic matrix (Figure 12.2) [6–9]. In systems where the elastomer is already in the dispersed state in the continuous thermoplastic phase, that is, the concentration of elastomeric component is relatively lower than the thermoplastic, no phase inversion take place; but, the rubber particles are more finely dispersed with dynamic crosslinking. Phase inversion happens only when the ratio of elastomer/thermoplastic during the dynamic vulcanization process is high [10].

Contrary to the low-yield strain nature of the thermoplastic matrix, the produced TPVs exhibit elastic properties upon tension or compression at moderate temperature. The deformation mechanism of the TPVs has been investigated intensively but it is still subject to controversies. Soliman *et al.* demonstrated that during the stretching process the matrix ligament (i.e., the thermoplastic inter-layer) between the deformed rubber particles acts as glue, and only a small fraction of it is irreversibly yielded. In the following recovery process, the matrix ligament can be pulled back again (partially) by the recovery of the rubber phase (Figure 12.3) [11]. Following this study, AbdouSabet gave another explanation for the deformation mechanism by using ethylene-propylene-diene

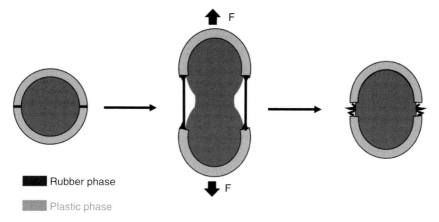

Figure 12.3 Sketch illustrating the deformation and recovery of dynamically vulcanized blends.

tripolymers/polypropylene (EPDM/PP) TPV as example. They claimed that the PP phase remains amorphous and deformable, because the limited region of the PP interstices between the rubber particles inhibits the crystallization [12]. Subsequently, Datta et al. studied the stress–strain behavior of phenolic resin-cured PP/EPDM TPVs by the application of an experimental design [13]. Boyce et al. used a series of micromechanical models to explain the micromechanisms of deformation and recovery in the TPVs [14].

Compared to simple blends without dynamic vulcanization, the properties of TPVs such as mechanical, permanent set, and elastic recovery, and resistance to the effect by oil and other organic solvents, are remarkably improved, while the melt processability is still preserved. Surveys indicate that currently TPVs constitute the fastest growing TPE market with a global annual growth rate of about 15% [15]. In this chapter, the recent developments of TPVs from immiscible blends and, especially, miscible blends will be reviewed.

12.2 Morphological Development of TPVs from Immiscible Polymer Blends

TPVs have at least two phases: the thermoplastic matrix providing the processability and the crosslinked rubber providing the elasticity. Therefore, TPVs are generally fabricated from immiscible plastic/rubber blends. Traditionally, the compositions of TPVs containing a very broad range of possible combinations of appropriate types of elastomers and thermoplastics were immiscible with each other. The rubbers included butyl rubber (IIR), EPDM, natural rubber (NR), butadiene rubber (BR), styrene-butadiene-rubber (SBR), ethylene vinyl acetate (EVA), acrylic rubber (ACM), chlorinated polyethylene (CM), polychloroprene (CR), and nitrile rubber (NBR). The thermoplastics included PP, polyethylene (PE), polystyrene (PS), acrylonitrile butadiene styrene (ABS), styrene acrylonitrile (SAN), polymethyl methacrylate (PMMA), polybutylene terephthalate

(PBT), polyamide (PA), and polycarbonate (PC) [16]. Only a few of them were commercialized, because the poor compatibility of most blends required one or more steps to make them compatible [17]. To improve the mechanical property and ensure large-scale fabrication of TVPs, the miscibility of the plastic and the rubber with relevance to dynamic vulcanization is a critical factor.

The most common and important industrial-scale TPVs are based on PP and EPDM. PP is chosen because of its high melting point (~165 °C) compared to PE and PS as well as its high crystallinity, resulting in good TPV properties even at elevated temperatures. EPDM with saturated main chains possesses stability against high temperature, oxygen and ozone resistance. The service temperature is range from −50 °C to 150 °C. The fraction of non-conjugated diene of commercial EPDM is 0 – 10 wt%. The significantly less oxidant attach points (unsaturation sites) provides EPDM inherently ozone resistance. Therefore, TPVs consist of EPDM also can perform good heat oxidation and ozone resistance. Moreover, the blend of EPDM and PP is characterized by a relatively low interfacial tension of about $0.3\,\mathrm{mN\,m^{-1}}$ [2].

Intensive studies were carried out on the interrelation between the morphology and the physical properties of PP/EPDM TPVs. It was deduced that the mechanical properties of the TPVs are primarily governed by (i) the particle size of rubber domains and (ii) the interfacial adhesion between the thermoplastic and rubber phases. Therefore, based on the compatibility of pure PP and EPDM, PP/EPDM TPV can be characterized by the small rubber particles of around 0.5–5 μm in the continuous PP matrix without the addition of any compatibilizer [18]. Radusch and Pham studied the morphology of PP/EPDM TPV with phenolic resin and concluded that only a fine dispersion of the relatively high amount of rubber into small particles (<0.5 μm) surrounded by the thermoplastic matrix guarantees an elastomer-typical and a high mechanical property level [19]. Asaletha *et al.* compared the mechanical properties of the TPVs based on NR and PP with different crosslinking agents. Compared to the TPVs prepared with sulfur, a higher value of tensile strength in peroxide-cured TPV was observed, which is attributed to the finer particle size of the dispersed rubber phase and higher value of crosslink density [10]. AbdouSabet *et al.* found that properties such as compression set and oil resistance as well as the processing characteristics of PP/EPDM TPVs could be improved by using a dimethyloloctyl phenol curing resin, because a small amount of *in situ* formation of PP/EPDM graft copolymer produced by the coupling of radicals increases the high interfacial activity of crosslinked blends [20].

Processing oil, in most cases paraffinic oil, is a common additive that is widely employed in PP/EPDM TPVs. Since the polarity difference between oil and PP/EPDM is almost the same, the oil is present in both PP and EPDM. The presence of a substantial amount of extender oil lowers the hardness and increases the melt processability. It is considered that oil also plays a role in rubber-like properties of PP/EPDM TPVs, because it can lead to a decrease of the dimension of the PP spherulites and thereby promote yielding in the PP matrix.

12.3 TPVs from Miscible PVDF/ACM Blends

The dynamic vulcanization of an immiscible rubber/plastic system leads to a two-phase material in which crosslinked rubber particles are dispersed in the plastic matrix [21]. Such morphology generation is prerequisite for TPE, that is, the plastic component should be the matrix to provide the melt processability. A static vulcanization of the single-phase mixture would result in a three-dimensionally crosslinked material that cannot be processed by melt processing methods [11, 12]. In contrast, the dynamic vulcanization of a miscible rubber/plastic system is expected to induce phase decomposition because (i) at the early stage of vulcanization, a microgel (loosely crosslinked domain) is usually formed, (ii) the gel particles would hardly coalesce with each other under high shear forces (dynamic state), (iii) the increase in molecular weight and the crosslinking are unfavorable for the polymer–polymer miscibility so that the non-crosslinkable plastic chains in the microgel particles would be forced to get out, and (iv) the phase decomposition would proceed during the dynamic vulcanization, which results in the crosslinked particles and the non-crosslinked matrix. This morphology may be desirable for TPEs.

Li et al. first used miscible thermoplastic/rubber system – poly(vinylidene fluoride) (PVDF)/acrylic rubber (ACM) system – to prepare TPVs. The hexamethylene diamine carbamate (HMDC) was chosen as the curative. They introduced a new concept in the preparation of TPVs based on the dynamic vulcanization, in which phase decomposition was induced in miscible blends of a semi-crystalline thermoplastic and an elastomer. Because both PVDF and ACM are highly polar and show high oil resistance, the TPVs from PVDF/ACM exhibit super oil resistance.

The CF_2 groups of PVDF have special interaction with the carboxyl groups of the ACM. This results in good compatibility between PVDF and ACM. The dynamic mechanical analysis (DMA) results (Figure 12.4) show that only one T_g (12.5 °C) of the PVDF/ACM blends with PVDF less than 60 wt%, indicating the miscible state between ACM and the amorphous part of PVDF. With increasing concentration of PVDF, two T_gs appeared, suggesting the two-phase structure of the blends. Such PVDF/ACM blends at the 50/50 component ratio provide the possible crosslinked-induced phase separation by dynamic vulcanization. [22].

The phase decomposition occurred in PVDF/ACM blends after the dynamic vulcanization processes. The produced PVDF/ACM TPVs possess a complex morphology, in which a small number of ACM nanodomains are dispersed

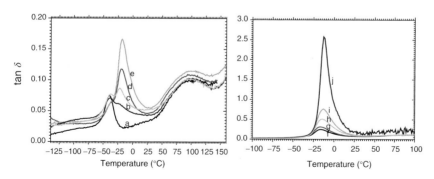

Figure 12.4 Tan δ versus temperature for the PVDF/ACM blends with various composition ratios: (a) neat PVDF, (b) PVDF/ACM 95/5, (c) PVDF/ACM 90/10, (d) PVDF/ACM 80/20, (e) PVDF/ACM 70/30, (f) PVDF/ACM 60/40, (g) PVDF/ACM 50/50, (h) PVDF/ACM 40/60, (i) PVDF/ACM 20/80, and (j) neat ACM.

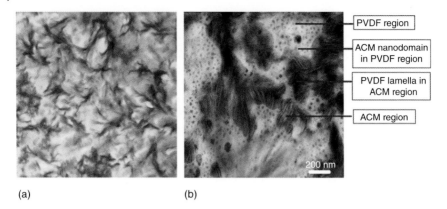

Figure 12.5 TEM images of the PVDF/ACM blends dynamically vulcanized with (a) 0 wt%, (b) 0.8 wt% curative (HMDC).

in the PVDF matrix and PVDF crystal lamella are located in the crosslinked ACM particles (Figure 12.5). The reasons for the phase decomposition and the special morphology were explained by selective crosslink reactions and partially preserved miscibility. On one hand, the elastomer precursor ACM phase was selectively crosslinked into rubber particles. They hardly coalesce with each other, since the vulcanization takes place under high shear force (dynamic state). The crosslinking disturbed the interaction between the CF_2 dipole of PVDF and the carbonyl group of ACM, as well as led to the significant increased molecular weight of ACM, which decreased the entropy upon mixing. The unfavorable PVDF–ACM miscibility caused the non-crosslinkable PVDF chains to be excluded from the crosslinked ACM phase, resulting in phase decomposition. On the other hand, miscibility was partially preserved after the dynamic vulcanization in the molten state. During the subsequent cooling to room temperature, PVDF chains crystallized and segregated out as crystal lamella in ACM particles, and a small amount of ACM was expelled during the crystallization of PVDF as the nanodomains in the PVDF matrix (Figure 12.6).

This special morphology provided PVDF/ACM TPVs excellent processability. Because of the PVDF-rich matrix, these TPVs could be shaped into dumbbell-shaped samples by thermoplastic injection molding and reprocessed for many times: the molded sheet was ground and then melt-pressed to sheet (Figure 12.7).

Figure 12.8 shows the physical properties of PVDF/ACM TPVs with different curative content. The TPV with highest amount of curative (0.8 wt%) displayed highest tensile strength and elongation as well as lowest residual strain. The authors also noted that increasing the curative does lead to a higher crosslink density, which results not only in better mechanical properties but also in poorer processability of the TVPs.

Li *et al.* claimed that the miscibility of PVDF and ACM can be the main reason for the excellent physical properties displayed by PVDF/ACM TPVs. After phase

12.3 TPVs from Miscible PVDF/ACM Blends

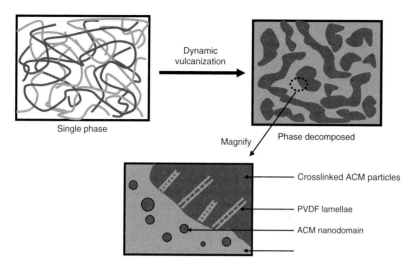

Figure 12.6 Schematic diagram of dynamic vulcanization induced phase decomposition and the phase structure for the TPVs (dark gray part indicates ACM and light gray part indicates PVDF).

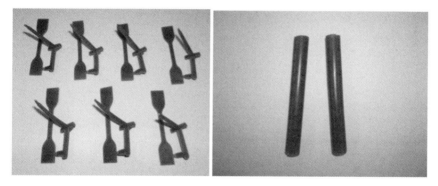

Figure 12.7 Photograph of the injection molded dumbbell-shaped and tube-shaped TPV samples.

separation, a small amount of one polymer even partially exist in the other phase with nanoscale structure because of the miscibility. This partial miscibility provided by the high surface area largely increased the tensile strength as well as the elongation at break of the TVPs. Moreover, the small amount of ACM nanoparticles in the matrix decreased the regularity of PVDF spherulites. It contributed to a significant improvement in the residual strain of TVPs, which was much smaller than that of the commercialized PP/EPDM TPE with extension oil [23]. The novel TVPs also exhibit excellent heat and oil resistance, which was confirmed by measuring the retention of physical properties after 2 weeks exposure to 140 °C air or toluene swelling.

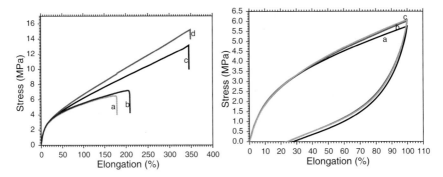

Figure 12.8 Strain–stress curves and strain recovery curves for the PVDF/ACM TPVs dynamically vulcanized with (a) 0 wt%, (b) 0.2 wt%, (c) 0.4 wt%, and (d) 0.8 wt% curative (HMDC).

12.4 TPVs from Highly Compatible EVA/EVM Blends

Crosslink-induced phase separation paves a new avenue for the preparation of TPVs [24]. EVA copolymer is a kind of thermoplastic with different percentages of vinyl acetate (VA); a certain amount of polar groups included in it can offer superior oil resistance property. EVAs with different VA content have similar chemical structure, as the VA contents affect the properties of the copolymer significantly. Therefore, the blends of EVAs show different compatibilities that are dependent upon VA contents in the EVA components. In order to select a good EVA matrix to prepare a blend with EVA rubber (VA content = 50 wt%) (EVM) for dynamic vulcanization, three types of EVAs with different VA contents (VA content = 9, 28, 40 wt% termed as *EVA9*, *EVA28*, and *EVA40*, respectively) have been blended with EVM (50/50) without a curing agent initially (Figure 12.9). The VA content in EVA affects the miscibility of EVA/EVM significantly. The EVM/EVA9 blend (Figure 12.9a) shows a very coarse and sea-island structure where the EVM is dispersed in an EVA matrix with the domain size of 10–20 μm. The EVM/EVA28 blend (Figure 12.9b) has a much finer co-continuous morphology with a characteristic length scale in the order of 1–3 μm. On the other hand, the EVM/EVA40 blend (Figure 12.9c) shows a homogeneous phase morphology, which indicates that EVM and EVA40 are thermodynamically miscible and form a one-phase mixture. In Figure 12.10 obtained by DMA, the T_g of neat EVM is at −30 °C and that of neat EVA 28 is at −15 °C, while the EVM/EVA28 (50/50) blend shows only one relaxation peak at about −18 °C with a broad relaxation peak. From SEM measurements that show the phase separated structure for the EVM/EVA28 blends, the wider single relaxation peak can be attributed to the overlapping of the molecular relaxation of the components. The intermediate relaxation peak temperature of the blend originates from the shifting of the relaxation peaks of each of the components due to the compatibility. The DMA investigation again means that EVM and EVA28 are phase separated, but are highly compatible.

The dynamic vulcanization has been carried out for the all above mentioned three blends. It was found that the mechanical properties of the dynamic vulcanized EVM/EVA9 blend were unsatisfactory due to the poor compatibility between the components. On the other hand, the dynamic vulcanization of

12.4 TPVs from Highly Compatible EVA/EVM Blends | 339

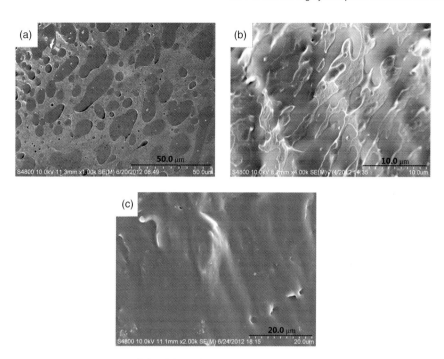

Figure 12.9 SEM images of blends based on (a) EVM/EVA9, (b) EVM/EVA28, and (c) EVM/EVA40 at the constant ratio of 50/50. The dark regions represent the EVM phase and the light regions the EVA phase.

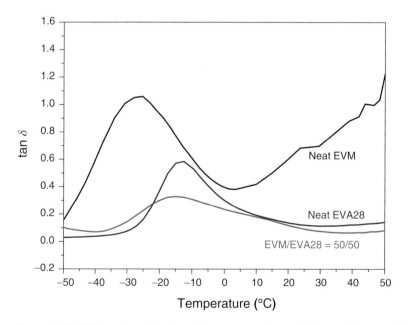

Figure 12.10 DMA–tanδ as a function of temperature for neat EVM, neat EVA28, and the EVM/EVA28 (50/50) blend.

EVM/EVA40 resulted in a crosslinked rubber with no processability. In other words, no crosslink-induced phase separation was observed for the EVM/EVA40 blends. EVM/EVA28 blend is not miscible, but highly compatible. This blend was appropriate for dynamic vulcanization [24].

The TPVs from EVM/EVA28 blends exhibit wonderful mechanical properties and excellent oil resistance by dynamic vulcanization with dicumyl peroxide (DCP) as curative [24]. Before that, EVA has been widely investigated as an elastomeric component in PE/EVA TPVs systems, because of its excellent aging, weather resistance and mechanical properties [25]. However, those TPVs unfortunately are far from the real application, because the rubber material EVA has poor compatibility with the other thermoplastic material PE, resulting in large size of dispersed EVA rubber particles.

This work cleverly employed EVM and EVA28 that had similar chemical structure that allowed the EVM with higher VA content take the character of an elastomer, that is, the lower VA content in EVA28 of the thermoplastic. The differences between the VA content in EVM and EVA28 enable the EVM phase to selectively crosslink in the EVM/EVA28 blend during dynamic vulcanization with appropriate dose of curative and with appropriate processing time. The vulcanization speed of the EVM was much faster than that of EVA28 (Figure 12.11), since the EVM with more VA is easily crosslinked by the radical generated from DCP. By utilizing the time lag between the crosslinking processes of the two components (EVM and EVA28) in the EVM/EVA28 (50/50) blends, the EVM phase was selectively crosslinked and embedded in the EVA28 matrix as rubber particles. It was strongly improved by the drastic transition of the morphology before and after dynamic vulcanization (Figure 12.12). The unique

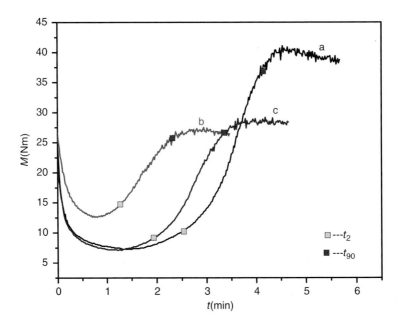

Figure 12.11 Vulcanization curves of (a) pure EVA28, (b) pure EVM, and (c) the EVM/EVA28 (50/50) blends containing 0.2 wt% DCP.

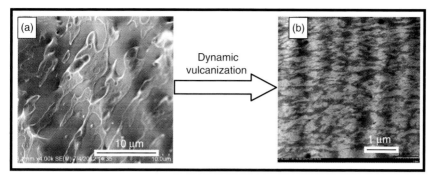

Figure 12.12 Morphology transition of the EVM/EVA28 (50/50, wt/wt) after dynamic vulcanization. Panel (a) is the SEM image of non-vulcanized blend and (b) is the TEM image of dynamically vulcanized blend with 0.1 wt% DCP.

nanostructures of the TPV materials originated from both the compatibility and the viscosity change during the dynamic vulcanization. The similarity in the chemical structure for EVM and EVA28 results in a very low interfacial tension between the EVA and EVM phase. It was claimed that the reaction between EVA28 and EVM would occur during the dynamic vulcanization. The formation of certain copolymers leads to the reduction of the interfacial tension, which could help achieve fine morphology of the TPVs.

Because of the nanoscale size of rubber particles and the co-continuous phase morphology, the deformation and retention behaviors upon the tensile deformation of EVM/EVA28 blend had significant improvement after dynamic vulcanization. The TPV with 0.1 wt% DCP shows an elongation at break of about 920%, which is higher than the elongation at break of the commercialized PP/EPDM TPVs (Figure 12.13) [26]. The remnant strain was as low as 19% after 100% stretching for the EVM/EVA28 vulcanized sample with 0.1 wt% DCP, indicating the excellent elastic recovery of the TPV (Figure 12.14). Moreover, the physical properties of TPVs after an oil swelling test nearly did not change (Table 12.1).

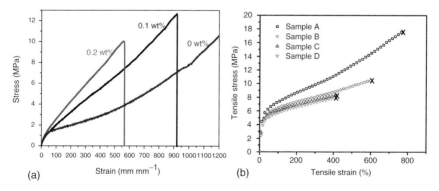

Figure 12.13 (a) Strain–stress curves for the EVM/EVA28 (50/50) blends with addition of the indicated DCP content. (b) Strain–stress curves of EPDM/PP (60/40) TPV samples A, B, C, D with same degree of resol and $SnCl_2$ with 3, 10, 15, 20 min dynamic vulcanization time.

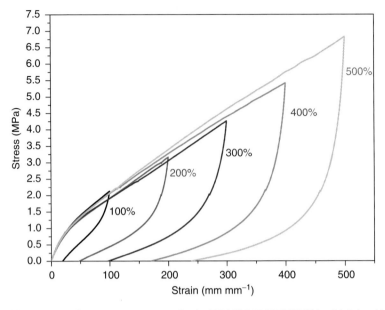

Figure 12.14 Strain recovery curves for the EVM/EVA28 (50/50) TPVs with 0.1 wt% DCP at the indicated stretching.

Table 12.1 Comparison of physical properties of the EVM/EVA28 (50/50, 0.1 wt% DCP) TPV before and after oil aging at room temperature for 24 h.

	Elongation at break	Tensile set at break	Mass	Volume	Thickness	Width	Length
Rate of change (%)	−44.15	−31.0	1.36	0.40	1.42	0.31	0.50

12.5 Conclusions and Future Remarks

TPVs are multi-phase materials. Therefore, the miscibility property of the components is critically important. Obviously, the dynamical vulcanization process induces a drastic change in the phase structure of the components. The miscible polymer blends can also be used for fabrication of high-performance TPVs because of the crosslink-induced phase separation during melt mixing. At the same time, highly compatibile blends, such as PP/EPDM, EVA/EVM, are accessible to prepare TPVs with good flexibility, ductility, and elasticity.

References

1 Coran, A.Y. and Patel, R.P. (1981) Rubber-thermoplastic compositions. Part IV. Thermoplastic vulcanizates from various rubber-plastic combinations. *Rubber Chem. Technol.*, **54** (4), 892–903.

2 van Duin, M. and Machado, A.V. (2005) EPDM-based thermoplastic vulcanisates: crosslinking chemistry and dynamic vulcanisation along the extruder axis. *Polym. Degrad. Stab.*, **90** (2), 340–345.
3 Coran, A.Y. and Patel, R.P. (1980) Rubber–thermoplastic compositions. Part 1: EPDM-polypropylene thermoplastic vulcanizates. *Rubber Chem. Technol.*, **53** (1), 141–150.
4 Shanks, R. and Kong, I. (2012) *Thermoplastic Elastomers*, In Tech, p. 416.
5 Spontak, R.J. and Patel, N.P. (2000) Thermoplastic elastomers: fundamentals and applications. *Curr. Opin. Colloid Interface Sci.*, **5** (5–6), 333–340.
6 Goharpey, F., Katbab, A.A., and Nazockdast, H. (2001) Mechanism of morphology development in dynamically cured EPDM/PP TPEs. I. Effects of state of cure. *J. Appl. Polym. Sci.*, **81** (10), 2531–2544.
7 Naskar, R.R.B.K. (2010) *Recent Developments on Thermoplastic Elastomers by Dynamic Vulcanization*, vol. 239, Springer, Berlin, Heidelberg.
8 Coran, A.Y. and Patel, R.P. (1983) Rubber-thermoplastic compositions. Part VIII. Nitrile rubber-polyolefin blends with technological compatibilization. *Rubber Chem. Technol.*, **56** (5), 1045–1060.
9 Coran, A.Y., Patel, R.P., and Williams-Headd, D. (1985) Rubber-thermoplastic compositions. Part IX. Blends of dissimilar rubbers and plastics with technological compatibilization. *Rubber Chem. Technol.*, **58** (5), 1014–1023.
10 Asaletha, R., Kumaran, M.G., and Thomas, S. (1999) Thermoplastic elastomers from blends of polystyrene and natural rubber: morphology and mechanical properties. *Eur. Polym. J.*, **35** (2), 253–271.
11 Soliman, M., Van Dijk, M., Van Es, M., and Shulmeister V. (1999) *SPE/ANTEC*. New York, p. 1947.
12 AbdouSabet, S. (2000) *ACS Rubber Division Meeting*. Cincinnati, OH, USA.
13 Datta, S. (2000) *ACS Rubber Division Meeting*. Dallas, TE, USA.
14 Boyce, M.C., Socrate, S., Kear, K., Yeh, O., and Shaw, K. (2001) Micromechanisms of deformation and recovery in thermoplastic vulcanizates. *J. Mech. Phys. Solids*, **49** (6), 1323–1342.
15 Datta, S., Naskar, K., Jelenic, J., and Noordermeer, J.W.M. (2005) Dynamically vulcanized PP/EPDM blends by multifunctional peroxides: characterization with various analytical techniques. *J. Appl. Polym. Sci.*, **98** (3), 1393–1403.
16 Coran, A.Y. (1995) Vulcanization – conventional and dynamic. *Rubber Chem. Technol.*, **68** (3), 351–375.
17 Naskar, K. (2004) *Dynamically vulcanized PP/EPDM thermoplastic elastomers: exploring novel routes for crosslinking with peroxides*, University of Twente, Calcutta.
18 Kresge, E. N., *Polymer Blends*, vo. 1. Academic Press: New York, 1978.
19 Radusch, H.J. and Pham, T. (1996) Morphology formation in dynamic vulcanized PP/EPDM blends. *Kaut Gummi Kunstst*, **49** (4), 249–257.
20 AbdouSabet, S., Puydak, R.C., and Rader, C.P. (1996) Dynamically vulcanized thermoplastic elastomers. *Rubber Chem. Technol.*, **69** (3), 476–494.

21 Li, Y.J., Kadowaki, Y., Inoue, T., Nakayama, K., and Shimizu, H. (2006) A novel thermoplastic elastomer by reaction-induced phase decomposition from a miscible polymer blend. *Macromolecules*, **39** (12), 4195–4201.
22 Li, Y.J., Oono, Y., Nakayama, K., Shimizu, H., and Inoue, T. (2006) Dual lamellar crystal structure in poly(vinylidene fluoride)/acrylic rubber blends and its biaxial orientation behavior. *Polymer*, **47** (11), 3946–3953.
23 Yang, Y., Chiba, T., Saito, H., and Inoue, T. (1998) Physical characterization of a polyolefinic thermoplastic elastomer. *Polymer*, **39** (15), 3365–3372.
24 Tang, Y.C., Lu, K., Cao, X.J., and Li, Y.J. (2013) Nanostructured thermoplastic vulcanizates by selectively cross-linking a thermoplastic blend with similar chemical structures. *Ind. Eng. Chem. Res.*, **52** (35), 12613–12621.
25 Radhakrishnan, C.K., Sujith, A., Unnikrishnan, G., and Thomas, S. (2004) Effects of the blend ratio and crosslinking systems on the curing behavior, morphology, and mechanical properties of styrene-butadiene rubber/poly(ethylene-*co*-vinyl acetate) blends. *J. Appl. Polym. Sci.*, **94** (2), 827–837.
26 Wu, H.G., Tian, M., Zhang, L.Q., Tian, H.C., Wu, Y.P., Ning, N.Y., and Hu, G.H. (2016) Effect of rubber nanoparticle agglomeration on properties of thermoplastic vulcanizates during dynamic vulcanization. *Polymers*, **8** (4), 127.

Part VI

Selected Examples of Processing

13

Reactive Extrusion of Polyamide 6 with Integrated Multiple Melt Degassing

Christian Hopmann[1], Eike Klünker[2], Andreas Cohnen[1], and Maximilian Adamy[1]

[1] *RWTH Aachen University, Institute of Plastic Processing (IKV), Seffenter Weg 201, D-52074 Aachen, Germany*
[2] *3M Deutschland GmbH, Carl-Schurz-Str. 1, D-41453 Neuss, Germany*

13.1 Introduction

Polyamide 6 (PA6) can be produced by reactive extrusion in a twin-screw extruder based on anionic polymerization of ε-caprolactam. Reactive extrusion provides advantages such as flexible production of specialities compared to production in large-scale reactors. So far, reactive extrusion of PA6 is not common in industry. This is due to the high amount of residual monomer that remains in the polymer, which has a negative effect on product properties. An in-line removal of residual monomers during reactive extrusion of PA6 in a twin-screw extruder is presented in this chapter. Multiple melt degassing and use of entrainers are investigated. The influence of process parameters on both, the amount of residual monomer and the molecular weight of the polymerized PA6, is shown.

13.2 Synthesis of Polyamide 6

Two industrially relevant production processes exist for the production of PA6: hydrolytic and anionic polymerization. Both chemical reactions are based on a chain polymerization of lactams [4]. The annular structure of the monomer ε-caprolactam (CPL) is opened during the reaction and initiates the synthesis of PA6. As an important representative of the engineering thermoplastics, PA6 is mainly produced by conventional reactor technology in large-scale reactors in the chemical industry based on hydrolytic polymerization of ε-caprolactam.

13.2.1 Hydrolytic Polymerization of Polyamide 6

The industrial synthesis of PA6 based on hydrolytic polymerization takes place in the presence of water. The production method was developed by Schlack at IG Farben AG, Frankfurt, Germany, in 1938 [46], and covers most of the polyamide

---

Reactive Extrusion: Principles and Applications, First Edition. Edited by Günter Beyer and Christian Hopmann.
© 2018 Wiley-VCH Verlag GmbH & Co. KGaA. Published 2018 by Wiley-VCH Verlag GmbH & Co. KGaA.

market [3]. Polymerization takes place at high temperatures of 260–280 °C, a pressure of about 10 bar, and by using 1–5 wt% water as activator [3, 13, 30]. Water causes the opening of the monomer CPL ring. Over a period of 8–10 h, the chain growth occurs until the desired molecular weight is achieved. Then, the produced PA6 is granulated. With increasing molecular weight, melt viscosity also increases and melt flow inside the reactor is limited. Thus, high molecular weights, which are usually needed for further processing in extrusion, can only be achieved through a solid-phase condensation of the pelletized PA6 after the polymerization step. This energy- and time-intensive process step is conducted intermittently at temperatures below the melting temperature of polyamide over several hours under vacuum [3]. Due to polymerization temperature an equilibrium between the product PA6 and low molecular residuals with a content of around 10 wt% occurs. The dominant amount of low molecular residuals is represented by monomeric CPL. Approximately 2 wt% are higher oligomers, for example, di-, tri-, and tetramers [3, 21, 29, 41, 55, 59]. The low molecular residuals, especially monomeric CPL, significantly influence the mechanical properties of the PA6 material such as mechanical strength and rigidity. Due to the softening effect of these components, a total content higher than about 3 wt% leads to a significant decrease in rigidity [58]. In addition, the low molecular residuals show a negative impact on further processing of the material, as blistering through off-gassing of residual monomer can occur. The evaporation of low molecular components leads to depositions in injection molds or on rolls in PA6 film production [3].

As a consequence, different applications tolerate different specific maximum levels of low molecular weight components. Becker et al. give a maximum content of low molecular weight components of less than 0.6 wt% for applications such as textile filaments, carpet yarns, industrial yarns, tire cord, and films [3]. In addition, there are legal regulations specifying the permissible limits of low molecular ingredients that can migrate out of plastics and can come in contact with food [43]. For CPL, a specific migration limit of 15 mg kg^{-1} packaged food is given to restrict contamination of food due to residual monomers [43]. In technical applications, for example, with relevance to engine parts of automobiles, the tolerable content of low molecular residuals relates to the required mechanical properties of the parts.

For hydrolytic polymerization of PA6, the extraction of volatiles is performed in extraction columns with water. Residual monomer, dimers, and trimers are dissolved out from PA6 granules at temperatures of 90–100 °C over a period of up to 24 h. Further processing of hydrolytic polymerized PA6 at temperatures above the melting temperature can cause a regression of residual monomers due to the thermodynamic equilibrium at such high temperatures [3].

13.2.2 Anionic Polymerization of Polyamide 6

In contrast to the previously described hydrolytic polymerization, a catalyst and an activator are used in the activated anionic polymerization of PA6 in addition to the monomer CPL. The reaction sequence can be divided into three steps (Figure 13.1) and is described in detail in literature [3, 15, 49, 53, 58, 63].

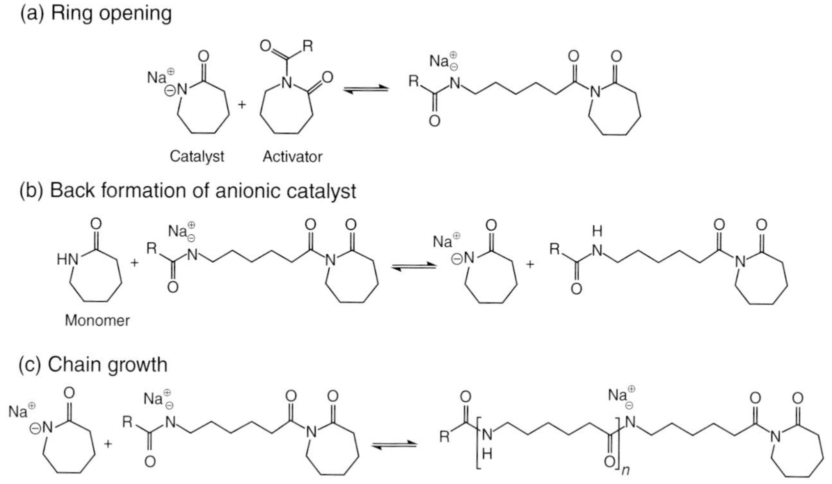

13.2 Synthesis of Polyamide 6

(a) Ring opening

(b) Back formation of anionic catalyst

(c) Chain growth

Figure 13.1 Reaction scheme of anionic polymerization of PA6.

In the first step, the catalyst dissociates to a lactam anion and a metal cation. The lactam anion opens the ring structure of the activator to form an intermediate. The back formation of lactam anion that is needed for further polymerization is carried out in the subsequent step 2. The chain growth to high molecular weight PA6 occurs in step 3. After each step of chain growth the lactam anion is formed back. It remains in the polymerized material at the end of the polymerization reaction.

Due to the weakened ring structure of the activator molecule, less energy is required to open the ring by the lactam anion as compared to monomeric CPL. This allows performing the polymerization reaction at temperatures below the mass temperature of PA6. An ideal reaction temperature is 140 °C [15]. The solubility of growing PA6 chains in the remaining low molecular weight reaction mixture decreases. Thus, the growing PA6 chains segregate from the reaction mixture which has not yet reacted and a direct transition from the monomer to the partially crystalline solid results. Compared to hydrolytic polymerization, the anionic polymerization of PA6 provides a fast reaction kinetic even at low reaction temperatures. A higher reaction conversion is achieved in much shorter time (Figure 13.2) [61].

The formulation of the anionic polymerization of PA6 is defined substantially by the types of activator and catalyst. Alkali metal salts of the CPL have proven to be successful as catalysts (sodium or magnesium lactamate), and acylated compounds by reaction with CPL come into consideration. Of practical importance is mono- or polyisocyanates, polycarbodiimides, or acetylcaprolactam [3]. Different activators and catalysts can be used, but not every combination of activator and catalyst has a sufficiently high reactivity and is suitable for polymerization [15, 53]. By using cyanates or amines (called carbamoyllactams) side reactions can occur at reaction temperatures already below the melting temperature of PA6 leading to a branching or crosslinking of the polymerized PA6 [3]. The attainable molecular weight of PA6 is influenced by the amount of added activator. Because

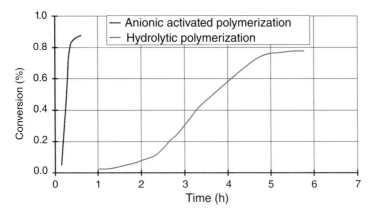

Figure 13.2 Conversion as a function of time for the polymerization of ε-caprolactame [61].

of the weakened structure of the activator ring, it is opened preferably by a catalyst, contrary to monomeric CPL. Consequently, the activator molecules are the starting point of growing PA6 chains. Therefore, a high proportion of activator leads to a low molecular weight and a small amount of activator causes a high molecular weight. Narrow molecular weight distributions can be obtained [3, 20]. The amount of activator also influences the kinetics of the reaction. The reaction rate decreases with decreasing content of activator leading to a longer polymerization time required for production of high molecular weight PA6 [12, 53]. The reaction rate can be further accelerated by higher catalyst contents and reaction temperature, with increasing reaction temperatures leading to shorter reaction times.

The anionic polymerization is exothermic with an energy release of 125 kJ kg^{-1}. The catalyst is frequently deactivated by proton donators, for example, water (Figure 13.3). Water forms a metal hydroxide and CPL with the catalyst sodium lactamate. Depending on the nature of the catalyst the deactivation by water is irreversible. In case of sodium lactamate as catalyst, a reverse reaction at high temperature of sodium hydroxide and CPL to water and sodium lactamate can occur [6]. Thus, the anionic polymerization of PA6 is very sensitive to moisture as small amounts of moisture can deactivate the catalyst and stop the polymerization reaction totally. A temperature-dependent equilibrium between CPL and PA6 also occurs for the anionic polymerization and residual monomers remain

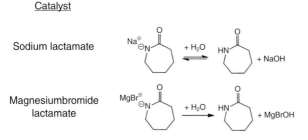

Figure 13.3 Inhibition of active catalyst by addition of water.

in the polymerized PA6. Similarly to the hydrolytically polymerized PA6, higher oligomers in a proportion between 0 and 1 wt% occur in anionically polymerized PA6 besides the actual residual monomer CPL [3, 42, 59]. The reaction sequence for the formation of high-order oligomers is described by Ueda et al. [59]. It is stated that the formation of oligomers is supported by increasing temperature [3, 59].

The main advantage of the anionic polymerization compared to the hydrolytic polymerization is the significantly higher reaction rate. At reaction temperatures below the melting temperature of PA6, a higher molecular weight, higher degree of crystallization and faster reaction conversions can be achieved for the polymerized PA6. Due to the fast reaction conversion at temperatures below the mass temperature of PA6 only a small amount of low molecular residuals is present in the product. Thus, for anionic polymerization at temperatures below the melting temperature of PA6 an additional extraction step to remove residual monomers is not necessary [1, 53, 58].

Every year, approximately 5% of the amount of PA6 produced worldwide is by anionic activated polymerization [6]. It is used in monomer casting for the batch production of thick-walled preforms of high molecular PA6. Furthermore, large-volume hollow body, gears, or long pipes are produced by rotational molding or centrifugal casting [3, 58]. Due to the high reaction rate, the anionic polymerization of PA6 can also be performed continuously in a twin-screw extruder at temperatures above the melting temperature of PA6.

In literature, there is a lot of information on investigations carried out on the influence of formulation and process parameters on the kinetics of the reaction and the properties of anionically polymerized PA6 based on different activator/catalyst systems in the twin-screw extruder. In addition, several approaches for modeling the polymerization process are described. For detailed information the following references may be useful [11, 12, 52, 53, 57]. The qualitative influence of parameters such as formulation and temperature on the reaction kinetics and the molecular weight are explained hereinafter.

In the anionic polymerization of PA6 performed in a twin-screw extruder at temperatures above the melting temperature of PA6, a residual monomer content of around 10 wt% remains in the polymerized material. Similar to hydrolytic polymerization of PA6 the amount of residual monomers has to be removed from the PA6 to allow sufficient product properties. In addition, processing at temperatures above the melting temperature can cause different side reactions that can result in a change of the molecular weight. If the alkaline catalyst is not deactivated after the polymerization reaction, side reactions of individual PA6 chains can occur (branching, claisen condensation, transamidation) [12, 15, 52]. A deactivation of the catalyst, for example, by means of adding water can cause a hydrolysis of the lactam at the end of the polymerized PA6 chains. The hydrolyzed ends of the PA6 chains on the other hand can perform condensation reactions with amine groups of nearby PA6 chains. This condensation reaction also leads to branching of individual PA6 chains and the molecular weight increases. In the absence of a catalyst, a thermal degradation of PA6 occurs only at elevated temperatures greater than 350 °C[49]. But degradation by hydrolysis and catalyzed depolymerization may occur even at lower temperatures if an active catalyst of the anionic polymerization reaction is present [60]. Hydrolysis of the amide groups of PA6 is accelerated by strong bases such as sodium hydroxide or acids [10, 16]. Without catalyzing effects the hydrolysis of PA6 is slow and not noticeable at usual residence times in extrusion [16]. A good overview of the described reaction mechanisms can be found in the works of Puffr and Kubanek [49] and Ueda et al. [60]. The presented reactions must be taken into account after considering the possible changes of molecular weight that may occur during reactive extrusion processing of PA6.

13.3 Review of Reactive Extrusion of Polyamide 6 in Twin-Screw Extruders

For the first time Illing used a twin-screw extruder for reactive extrusion of PA6 on the basis of anionically activated polymerization in 1968 [27, 28]. Monomer, caprolactam, activator, and catalyst were melted in two extruders separately and fed to the reaction extruder. The reaction could be successfully performed within a few minutes.

At the Institute of Plastics Processing (IKV) at RWTH Aachen University the reactive extrusion and anionic polymerization of PA6 in a twin-screw extruder was further developed in the following years [1, 5, 20, 36, 37]. Michaeli, Bartilla, and Berghaus used models for the theoretical description and interpretation of the reactive extrusion process. A twin-screw extruder with 30 mm screw diameter was used in the lab. The polymerization was performed at a throughput of 8 kg h^{-1}. An adduct of hexamethylene diisocyanate and CPL was used as activator, which was produced by dissolving hexamethylene diisocyanate in CPL at temperatures above 110 °C. Sodium lactamate was used as a catalyst. Average molecular weights of 50 000–120 000 g mol^{-1} were achieved depending on the formulation and the shear stress during processing [1, 37].

Hereinafter, Michaeli and Berghaus examined the possibility of direct extrusion of PA6 profiles starting from the monomer [36]. Based on previous work, a machine concept for the coupling of polymer synthesis, purification of the crude polymer, and further processing into solid bar profiles based on the reactive extrusion of PA6 was developed and studied. Correlations between process parameters of the extrusion, the average molecular weight and the residual monomer content were examined. The adduct of hexamethylene diisocyanate and CPL was used as activator again (carbamoyllactam) and sodium lactamate was used as catalyst.

To reduce the residual monomer content directly during reactive extrusion in the twin-screw extruder, Michaeli and Berghaus investigated different entrainers in a single-step vacuum degassing process [36]. Materials were chosen that do not deactivate the catalyst, such as chain-like hydrocarbons of alkanes, n-heptane, and n-decane. In addition, materials were studied, which can deactivate the catalyst specifically, such as water, dibutyl ether, and diethylene glycol dimethyl ether. By use of a catalyst deactivating material minimal residual monomer levels just below 2 wt% have been realized. The use of entrainers caused a reduction in molecular weight in all cases. The best result was achieved with diethylenglycoldimethylether [36]. However, this substance is classified as hazardous [45]. The decrease of the molecular weight caused by the addition of an entrainer can be reduced by adding a copper complex and phosphonates [20]. An influence of the content of entrainer in the range of 0.2–2 wt% on the amount of residual monomer for a single-step vacuum degassing process could not be observed [36]. Michaeli and Berghaus showed a decrease of stiffness of directly produced solid rods with increasing residual monomer content.

Grefenstein *et al.* used acetylcaprolactam as activator and magnesium lactamate as catalyst and performed trials to reduce the monomer content [20, 35]. Using a single-step degassing process minimal residual monomer levels of about 1.5–2 wt% were achieved by addition of water as an entrainer. The experiments were performed on a twin-screw extruder with 30 mm screw diameter at a maximum throughput of 3 kg h^{-1}.

Kye and White studied the influence of screw configuration and process conditions on product quality [33]. The experiments were performed at a very low throughput of 2 kg h^{-1} in a twin-screw extruder having 30 mm screw diameter. To reduce the residual monomer content directly during the production process, a single-step vacuum degassing was applied at the end of the extruder. Residual monomer content of 2–5 wt% was achieved. Average molecular weights in a wide range of 14 000–32 000 g mol^{-1} were obtained for a constant amount of activator, depending on screw configuration and the average residence time in the process. A slight increase in the mean molecular weight was also observed by application of vacuum degassing. An explanation for the increase of the mean molecular weight is not delivered. An acetylcaprolactam was used as activator, and sodium lactamate as catalyst.

Hornsby *et al.* characterized the material properties of PA6, prepared by reactive extrusion in a twin-screw extruder. The polymerization was performed in an extruder having a 40 mm screw diameter at a low throughput of 4 kg h^{-1}. As performed by Kye and White a single-step vacuum degassing was implemented in the process. The achievable monomer content was 4–6 wt%. Average molecular weights from 50 000–150 000 g mol^{-1} could be obtained depending on the extrusion parameters. A carbamoyllactam was used as activator and sodium lactamate as catalyst [26].

Lee and White compared different types of twin-screw extruders and an internal mixer as chemical reactors for the anionic polymerization of PA6 in 2001. Better turnovers were found for twin-screw extruders at same residence times equipped with adequate mixing elements. A commercially available reaction system was used for the monomer casting of PA6, with Carbamoyllactam as activator and sodium lactamate as catalyst [34].

Michaeli and Rothe for the first time used a modified reaction system. Acetylcaprolactam as activator was modified to provide lower volatility at high temperatures. Magnesium lactamate was used as catalyst. The reaction system was easy to control and showed a high reactivity for polymerization of PA6 with a limited residence time in the twin-screw extruder [16, 40].

Hopmann and Klünker used the same modified reaction system and investigated on the addition of cellulose fibers to the low viscous reaction system in order to improve wetting and distribution of the fibers during extrusion [22, 23]. Poindl and Bonten improved impact resistance of PA6 in presence of a polyethertriol [48].

13.4 Recent Developments in Reactive Extrusion of Polyamide 6 in Twin-Screw Extruders

Scientific publications have shown the technical potential of reactive extrusion of PA6 using twin-screw extruders. The modular design of twin-screw extruders allows the combination of different process steps, such as polymer synthesis, removal of residual monomer, and addition of fillers in a single step without re-melting the material. The integration of polymer synthesis to the extrusion step enables adapting and fine tuning the polymer properties to product requirements. Thus, reactive extrusion is an ideal process for the production of tailor-made products in small volumes.

At IKV, recent developments have focused on reactive extrusion of PA6 with integrated melt degassing to minimize the residual monomer content allowing an industrial application of the process. A quantitative investigation of the influence of melt degassing on the achievable residual monomer content and the molecular weight has been carried out for anionic polymerization of PA6 in a twin-screw extruder [20, 35].

13.4.1 Reaction System and Experimental Setup

CPL of the type AP-Nylon® Caprolactam from Brüggemann Chemical KG, Heilbronn, Germany, is used as monomer for the anionic polymerization of PA6. Due to its low initial content of moisture of less than 0.01 wt% it is preferably used for anionic polymerization of PA6. Acetylcaprolactam (activator) and magnesium lactamate (catalyst) are supplied by Brüggemann Chemical KG.

The anionic polymerization of PA6 is performed continuously in a twin-screw extruder ZSK 26MC of Coperion GmbH, Stuttgart, Germany. The screw diameter is 26 mm and the L/D ratio is 57. A schematic overview of the experimental setup is shown in Figure 13.4. In the first zone of the extruder the polymerization reaction takes place, followed by the zone for multiple melt degassing. The solid CPL is filled into two heated stirred tanks and melts under dry nitrogen atmosphere. Activator and catalyst are held separately in the associated tanks and mixed with liquid CPL. The liquid reaction mixture is metered at a temperature of 100 °C by two gear pumps without counter pressure through heated pipes into the first zone of the extruder. Dry nitrogen is added to the feeding zone during the extrusion run. Due to the melting of CPL under nitrogen atmosphere and the addition of nitrogen to the feeding zone, the effect of moisture from the environment on polymerization reaction is minimized. For multiple melt degassing, two different vacuum pumps from Dr Ing. K. Busch GmbH, Maulburg, Germany are used. A liquid ring vacuum pump Dolphin type LB 0113 with a maximum volume flow of 32 $m^3 h^{-1}$ and a final pressure of 30 mbar can be operated for vacuum degassing at two locations along the extruder. In addition, a combi vacuum pump with a maximum volume flow of 280 $m^3 h^{-1}$ and a final pressure of 30 mbar can be used at a third vacuum port.

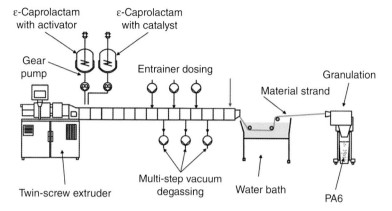

Figure 13.4 Schematic process set-up for the continuous anionic polymerization of polyamide 6 and multiple degassing of residual monomer in a twin-screw extruder.

Table 13.1 Used process parameters for reactive extrusion of PA6.

Parameter			
Screw speed (rpm)	400		
Throughput (kg h^{-1})	5, 10, 20		
Catalyst content (wt%)	3		
Activator content (wt%)	0.5, 2		
Barrel temperature (L/D) (°C)	0–4	4–28	28–56
	170	260	220
Die temperature (°C)	260		
Degassing pressure (mbar)	Degassing 1	Degassing 2	Degassing 3
	40	40	10
Entrainer throughput, H$_2$O (kg h^{-1})	0.6	0.6	0.6

Liquid entrainers are injected into the melt by an HPLC pump from ERC GmbH, Riemerling, Germany. The used entrainers donate protons and can deactivate the active catalyst after the polymerization reaction [20]. Thus, it is assumed that a back formation of monomer can be avoided during residual monomer degassing or following melting of the material, for example, in further processing. If not mentioned otherwise, distilled water is used as entrainer.

The process parameters of the reactive extrusion process are listed in Table 13.1.

The low molecular residual content is determined by measuring the weight loss after extracting the water-soluble residuals by means of hot water. With this method water-soluble high-order oligomers are also detected. To determine only the content of the monomeric residuals in PA6 gas chromatography, analysis is performed for selected samples. The relative viscosity is determined by solution viscometry. For this purpose the polymer is dissolved in formic acid (90%). Then

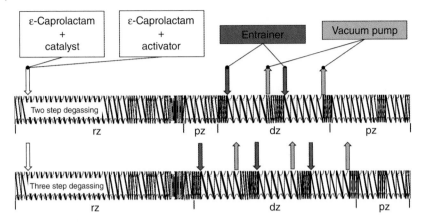

Figure 13.5 Screw configurations for multiple melt degassing.

the ratio of the flow times of the solution and the pure solvent in the capillary of a viscosimeter (z. B. Ubbelohde viscometer) is measured. An overview of relevant methods used to determine the amount of low molecular residuals in PA6 is given by Klünker [24, 31].

Figure 13.5 shows the screw configurations used for multiple melt degassing. For both, step 2 and step 3 of melt degassing, the length of the reaction zone is 24 L/D. Each of the individual degassing zones has a length of 6 L/D. The degassing zones are separated from each other and from the reaction zone by fully filled tooth mixing elements providing a sufficient melt sealing. This allows running the process with different vacuum pressures in the degassing zones. At the same time it is ensured that the injected entrainer is only effective in the designated degassing zone. The entrainer is injected directly into the polymer melt at the beginning of each degassing zone. At the end of the degassing zones the vacuum pumps are connected.

13.4.2 Influence of Number of Degassing Steps and Activator Content on Residual Monomer Content and Molecular Weight

The influence of the number of degassing steps on residual monomer content is shown in Figure 13.6 (2 wt% activator content, and entrainer not used).

The residual monomer content decreases with increasing number of degassing zones. Without monomer degassing the initial residual monomer content is about 10 wt%. It can be reduced to about 6 wt% using a two-step melt degassing without entrainer addition.

During the first degassing the thermodynamic reaction equilibrium shifts toward the monomer. As no entrainer is injected the catalyst cannot be deactivated. Thus, a partial back formation of monomer during or after the first degassing zone is assumed. The resulting residual monomer content can be removed from the melt in the second degassing step. As a consequence, there

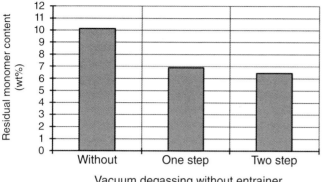

Figure 13.6 Residual monomer content for vacuum degassing without an entrainer (amount of activator: 2 wt%, polymer throughput: 10 kg h^{-1}, screw speed: 400 rpm).

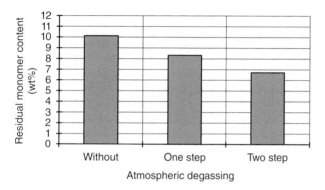

Figure 13.7 Residual monomer content for atmospheric degassing using an entrainer (amount of activator: 2 wt%, polymer throughput: 10 kg h^{-1}, screw speed: 400 rpm).

is no significant difference in the final residual monomer content for one- or two-step melt degassing.

Figure 13.7 shows the influence of addition of water as entrainer for multiple atmospheric degassing.

With additional steps of entrainer the residual monomer content decreases for atmospheric degassing without applying vacuum. By injecting a low-boiling entrainer to the polymer melt prior to the degassing zone, the degassing efficiency can be improved. The entrainer causes a decrease of the partial pressure of the volatiles without lowering the absolute pressure in the gas phase. Thus, the concentration of the volatiles in the melt, which is proportional to the partial pressure of the volatiles in the gas phase, can be reduced and the residual monomer content can be decreased. In addition, the entrainer causes a foaming of the melt and the surface of the interphase between melt and gas phase can be increased. This also improves the degassing efficiency. By adding water as entrainer, the catalyst of anionic PA6 polymerization can be deactivated and a back formation of residual monomer can be reduced during and after the respective degassing steps.

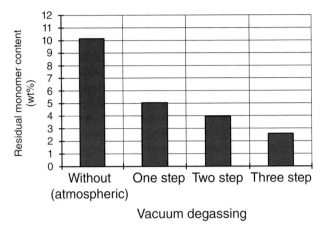

Figure 13.8 Residual monomer content depending on number of degassing steps for 2 wt% activator and use of an entrainer in each degassing zone (polymer throughput: 10 kg h^{-1}, screw speed: 400 rpm).

Thus, in contrast to Figure 13.6 a significant difference in the achievable residual monomer content can be observed between one- and two-step degassing in Figure 13.7. This demonstrates the tremendous importance of a sufficient deactivation of the catalyst after the polymerization reaction and before or during the monomer degassing.

The dependence of the residual monomer content on the number of vacuum degassing zones in combination with an addition of water as entrainer is shown in Figure 13.8 (2 wt% activator content, addition of water in each degassing zone).

For a three-step vacuum degassing, the residual monomer content can be reduced from 10 to 2.6 wt% by a throughput of 10 kg h^{-1}, a screw speed of 400 rpm, and an activator content of 2 wt%. The concentration gradient between the residual monomer in the melt and the degassed monomer in the gas phase is the driving force for melt degassing. The concentration gradient decreases with decreasing content of the volatile component in the melt. Therefore, it becomes more difficult to remove the remaining small portions of volatiles from the melt. The positive effect of adding an entrainer is higher in case of multiple addition of entrainers and application of vacuum.

The residual monomer content that is dependent on the number of degassing zones for different contents of activator is shown in Figure 13.9. In addition, the mass temperature of PA6 measured at the exit of the extrusion is listed.

For both amounts of activator the residual monomer content decreases with increasing number of vacuum degassing steps. Lowest levels of less than 1 wt% can be obtained for three-step degassing of PA6 with 0.5 wt% activator. The initial residual monomer content thereby is higher for 0.5 wt% activator, whereas for the two- and three-step degassing lower amounts of residual monomer can be achieved compared to the 2 wt% activator. These results contradict the theory for degassing: The molecular weight of PA6 increases with decreasing activator content. A higher melt viscosity is to be expected at the same mass temperature as compared to the low molecular weight PA6 based on 2 wt% activator. A high melt

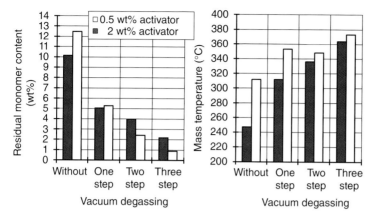

Figure 13.9 Residual monomer content and resulting mass temperature depending on number of degassing steps for different amounts of activator (polymer throughput: 10 kg h^{-1}, screw speed: 400 rpm).

viscosity has a negative effect on the degassing, as the diffusion of the volatiles is slowed down.

The improved degassing efficiency observed for low activator content with relevance to the mass temperature measured at the extruder exit can be explained by the following two general statements:

1) The mass temperature increases with increasing number of degassing zones for both activator contents.
2) High mass temperatures of 310 °C are observed for low activator content of 0.5 wt% even without melt degassing. For a three-step degassing, higher temperatures of up to 370 °C are measured. For 2 wt% activator mass temperatures are lower and highest values of 360 °C are only reached for three-step melt degassing.

Based on the measured temperature of PA6 melt at the die of the extruder, the temperature at the end of the reaction zone, and the start temperature of the reaction system as added in the feeding zone, a course of mass temperature can be assumed over the process length. It is a function of the activator content for multiple melt degassing with addition of entrainer (Figure 13.10). A linear course of mass temperature is assumed. A lowering of the mass temperature in the degassing zone by adding the entrainer is not taken into account.

It is clearly recognizable that the mass temperature is significantly higher at 0.5 wt% activator content in comparison to 2 wt% activator content over the entire length of the process. The temperature difference decreases with an increasing number of degassing zones. A high mass temperature of around 350 °C occurs for 0.5 wt% activator content already in the first degassing zone, whereas a similar temperature level for 2 wt% activator content is first present in the second degassing zone. The low activator content leads to higher mass temperature, which stresses the PA6 melt for a significantly longer time in the process. A high mass temperature leads to an increase of the diffusion coefficient and the partial pressure in equilibrium state of the volatiles in the

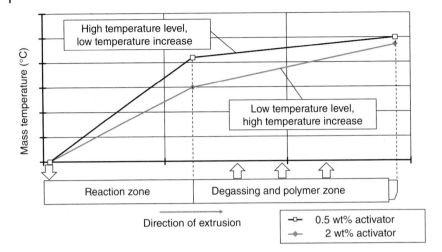

Figure 13.10 Assumed trend of mass temperature along the multiple degassing process depending on amount of activator.

melt. Further, melt viscosity is decreased. Both improve degassing efficiency, which explains the observed lower amounts of residuals for PA6 with 0.5 wt% activator compared to the process with higher activator content.

The different trend of mass temperatures for varied activator content can be explained by energy dissipation to the melt, which among others depends on melt viscosity. The low activator content causes a PA6 melt with a higher initial molecular weight. Compared to high activator content more shear energy can be dissipated in the melt. Thus, initial mass temperature for a 0.5 wt% activator is higher compared to a 2 wt% activator. Residual monomer acts as a lubricant and as a softening agent in the PA6 melt. By removing the residual monomer from the melt the lubricating and softening effect decreases, which causes an increase of melt viscosity. With increasing degassing steps more residual monomer is removed and more shear energy can be dissipated. As a consequence, mass temperature rises with increasing number of degassing steps for contents of both activator and 0.5 wt% activator leading to a significant higher initial mass temperature.

The mass temperature during degassing is limited by the polymer to be degassed: At high temperatures reactions can occur, which have an effect on the molecular weight of the degassed polymer. Figure 13.11 shows the influence of residual monomer degassing on the relative viscosity of PA6 with 2 wt% activator and the corresponding mass temperatures. Relative viscosity correlates with molecular weight of the polymer, with high molecular weights having high relative viscosities.

With increasing amount of degassing steps the relative viscosity increases from 2.3 up to 2.9 for two-step vacuum degassing without entrainer and 2.7 to 2.8 for two and three-step vacuum degassing with entrainer. A change of relative viscosity and molecular weight, respectively without varying the amount of activator can only be explained by side reactions, which occur after the polymerization reaction (see Section 13.2.2). For a melt degassing without using water as

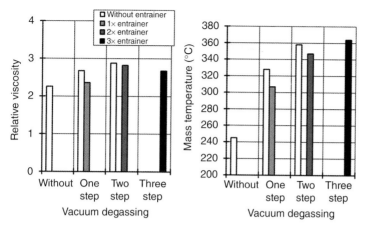

Figure 13.11 Relative viscosity and resulting mass temperature depending on number of degassing steps for 2 wt% activator (polymer throughput: 10 kg h^{-1}, screw speed: 400 rpm).

entrainer, the catalyst remains active in the melt after the polymerization reaction. Side reactions such as branching and claisen condensations or transamidition can be favorable at elevated temperatures causing an increase in molecular weight and relative viscosity, respectively. Adding water to the anionic PA6 melt can deactivate the catalyst, but can also cause a hydrolysis of lactam at the end of the polymerized PA6 chains. The hydrolyzed ends of the PA6 chains can perform condensation reactions with amine groups of nearby PA6 chains and molecular weight and relative viscosity can increase. For multiple melt degassing of PA6 with 2 wt% activator it can be assumed that the described side reactions occur which influence thermal degradation reactions even at the observed elevated mass temperatures leading to an increase of molecular weight of PA6. The influence of residual monomer degassing on the relative viscosity of PA6 prepared with 0.5 wt% activator, and the corresponding mass temperature are shown in Figure 13.12.

In contrast to PA6 with 2 wt% activator, a decrease in relative viscosity with increasing number of degassing zones is found with 0.5 wt% activator. A two-step degassing without the addition of water leads to a reduction of relative viscosity from 3.1 to a lowest value of 2.5. By adding water as entrainer, the decrease in the relative viscosity can be reduced. A relative viscosity of 2.76 is reached for two-step degassing, whereas in three-step degassing a slightly lower relative viscosity of about 2.7 is measured, when water is used as entrainer.

Again, the change in relative viscosity and molecular weight at constant activator content can only be explained by reactions that take place after the initial polymerization reaction. Higher mass temperatures are present compared to the process with 2 wt% activator and these higher mass temperatures stress the PA6 directly after the reaction zone until the end of the process. For the melt with 0.5 wt% activator prone to the measured high mass temperatures over the entire length of the process it can be assumed that thermal degradation reactions are dominant and consequently the molecular weight and relative viscosity of the material decrease.

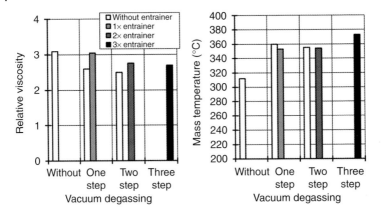

Figure 13.12 Relative viscosity and resulting mass temperature depending on number of vacuum degassing steps for 0.5 wt% activator (polymer throughput: 10 kg h^{-1}, screw speed: 400 rpm).

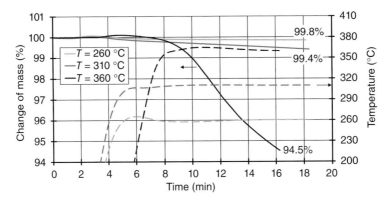

Figure 13.13 Thermal degradation of polyamide 6 with 2 wt% activator based on anionic polymerization depending on temperature and time.

An impression of the assumed thermal degradation can be obtained by isothermal thermogravimetric analysis (TGA). Figure 13.13 shows the detected thermal degradation of PA6 based on anionic polymerization as a function of temperature and time in the TGA. Temperatures have been chosen to be close to the mass temperatures observed in the process. The measurement is performed under inert gas atmosphere (nitrogen).

At temperatures around 360 °C, which are comparable to the mass temperatures arising during processing of 0.5 wt% activator, a thermal degradation already takes place after a short time (~1.5% weight loss per 2 min). For lower temperatures nearly no degradation occurs. The slightly lower decrease of the relative viscosity of PA6 with 0.5 wt% activator when using water as an entrainer can be explained by a local cooling of the melt. Heat energy is removed from the melt by vaporization of the entrainer and the mass temperature decreases. Thus, less strong thermal degradation effects occur.

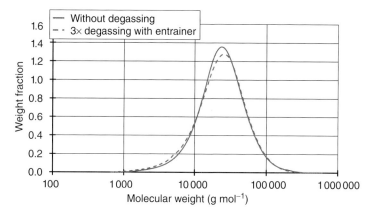

Figure 13.14 Change of molecular weight distribution for low amount of activator due to monomer degassing (polymer throughput: 10 kg h^{-1}, screw speed: 400 rpm).

No uniform trend is found for the influence of multiple melt degassing on the relative viscosity of PA6 based on anionic polymerization and different amounts of activator. Side reactions are assumed in the case of 2 wt% activator content, which result in branching and increase of relative viscosity and molecular weight. For 0.5 wt% activator content, thermal degradation reactions are assumed to be dominant, which causes a decrease in relative viscosity and molecular weight.

In addition to the measured values of relative viscosity, selected materials are analyzed by gel permeation chromatography (GPC) for the average molecular weight and molecular weight distribution. The molecular weight distribution of PA6 with 0.5 wt% activator measured in GPC shows a slightly higher fraction of polymer chains with lower molecular weights for the three-step degassing compared to the process without degassing (Figure 13.14).

This can be explained by the previously observed and discussed thermal degradation of the PA6 melt prone to high mass temperatures for the entire length of the process. The thermal degradation leads to a reduction of the molecular weight of long polymer chains and to an increasing amount of polymer chains with lower molecular weight.

For 2 wt% activator content, GPC analysis shows significantly higher molecular weights with increasing number of degassing steps (Figure 13.15). This is in good accordance with the assumed side reactions leading to branching and thus, to a higher number of polymer chains with high molecular weights.

The change of average molecular weights for a three-step vacuum degassing as a function of the investigated activator content is shown in Figure 13.16.

No change of average molecular weight M_w can be observed in the three-step degassing at a low activator content of 0.5 wt%, but the number average molecular weight M_N decreases slightly. At activator content of 2 wt% an increase in the weight and number average molecular weight is measured. Again, this is in good correlation with the observed changes in relative viscosity due to melt degassing and clearly indicates thermal degradation reactions for PA6 with 0.5 wt% activator and branching side reactions for 2 wt% activator.

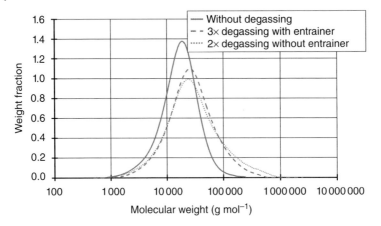

Figure 13.15 Change of molecular weight distribution for high amount of activator due to monomer degassing (polymer throughput: 10 kg h^{-1}, screw speed: 400 rpm).

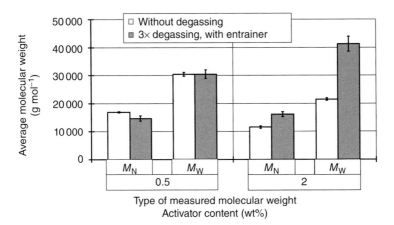

Figure 13.16 Change of molecular weight due to triple degassing for different contents of activator (polymer throughput: 10 kg h^{-1}, screw speed: 400 rpm).

The results show that residual monomer can be removed more effectively from the melt with an increasing number of degassing zones. The use of an entrainer in each degassing zone is essential: On one hand, the entrainer improves the removal of residual monomer from the melt, and on the other hand, the catalyst is deactivated sufficiently by the entrainer and a back formation of residual monomer can be avoided during and after degassing. A lower residual monomer content is realized against expectations when degassing PA6 with 0.5 wt% activator content. This can be explained by the initially higher mass temperatures during processing compared to 2 wt% activator. On the other hand, the high mass temperatures show an influence on the molecular weight as well as on the relative viscosity of the material produced. Thus, the multiple melt degassing of PA6 based on anionic polymerization in a twin-screw extruder is a trade-off between low levels of residual monomers and achievable molecular weight.

13.4.3 Influence of Amount and Type of Entrainer on Residual Monomer Content and Molecular Weight

Starting from an entrainer content of 6 wt% per degassing zone the influence of higher amounts of entrainer and different types of entrainers to the residual monomer content and the relative viscosity is investigated. For a two-step degassing and an activator content of 0.5 wt% an increased amount of water as entrainer leads to less efficient degassing characterized by an increase of residual monomer content (Figure 13.17). At the same time the measured mass temperature at the extruder outlet decreases with increasing amount of water added to the melt. It can be assumed that with higher amounts of water the local cooling of the melt in the degassing zone increases, leading to both, an increased melt viscosity and a decrease of diffusion coefficient. Both have significant negative effects on the degassing results.

The increased local cooling of the melt has a positive effect on the observed thermal degradation of molecular weight for PA6 with 0.5 wt% activator. With increasing amount of entrainer added to the melt, the relative viscosity approaches the initial value of 3.1 for PA6 based on 0.5 wt% activator and no degassing applied (Figure 13.18).

In addition to water, liquid ethanol and gaseous carbon dioxide (CO_2) are used as entrainers in the two-step degassing process. Ethanol is metered similarly to water with HPLC pumps. The CO_2 dosing takes place via a CO_2 metering station, which is connected to a liquid CO_2 bottle. The CO_2 is injected to the polymer melt in a gaseous state. Accordingly, no change of aggregation state takes place within the extruder. Thus, less heat is removed from the melt compared to using water or ethanol as entrainer. With the given metering system CO_2 can only be added in one degassing zone. Therefore, it is combined with the addition of water to the remaining degassing zone.

Ethanol and CO_2 display different properties compared to water (Table 13.2). The enthalpy of vaporization for ethanol is significantly lower compared to water.

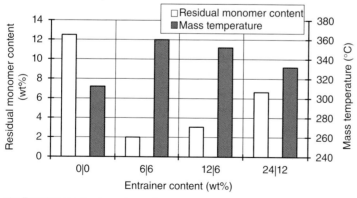

x|x: Entrainer content per degassing zone

Figure 13.17 Influence of amount of water as entrainer on residual monomer content for the two-step degassing of PA6 based on 0.5 wt% activator (polymer throughput: 10 kg h^{-1}, screw speed: 400 rpm).

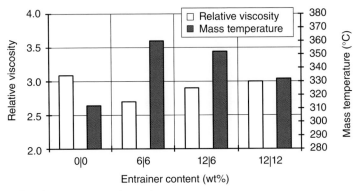

x|x: Entrainer content per degassing zone

Figure 13.18 Influence of amount of water as entrainer on relative viscosity for the two-step degassing of PA6 based on 0.5 wt% activator (polymer throughput: 10 kg h^{-1}, screw speed: 400 rpm)

Table 13.2 Selected material properties of used entrainers [39].

Entrainer	Enthalpy of evaporation (kJ kg^{-1})	pK_a-value
Water	2200	14
Ethanol	870	16
CO$_2$	Gaseous	Carbonic acid: 6.5

Thus, less heat can be removed from the melt during vaporization. The pK_a-value can be used to characterize the ability to donate protons with lower values equal to a higher rate of donation [39]. An entrainer with lower pK_a-values is assumed to better deactivate the catalyst of anionic polymerization of PA6. For the combination of CO$_2$ and water it can be assumed that there is a reaction to carbonic acid to some extent during the process. Carbonic acid (H$_2$CO$_3$) can also be used to deactivate the catalyst.

The influence of different entrainers on the residual monomer content, the relative viscosity, and the resulting mass temperature in the two-step degassing of PA6 with 0.5 wt% activator content is shown in Figure 13.19.

Due to the smaller enthalpy of vaporization of ethanol and water/CO$_2$ compared with two-step addition of water higher mass temperatures are observed for ethanol and water/CO$_2$ at the outlet of the extruder. Despite the approximately 10 K higher mass temperature obtained by addition of ethanol, around 20% higher residual monomer content occurs compared to two-step addition of water. As ethanol donates slightly less protons compared to water, it can be assumed to be less effective in deactivation of the catalyst. However, it is not possible to distinguish whether the lower degassing efficiency is a result of a lesser

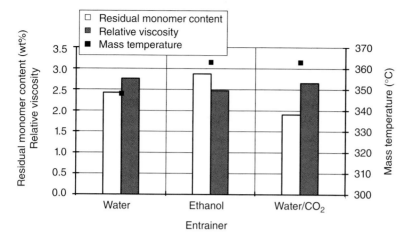

Figure 13.19 Influence of different entrainers on residual monomer content, relative viscosity and mass temperature for PA6 based on 0.5 wt% activator and two-step degassing (polymer throughput: 10 kg h^{-1}, screw speed: 400 rpm).

amount of good deactivation of the catalyst or whether it is due to a lower entraining effect of compared to water.

The combination of water and CO_2 also results in an approximately 10 K higher mass temperature compared to the two-step addition of water as entrainer. In this case, the residual monomer content can be lowered by around 20%, that is, to 1.9 wt% compared to water. It can be assumed that the addition of CO_2 gas provides an improved foaming of the melt and thus has a positive effect on the degassing by enhanced surfaces.

In good correlation with the measured higher mass temperatures by using ethanol and water/CO_2, there is an additional decrease of the relative viscosity compared to the two-step degassing process using water as entrainer. The relative viscosity decreases approximately by 10% to 2.5 using ethanol and by about 5% to 2.7 using water/CO_2. Taking into account the achievable residual monomer content and the change of relative viscosity water has proven to be the best entrainer.

13.4.4 Influence of Polymer Throughput on Residual Monomer Content

By varying the polymer throughput, the average residence time and the filling level of melt are influenced substantially in the degassing zone. A low mass throughput leads to an increased average residence time in the degassing zone, which has a positive effect on the degassing process limited by slow diffusion mechanism. From an economic point of view, it is aimed to operate processes at high throughputs and high machine utilization [7].

Figure 13.20 shows the influence of the throughput on residual monomer content.

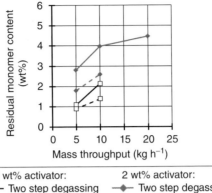

Figure 13.20 Residual monomer content depending on polymer throughput (screw speed: 400 rpm).

A reduction of polymer throughput from 20 to 5 kg h^{-1} results in a significant improvement in degassing performance. Lower residual monomer contents are obtained independently of the activator content for the two- and three-step degassing process.

The lowest residual monomer content including high-order oligomers of 0.9 wt% is achieved by three-step degassing with 0.5 wt% activator content at the lowest throughput of 5 kg h^{-1}. The associated value of pure monomeric CPL is approximately 0.4 wt% measured by gas chromatography. At a throughput rate of 10 kg h^{-1}, a residual monomer content of 1.4 wt% and of pure monomeric CPL of less than 1 wt% can be realized. At a throughput of 20 kg h^{-1}, which is feasible only with 2 wt% activator content under the given conditions, a residual monomer content of 4.5 wt% including high-order oligomers cannot be deceeded in a two-step degassing process.

13.5 Conclusion

The anionic polymerization can be performed continuously in a twin-screw extruder with integrated multiple residual monomer degassing. It features advantages concerning the flexible production of specialities compared to the conventional production process of PA6, hydrolytic polymerization. Multiple degassing leads to lower residual monomer levels with increasing number of degassing zones, decreasing throughput, and higher mass temperature. The latter is significantly influenced by the amount of activator used and the removal of monomer during each degassing step. In addition, the use of an entrainer is to achieve substantial low residual monomer contents. Thereby, water has proven as a sufficient entrainer, which also deactivates the catalyst of anionic polymerization of PA6. A deactivation is required to avoid a back formation of residual monomer during and after melt degassing.

Residual monomer contents of less than 1 wt% can be achieved by the use of a three-step monomer degassing at a PA6 throughput of 5 kg h^{-1} and an activator content of 0.5 wt%. For pure monomeric CPL that is mainly responsible for affecting the polymer properties, the lowest obtained value of 0.4 wt% is sufficient to use the produced PA6, for example, as polymer for technical applications [31]. Furthermore, it is possible to modify the reactive extrusion process and to integrate the addition of reinforcing fillers, such as glass fibers to the polymer melt. This allows industrial production of reinforced PA6 compounds starting from the monomer CPL with low levels of residual monomer. Thus, a subsequent removal of monomers in an additional process step is not required anymore. Additional and more detailed results can be found in [22, 24, 25, 31].

References

1 Bartilla, T. (1987) Polymerisation auf einem gleichlaufenden doppelschneckenextruder am beispiel des polycaprolactams. Dissertation at RWTH Aachen University.

2 Barth, U. (1989) Zweiwellige Schneckenreaktoren. Kontinuierliche Herstellung massiver Polyurethanprodukte. *Plastverarbeiter*, **40** (1), 100–106.

3 Becker, G., Braun, D., Bottenbruch, L., and Binsack, R. (1998) *Polyamide Kunststoff-Handbuch 3/4. Technische Thermoplaste*, Carl Hanser Verlag, Munich, Vienna.

4 Braun, D., Cherdon, H., Rehahn, M., Ritter, H., and Voit, B. (2005) Synthesis of macromolecules by chain growth polymerisation, in *Polymer Synthesis: Theory and Practice, Fundamentals, Methods, Experiments*, Springer-Verlag, Berlin, Heidelberg, New York.

5 Berghaus, U. (1991) Der Gleichdralldoppelschneckenextruder als kontinuierlicher Reaktor bei der reaktiven Extrusion von Polyamiden und Styrolpolymerisaten. Dissertation at RWTH Aachen University.

6 Bongers, J. (2014) *Personal Message*, Brüggemann Chemical KG, Heilbronn.

7 Barth, U. and Schuler, W. (1992) Restentgasen von Kunststoffen unter Berücksichtigung der Rückbildung von Monomeren, in *Entgasen beim Herstellen und Aufbereiten von Kunststoffen*, VDI-Verlag, Düsseldorf.

8 Coughlin, R.W. and Canevari, G.P. (1969) Drying polymers during screw extrusion. *AIChE J.*, **15** (4), 560–564.

9 Collins, G.P., Denson, C.D., and Astarita, G. (1985) Determination of mass transfer coefficients for bubble-free devolatilization of polymeric solutions in twin screw extruders. *AIChE J.*, **31** (8), 1288–1296.

10 Czernik, S., Elam, C.C., Evans, R.J., Meglen, R.R., Moens, L., and Tatsumoto, K. (1998) Catalytic pyrolysis of nylon-6 to recover caprolactam. *J. Anal. Appl. Pyrolysis*, **46** (1), 51–64.

11 Cimini, R.A. and Sundberg, D.C. (1986) A mechanistic kinetic-model for the initiated anionic-polymerisation of epsilon-caprolactam. *Polym. Eng. Sci.*, **26** (8), 560–568.

12 Dave, R.S., Kruse, R.L., Stebbins, L.R., and Udipi, K. (1997) Polyamides from lactams via anionic ring-opening polymerisation 2. Kinetics. *Polymer*, **38** (4), 939–947.
13 Elias, H.-G. (1999) *Makromoleküle*, Wiley-VCH Verlag, Weinheim.
14 Foster, R.W. and Lindt, J.T. (1990) Twin screw extrusion devolatilization – from foam to bubble free mass-transfer. *Polym. Eng. Sci.*, **30** (11), 621–634.
15 van Geenen, A. (2010) *Anionic Polyamide, Chemistry and Processing*, Personal Documentation, Brüggemann Chemicals, Heilbronn.
16 van Geenen, A. (2011) *Personal Message*, AvG Consultancy: Caprolactam Polymerisations and Polyamide-6 Processing Chemistry, Sittard.
17 Gestrig, I. (2002) Entgasen von polymeren. Dissertation at University Hannover.
18 Gestring, I. and Mewes, D. (2002) Degassing of molten polymers. *Chem. Eng. Sci.*, **57** (16), 3415–3426.
19 Grefenstein, A. (1994) Rechnergestütze Auslegung von Schneckenreaktoren am Beispiel des dichtkämmenden Gleichdralldoppelschneckenextruders. Dissertation at RWTH Aachen University.
20 Grefenstein, A. (1996) *Reaktive Extrusion und Aufbereitung: Maschinentechnik und Verfahren*, Carl Hanser Verlag, Munich, Vienna.
21 Guaita, C. (1984) HPLC analysis of cycol-oligoamides-6 and cyclo-oligoamides-66. *Makromol. Chem. Macromol. Chem. Phys.*, **185** (3), 459–465.
22 Hopmann, C. and Klünker, E. (2012) *Reaktive Extrusion von Zellulosefaser gefülltem Polyamid 6*, Abschlussbericht zum Forschungsvorhaben IGF-Nr. 16347N, Institute of Plastics Processing (IKV) at RWTH Aachen University.
23 Hopmann, C. and Klünker, E. (2012) Naturfasern im Extruder imprägnieren. *Kunststoffe*, **112** (8), 64–67.
24 Hopmann, C. and Klünker, E. (2014) Reaktive Extrusion und integrierte Compoundierung von Polyamid 6-Glasfaser Compounds: Reduzierung des Restmonomergehalts und flexibles Einstellen des Molekulargewichts. Final report on the research project IGF-Nr. 16775N. Institute of Plastics Processing (IKV) at RWTH Aachen University.
25 Hopmann, C. and Klünker, E. (2015) Polyamid 6 direkt compoundieren. *Kunststoffe*, **112** (2), 68–71.
26 Hornsby, P.R., Tung, J.F., and Tarverdi, K. (1994) Characterization of polyamide 6 made by reactive extrusion. 1. Synthesis and characterization of properties. *J. Appl. Polym. Sci.*, **53** (7), 891–897.
27 Illing, G. (1968) Herstellung von Fäden, Bändern, Folien aus Polyamiden durch alkalische Schnellpolymerisation von Lactamn im Extruder. *Kunststofftechnik*, **7** (10), 351–356.
28 Illing, G. (1969) Direct extrusion of nylon products from lactam. *Mod. Plast.*, **8** (2), 70–76.
29 Illing, G. (1989) Energiesparendes Herstellen von PA 6 durch direkte polymerisation. *Kunststoffe*, **79** (10), 967–973.

30 Kaiser, W. (2007) *Kunststoffchemie für Ingenieure*, Carl Hanser Verlag, Munich.

31 Klünker, E. (2015) Restmonomerentgasung und Direktcompoundierung bei der reaktiven Extrusion von Polyamid 6. Dissertation at RWTH Aachen University. ISBN 978-3-95886-031-5.

32 Keum, J. and White, J.L. (2004) Engineering analysis of devolatilization of various additives in intermeshing co-rotating twin-screw extruders. *Int. Polym. Process.*, **19** (2), 101–110.

33 Kye, H. and White, J.L. (1994) Continuous polymerisation of caprolactam in a modular intermeshing corotating twin-screw extruder integrated with continuous melt spinning of polyamide 6 fiber: influence of screw design and process conditions. *J. Appl. Polym. Sci.*, **52** (9), 1249–1262.

34 Lee, B.H. and White, J.L. (2001) Comparison studies of anionic polymerisation of caprolactam in different twin-screw extruders. *Int. Polym. Process.*, **16** (2), 172–182.

35 van Marwick, J. (1992) *Stabiliserung von im Extruder Polymerisierten Polyamid-Formmassen*, Institute of Plastics Processing (IKV) at RWTH Aachen University, unpublished study work.

36 Michaeli, W. and Berghaus, U. (1991) Kontinuierliche Herstellung von Polyamidprofilen. Final report on the research project AiF-Nr. 7362, Institute of Plastics Processing (IKV) at RWTH Aachen University.

37 Michaeli, W., Berghaus, U., and Bartilla, T. (1990) Prozeßanalyse und Auslegung von Gleichdralldoppelschneckenextrudern zur Polymerisation. Final report on the research project AiF-Nr. 7019, Institute of Plastics Processing (IKV) at RWTH Aachen University.

38 Michaeli, W. and Grefenstein, A. (1995) Engineering analysis and design of twin screw extruders for reactive extrusion. *Adv. Polym. Technol.*, **14** (4), 263–276.

39 Mortimer, C. and Müller, U. (2010) *Chemie: Das Basiswissen der Chemie*, Georg Thieme Verlag, Stuttgart.

40 Michaeli, W. and Rothe, B. (2010) Innovatives Herstellungsverfahren von Polyamid 6-Nanocompounds auf Basis von Schichtsilikaten und ε-Caprolactam. Final report of the BMBF research project no. 01RI0624A Part1, Institute of Plastics Processing (IKV) at RWTH Aachen University.

41 Mori, S. and Takeuchi, T. (1970) Gel permeation chromatography of linear monomer and oligomers in polyamides. *J. Chromatogr.*, **50** (3), 419–428.

42 N.N. (2004) Process for preparing a melt-processable polyamide composition. EP Patent 1,594,910.

43 N.N (2011) *Verordnung (EU) Nr.10/2011 der Kommission über Materialien und Gegenstände aus Kunststoff, die dazu bestimmt sind, mit Lebensmitteln in Berührung zu kommen*, Official Journal of the European Union.

44 N.N (2014) *AP-Nylon Caprolactam*, Produktinformation, Brüggemann Chemical KG, Heilbronn.

45 N.N (2016) *Diethylenglykoldimethylether*, Sicherheitsdatenblatt, Bernd Kraft GmbH, Duisburg.

46 N.N. (1994) Verfahren zur Herstellung verformbarer hochmolekularer Polyamide. DE Patentschrift 748,253.
47 N.N (1992) *Entgasen beim Herstellen und Aufbereiten von Kunststoffen*, VDI-Verlag, Düsseldorf.
48 Poindl, M. and Bonten, C. (2014) Reactive Extrusion of Cross Linked Block Copolymers – Structures and Potential as Impact Modifier for PA6. Conference Proceedings Society of Plastics Engineers ANTEC 2014. Las Vegas, USA.
49 Puffr, R. and Kubanek, V. (1991) *Lactam-Based Polyamides: Polymerisation, Structure, and Properties*, CRC Press, Boca Raton.
50 Pramoda, K.P., Liu, T.X., Liu, Z.H., He, C.B., and Sue, H.J. (2003) Thermal degradation behavior of polyamide 6/clay nanocomposites. *Polym. Degrad. Stab.*, **81** (1), 47–56.
51 Prusty, M. (2014) *Personal Message*, BASF SE, Ludwigshafen.
52 van Rijswijk, K., Bersee, H.E.N., Beukers, A., Picken, S.J., and van Geenen, A.A. (2006) Optimisation of anionic polyamide-6 for vacuum infusion of thermoplastic composites: influence of polymerisation temperature on matrix properties. *Polym. Test.*, **25** (3), 392–404.
53 van Rijswijk, K., Bersee, H.E.N., Jager, W.F., and Picken, S.J. (2006) Optimisation of anionic polyamide-6 for vacuum infusion of thermoplastic composites: choice of activator and initiator. *Composites Part A*, **37** (6), 949–956.
54 Reinecke, M. (2003) Polyamide im Motorraum. *Kunststoffe*, **93** (3), 20–21.
55 Reimschuessel, H.K. (1977) Nylon 6 chemistry and mechanisms. *J. Polym. Sci.*, **12** (1), 65–139.
56 Reimann, S. (1982) Entgasen von Polymeren auf gleichlaufenden, dichtkämmenden Doppelschneckenextrudern am Beispiel von Polystyrol. Dissertation at RWTH Aachen University.
57 Russo, S., Imperato, A., Mariani, A., and Parodi, F. (1995) The fast activation of epsilon-caprolactam polymerisation in quasi-adiabatic conditions. *Macromol. Chem. Phys.*, **196** (10), 3297–3303.
58 Rothe, B. (2010) Herstellung von Polyamid 6-Nanocompounds mit Schichtsilikaten durch reaktive Extrusion von Caprolactam. Dissertation at RWTH Aachen University.
59 Ueda, K., Hosoda, M., Matsuda, T., and Tai, K. (1998) Synthesis of high molecular weight nylon 6 by anionic polymerisation of epsilon-caprolactam. Formation of cyclic oligomers. *Polym. J.*, **30** (3), 186–191.
60 Ueda, K., Nakai, M., Hosoda, M., and Tai, K. (1996) Stabilization of high molecular weight nylon 6 synthesized by anionic polymerisation of epsilon-caprolactam. *Polym. J.*, **28** (12), 1084–1089.
61 Vernaleken, H. (1978) Die anionische polymerisation von epsilon-caprolactam. *Textil Praxis Int.*, **9**, 1021–1029.
62 Wang, N.H. (2001) Polymer extrusion devolatilization. *Chem. Eng. Technol.*, **24** (9), 957–961.

63 Wittmer, P. and Gerrens, H. (1965) Über die anionische Schnellpolymerisation von Caprolactam. *Die Makromol. Chem.*, **89** (1), 27–43.
64 White, J.L., Keum, J., Jung, H., Ban, K., and Bumm, S. (2006) Corotating twin screw extrusion reactive extrusion-devolatilization model and software. *Polym.-Plast. Technol. Eng.*, **45** (4), 539–548.
65 Yang, C.T. and Bigio, D.L. (1997) *Analysis and Modeling of Polymer Devolatilization in Screw Extrusion and Compounding Processes*. Conference Proceedings Society of Plastics Engineers ANTEC 1997. Toronto, Canada.

14

Industrial Production and Use of Grafted Polyolefins

Inno Rapthel[1], Jochen Wilms[2], and Frederik Piestert[2]

[1] BYK Chemie GmbH, Value Park Y 42, 06258 Schkopau, Germany
[2] BYK Chemie GmbH, Abelstrasse 45, 46483 Wesel, Germany

14.1 Grafted Polymers

Grafted macromolecules are defined by IUPAC as follows:
A macromolecule with one or more species of blocks connected to the main chain as side chains, these side chains having constitutional or configurational features that differ from those in the main chain [1].

This definition is very common but needs to be modified when talking about industrial grafted polymers. Industrially available grafted polymers are mostly produced based on non-polar polymers. Most often, polyolefins or polymers with a polyolefinic substructure are used.

These polymers are chemically modified via grafting with reactive monomers. Typically, unsaturated monomers containing additional functional groups are grafted onto the polymer by a radical addition mechanism. Common monomers are maleic acid anhydride and acrylic acid but styrene and acrylates are also used.

The grafted polymers differ in their properties compared to their non-grafted homologs. This extends the area of use of polyolefins by simple blending. The grafted polyolefins typically act as coupling agents in highly filled polyolefins, as compatibilizers between different incompatible polymers, as adhesion promotors in multilayer films and overmolding applications, and as impact modifiers.

A generic structural overview of a grafted polymer is shown in Figure 14.1.

Based on the choice of the unsaturated monomer grafted on the polyolefin, the resulting side chains can differ dramatically. When using acrylates or styrene, longer chains can be built up since these monomers are able to polymerize. The chain length and the amount of side chains can be controlled by the amount of radical starter used. Typical levels of grafting are 2–8 wt%, but higher levels can also be achieved. When using maleic acid anhydride as a monomer, the resulting product appears completely different than the products obtained with other grafted monomers since maleic acid anhydride cannot be homopolymerized. Some authors are of the opinion that only one molecule per grafting step is connected with the backbone [2] or that the grafting of maleic acid anhydride happens as a dimer [3]. However, the amount of maleic acid anhydride that can be

Reactive Extrusion: Principles and Applications, First Edition. Edited by Günter Beyer and Christian Hopmann.
© 2018 Wiley-VCH Verlag GmbH & Co. KGaA. Published 2018 by Wiley-VCH Verlag GmbH & Co. KGaA.

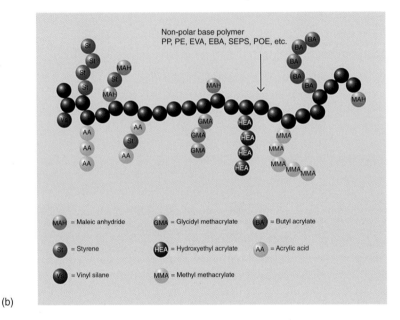

Figure 14.1 Chemical process of synthesis (a) and possible structures of grafted polymers achievable by grafting polar monomers onto a non-polar backbone (b).

grafted onto polyolefins is limited. Typical levels of maleic acid anhydride grafted onto polyolefins in an industrial process are in the range of 0.1–2.0 wt%. Higher levels of MAH-grafting can be achieved by using a second monomer like styrene and acrylates or methacrylates within the grafting process.

14.2 Industrial Synthesis of Grafted Polymers

Grafted polyolefins have become more and more common since the mid-1980s. Free radical grafting is the method of choice for the synthesis of these modified polymers. In general, there are four main technologies available for grafting monomers onto a polymeric backbone: melt phase grafting, solid state grafting, grafting in solution, and grafting in emulsions or suspensions.

A brief summary of the advantages and disadvantages of these technologies is given in Table 14.1.

A good overview on grafting technologies can be found in Rzayev [5].

Basic work for grafting of unsaturated monomers onto polyolefin backbones was already carried out in the 1960s [6, 7]. The first backbones used then

Table 14.1 Overview of the major technologies used for free radical grafting.

Technology	Advantage	Disadvantage	Main application
Melt phase	• Simple technology • No limit regarding melting point of backbone polymer	• Limited grafting level by short reaction time	Grafted polyolefins; low M_w SEBS
Solid state	• High grafting possible • Grafting of PP without β-scission possible • Grafting of high M_w SEBS possible • Grafting of various monomers possible	• Grafting only on partially crystalline polymers or very high molecular polymers	Grafted polyolefins High M_w SEBS
Solution	• Very homogeneous grafting • High grafting levels • No degradation	• High production cost • Waste solvent	Grafted polyolefins (gel-free)
Suspension/emulsion	• Use of sticky polymers possible • High grafting levels	• High production cost	Grafting of fibers High grafting of acrylates on crosslinked polyolefins

were polypropylene and polyethylene and, till today, they continue to be the main polymers to be used as basic polymeric backbones for grafting. Recent developments based on polymers like EVA, EBA, and SEBS have become more and more important.

For industrial synthesis, the most frequently used technology today is melt grafting. Currently, various companies offer state-of-the-art products using the melt grafting process. Solid state grafting has become more and more important over the years since products based on this technology show an improved performance compared to products based on melt grafting. In addition, products with unique performances can be made using solid state grafting since the variety of usable polymers for grafting is higher than for a melt grafting process.

Grafting in a solvent or onto a fine dispersed or emulsified polymer in water is based mostly on academic research, and only a few applications for the synthesis of grafted polymers using this technology are known. Bergström describe the grafting of a thermoplastic vulcanizate (TPV) with acrylic monomers [8]. Some grafted products using a technology based on solution grafting are used for multilayer film applications.

Since melt grafting and solid state grafting are the key technologies used, these technologies will be described more in detail.

14.2.1 Melt Grafting Technology

Many articles dealing with melt grafting can be found in the literature. Most of them deal with academic work performed in small kneaders and extruders. Some

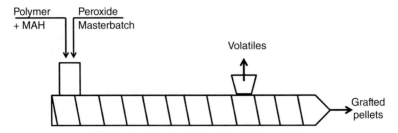

Figure 14.2 General principle of melt grafting via extrusion.

basic work was done by Rengarajan *et al.* [10] and Chandranupap *et al.* [35]. They show very interesting results pertaining to high grafting levels and low levels of side reactions. These results are often only possible due to very long residence times as well as to the possibility of using lower temperatures during a batch kneading process.

From an industrial point of view, it is mandatory that a continuous production process results consistently in obtaining a quality product and is also cost-effective. For this, an extrusion process is obviously the preferred method.

Figure 14.2 shows a typical configuration of an extruder used for melt grafting. The extruder is typically a co-rotating twin-screw extruder. The unsaturated monomer and the peroxide are fed into the main feeder along with the polymer. The feeding of the unsaturated monomer and the peroxide (when liquid peroxide is used) can also take place as liquids downstream by Sude stream feeding. In any case, the grafting takes place in the melt phase at the first parts of the extruder. Grafting can be influenced by:

- The choice of polymer
- Choice of peroxide and reaction temperature
- Extruder configuration (shearing rate, retention time) and screw design.

The influence of the retention time of the reaction mixture in the extruder is limited. Typical values for the retention times for industrial melt grafting processes are 30–180 s. It is important to note that process parameters, such as temperature, need to be chosen such that the grafting reaction needs to be completed within this short period of time. In addition, desorption of volatiles coming from unreacted monomers and from decomposition of the peroxide must be possible by venting. A general setup of an extruder for melt grafting including, for instance, the screw design and a temperature profile does not exist. For the desired grafting reaction, these parameters need to be evaluated individually.

14.2.2 Solid State Grafting Technology

Solid state grafting was described as an alternative method to melt grafting for many years [4, 9, 10]. The main attention paid to this technology was from an academic point of view. Up to now, only a few companies have used solid state grafting as a technology on an industrial scale. Initially, in the early 1990s, the company Montell started to commercialize some products made by solid state

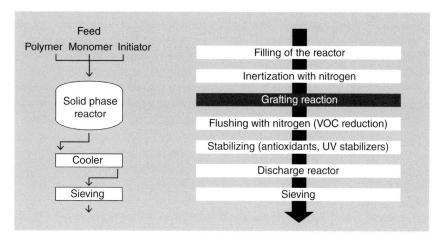

Figure 14.3 Working steps of solid state grafting in an industrial scale.

grafting [11]. At about the same time the former company BUNA GmbH started to produce a few products by this technology, based on polypropylene [12].

In 2001, KOMETRA GmbH started to offer grafted polyolefins synthesized by solid state grafting based on the technology from BUNA.

Solid state grafting is a batch process with long reaction times and very mild reaction conditions.

A brought range of polyolefins can be used as backbone for solid state grafting. They have to be partially amorphous or fully amorphous since the grafting only happens in the amorphous phase of the solid polymer.

Typically, polymeric powders or reactor grade flakes with particle diameters of 200–700 µm are used. The reaction step happens in special reaction vessels. The principle of solid state grafting is shown in Figure 14.3.

After inertization of the vessel with nitrogen and initiating heating, the monomers to be grafted and peroxides diffuse into the amorphous parts of the polymers (Figure 14.4).

The processing temperature for solid state grafting processes is in any case much lower than that required for a melt grafting process. The velocity of a reaction depends only on the kinetics of degradation of the peroxide. Depending on the polymer used a proper choice of the peroxide is eminent for sufficient control of reaction time for solid state grafting.

The use of reaction temperatures for solid state grafting below the melting point of the polymer also reduces the thermal degradation of the polymer used. In case of polypropylene as the polymer, the β-scission of the polymer radical is reduced to a minimum [15, 16]. If degradation in molecular weight is required, it can be easily controlled by the choice of mild reaction conditions combined with an extension of the reaction time. Figure 14.5 shows the different possible pathways in an example of grafting maleic acid anhydride onto polypropylene.

The possibility of controlling reaction conditions is one of the most important advantages of solid state grafting. The rate of degradation is dependent on the reaction conditions chosen; even at higher grafting rates the degradation can be

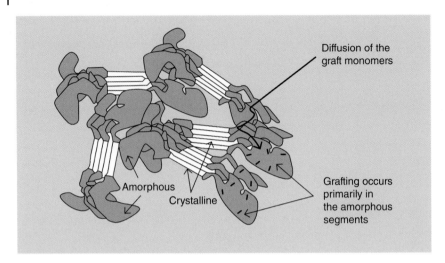

Figure 14.4 The principle of solid state grafting.

suppressed. In case of melt grafting, degradation is a significant side reaction and is dependent on the grafting rate: the higher the amount of grafted monomers, the lower is the molecular weight by degradation.

14.3 Main Applications

14.3.1 Use as Coupling Agents

One of the main applications of grafted polyolefins is its use as coupling agents for polar fillers and also for fibers to enable coupling of these fillers to a polymer matrix. This includes glass fiber reinforced polyolefins, especially polypropylene; wood plastic composites based on polypropylene and high density polyethylene; and the use of $Al(OH)_3$ or $Mg(OH)_2$ as fillers for flame retardant compounds, especially for cable compounds based on PP, PE, or EVA.

Polypropylene Glass Fiber Compounds The use of glass fibers as reinforcing materials in polypropylene compounds broadens the possible applications for polypropylene. The increased stiffness and heat deflection temperature (HDT) values of the compound compared to those of neat polypropylene allows the change of the polymeric matrix from the more expensive polyamide to fiber reinforced polypropylene in a number of different applications.

The main applications for glass fiber reinforced polypropylene compounds are:

- Parts for the car industry
- Parts in so-called white goods
- Pipes, tanks, and containers.

The fiber reinforced compounds can contain short or even long glass fibers, depending on the mechanical properties required. Nevertheless, a key topic in

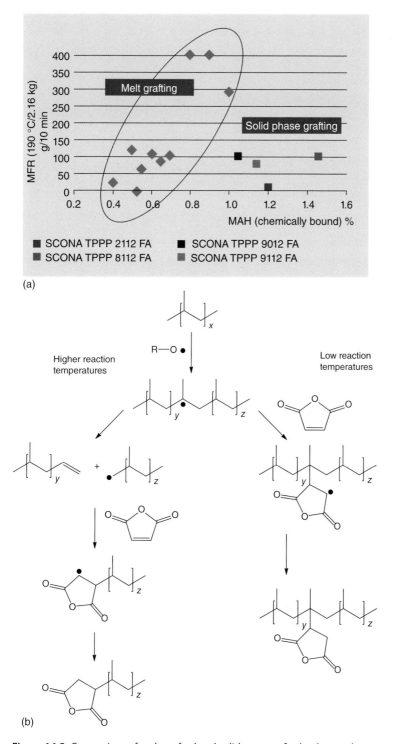

(a)

(b)

Figure 14.5 Comparison of melt grafted and solid state grafted polypropylene types (a) and reaction mechanism of grafting polypropylene with maleic acid anhydride using different reaction conditions (b).

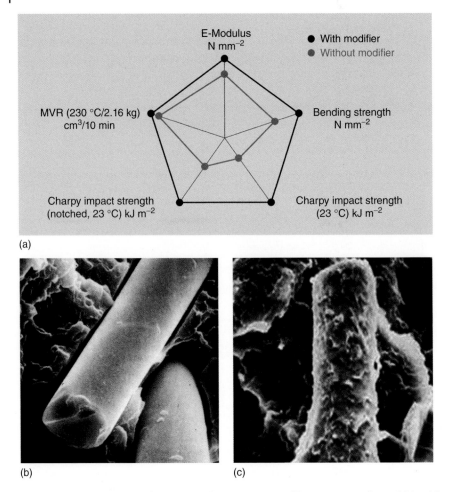

Figure 14.6 Improvement of properties of PP GF compound by using a coupling additive (a) and REM-figure of a polypropylene glass fiber compound without coupling agent (b) and with a solid state grafted polypropylene containing maleic acid anhydride groups (SCONA TPPP 9012 FA) as coupling agent (c). (Reproduced with permission of BYK-Chemie GmbH.)

these formulations is a strong coupling of the polymeric matrix to the fibers. This can be achieved by using coupling agents based on grafted polymers (Figure 14.6).

Two major factors have a significant impact on the coupling of glass fibers to polyolefins: the right choice of the coupling agent and the processing parameters of the compound.

In case of the processing parameters, the choice of the right screw configuration for the extruder is very important. The screw configuration should ensure a fast and complete dispersion of the glass fibers in the polymer. If the kneading is too strong, this will result in an overly strong distortion of the glass fibers. Figure 14.7 shows an exemplary screw design fulfilling this requirement, although, multiple other individual screw designs are also capable of fulfilling these requirements. There are a lot of other influences based on the compounding, such as by

Figure 14.7 Exemplary screw design used for compounding of glass in polyolefins.

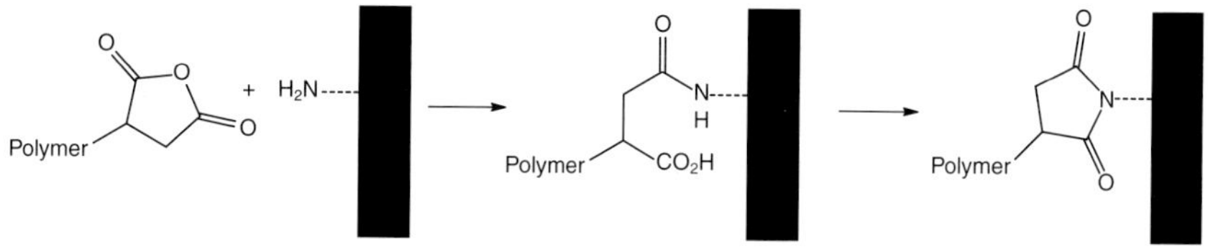

Figure 14.8 Reaction of the coupling agent with the sized glass fiber. (The black square illustrates a sized glass fiber containing primary amine groups.)

additional fillers, pigments, UV stabilizers, and lubricants that will not be mentioned here.

The right choice of coupling agent is eminent for the quality of the compound. For this, a better understanding of the coupling mechanism of grafted polymers to glass fibers is important.

Glass fibers for compounds with thermoplastic polymers are sized mostly with a functional silane [13]. For polypropylene, besides a wax the sizing contains an aminosilane. The silane group attaches to the glass surface, while the amine group stays free. The coupling will then take place by reaction with the acidic groups of the polypropylene grafted with acrylic acid or maleic acid anhydride to form amine or imide groups as illustrated in Figure 14.8. Acrylic acid grafted polypropylene was used successfully as a coupling agent in the past [14]. Today, maleic acid anhydride grafted polyolefins are used because of much better effectivity of coupling due to their higher reactivity properties.

Theoretically, for effective wetting of glass fibers, the best possible structure of an optimal coupling agent for the reaction with an aminosilane would be a structure which is built after β-Scission Regarding the grafting rate, a quite low grafting level of maleic acid anhydride is enough for an optimal coupling agent, if every polymer molecule is grafted by exactly one molecule of maleic acid anhydride. As an example, using a polymer with an melt flow index of 100 g/10 min (measured at 190 °C, 2.16 kg) has a molecular weight of about 100 000 Da. If exactly one molecule of maleic acid anhydride ($M_w = 98$ g mol^{-1}) is attached to the macromolecule, a grafting level of 0.1 wt% would be enough for an optimal coupling. Because the grafting process is a statistical process, the chance of matching every polymer molecule with at least one monomer is higher when using a higher level of grafting. The higher the grafting level, the better the fiber wetting and coupling to the glass fiber. A coupling agent with a higher grafting level at comparable MFR is therefore able to wet the surface more completely at lower dosage levels.

Comparing two formulations based on polypropylene with 30 wt% glass containing different coupling agents with small differences in grafting level demonstrates significant differences in performance. Figure 14.9 shows such a comparison in the performance of two products with different grafting levels.

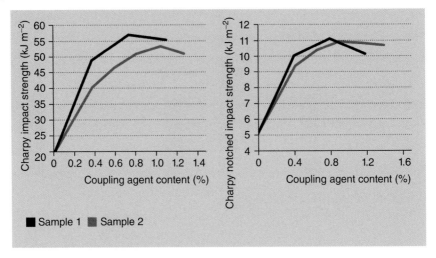

Figure 14.9 Comparison of performance of two grafted polyolefins with different grafting levels regarding impact strength.

Sample 1 (SCONA TPPP 9012 FA) has a grafting level of maleic acid anhydride of 0.9 wt%, Sample 2 of 0.7 wt%. The higher grafted product Sample 1 comes to the maximum level of Charpy impact strength at 0.8 wt% dosage, the lower grafted Sample 2 needs a dosage of 1.1 wt% for the same result. Overdosage of both coupling agents led to a little drop down of properties.

As long as the coupling agent is bound to the fiber it increases the Charpy impact strength. If the surface of the fibers is saturated with the coupling agent, an excess of coupling agent causes a little drop down in mechanical properties based on the amount of free coupling agent in the polymer matrix. As a consequence it is strictly recommended to determine the maximum concentration of the coupling agent required for covering the surface of the glass fiber.

An overdosage of the coupling agent makes sense only in a few cases. One of the more common exceptions is a glass fiber reinforced polypropylene compound used for contact with water (hot or cold) or soap suds. These compounds typically have to fulfill special stringent test procedures [17]. Here, the water or the soap suds can attack the sized surface of the glass fiber. This can cause a significant drop down of properties more than that is allowed for the recommended use for non-passing the requirements.

The reason is illustrated in Figure 14.10. The maximum mechanical properties that is required can already be achieved with a virtually total coverage of glass fibers. But there is still space for the attack of water or soap suds. For covering the complete surface, a higher dosage of coupling agent is needed. These phenomena are known from multilayer film application and cable compounds.

All these principles of grafting are more or less the same for short glass fiber compounds as for long glass fiber compounds. Depending on the grafting technology used, the same coupling agent can be chosen. The use of grafted polyolefins as coupling agents leads to the enhancement of the properties but the improvement is limited to a certain extent. If the glass fiber is wetted completely

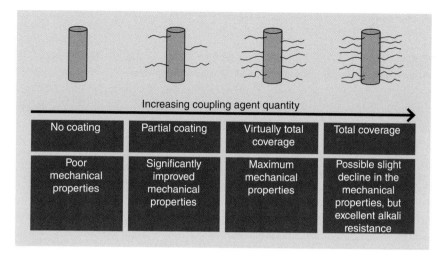

Figure 14.10 Influence of increasing coupling agent quantity on the properties of the PP/GF compound.

by a coupling agent no further improvement at the given concentration of glass fiber is possible. An additional improvement can then typically be achieved by increasing the content of glass. However, this increases the specific weight of finished parts.

Another possibility of getting better results without increasing the weight too much is by the use of a hybrid compound made of glass fiber reinforced polypropylene combined with polyethylenetherephthalate (PET). PET can be dispersed in PP by using special coupling agents as fillers in fiber form [18]. The fiber form is built *in situ* by using less than 10% PET in the compound. If more PET is used the morphology of PET in the PP matrix changes to spherical parts.

These principles can be used for the production of hybrid compounds with GF and PET. Table 14.2 shows the properties of a 30% short glass fiber reinforced polypropylene compound and a 30% long glass fiber reinforced polypropylene compound in Table 14.2.

The properties of the hybrid system (short GF and PET) achieves nearly the same properties by using short glass fibers like the compound using long glass fibers or an increased amount of 10% of short glass fibers. For this special compounding, a bimodal coupling agent is necessary.

Wood Plastic Composites (WPC) Wood plastic composites (WPC) is a fast growing market. The first trials to use wood flour as filler began a long time ago [19]. But since the beginning of the twenty-first century these compounds have become more and more interesting. Today, the use of wood-flour-filled thermoplastics is self-evident due to a number of aspects, not only due to sustainable requirements.

The choice of thermoplastics that can be used for this topic depends mostly on the processing temperature (see Figure 14.11) of the plastics material [19]. Most WPC products on the market are based on PP, HDPE, or PVC. But only for products based on HDPE and PP a coupling agent is necessary. Polypropylene has

Table 14.2 Comparison of short glass fiber reinforced polypropylene polyethylenetherephthalate hybrid compound.

Mechanical property	ISO test method	PP – 30% Short glass fiber +2.5% TPPP 6102 +10% PET (hybrid system)	PP – 30% Long glass fiber (commercially available compound from TICONA [33])	PP – 40% long glass fiber (commercially available compounds from A. Schulman [34])
Tensile modulus (MPa)	527	7000	6800	7800
Tensile strength (MPa)	527	100	110	84
Elongation at break (%)	527	4.5	2.2	4.3
Charpy impact strength at RT (kJ m^{-2})	179/1eU	55	50	57
Charpy impact strength (notched) at RT (kJ m^{-2})	179/1eAF	10	20	12
Heat deflection temperature HDT A (°C)	75	148	156	141
Density (g cm^{-3})	1183	1.14	1.12	1.23

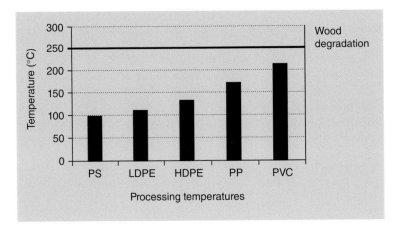

Figure 14.11 Processing temperatures for commonly used thermoplastics compared to the temperature of wood degradation [19].

better mechanical properties; high-density polyethylene has a better UV stability. So both polyolefins are typically used in the WPC market. Since both types of polymers are very non-polar, the adhesion between the matrix polymer and the natural filler is poor.

The more filler material used, the higher the E-modulus and lower the toughness of the compound. For improving these properties, again coupling agents are necessary.

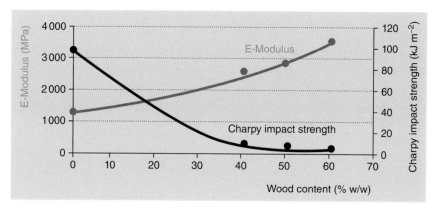

Figure 14.12 Influence of wood flour content on basic mechanical properties of PP-based WPC.

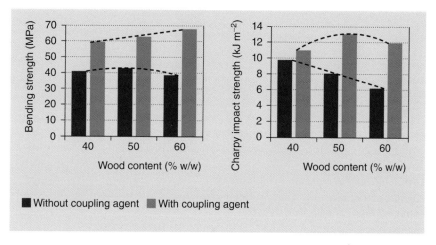

Figure 14.13 Bending strength and Charpy impact strength for PP/wood flour (60/40 wt%) with and without a coupling agent.

Figure 14.12 illustrates the improvements of WPC-based properties on polypropylene by using coupling agents. A rule of thumb for the use of coupling agents in WPC applications is: the higher the wood flour content in the formulation, the better the performance of the compound by using a grafted polymer as a coupling agent, as shown in Figure 14.13. The following properties are improved by using a coupling agent in WPC:

- Increased bending strength
- Increased impact strength
- Increased thermostability (HDT)
- Decreased water absorption with better dimension stability
- Better surface finish.

Multiple methods for coupling polyolefins to wood have already been tested. Bledzki and Gassan [21] described the use of maleic acid anhydride and silanes

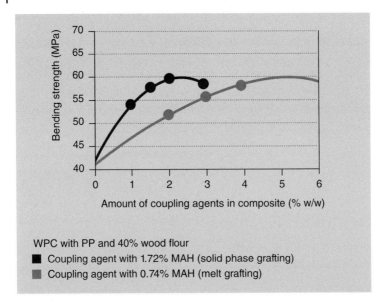

Figure 14.14 Comparison of the influence of coupling agents with different grafting levels onto bending strength of WPC based on polypropylene with 40 wt% wood flour [22].

for coupling of polypropylene to wood particles. Lu *et al.* [20] reviewed most of these trials: maleic acid anhydride, glycidylmethacrylates, isocyanates, peroxides, and other oxidizing systems were used with different success. Today, many well-working coupling agents based on maleic acid anhydride grafted polyolefins are available on the market.

The definition of an optimal modifier structure is difficult and should be discussed separately for PP and PE.

The first experiments toward a systematic understanding regarding the improvement of properties of WPC were carried out with coupling agents that were synthesized for polypropylene glass fiber compounds by Kreiter and Golombek [22]. They found out that the coupling to wood is similar to the coupling to glass: This is shown in Figure 14.14. Properties of the compound are increased as long as the coupling agent is interacting with the surface of the filler; However, an excess of coupling agent leads to decreasing properties. The amount of coupling agent needed is higher for wood than for glass fiber reinforced compounds. The reason is quite simple: the surface of the wood flour is much larger than the surface of a glass fiber.

Figure 14.15 illustrates the coupling mechanism of a maleic acid anhydride grafted polypropylene onto the surface of wood particles. The anhydride ring opens during esterification with one OH group of the cellulose. The second acid group is not able to build an ester group with another OH-group under the reaction conditions chosen.

Regarding the influence of the grafting rate on the coupling agents with similar molecular weight, products with a higher grafting rate perform better than products with a lower grafting rate. The higher the grafting rate, the lower the

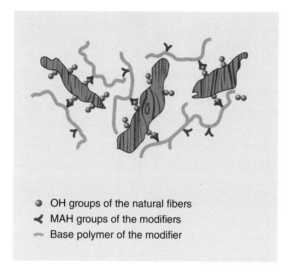

Figure 14.15 Coupling mechanism of maleic acid anhydride grafted polypropylene on wood (fibers).

concentration of the coupling agent needed. These results were mentioned by Specht [23] when comparing different grafted polypropylenes for a compound containing 40% wood flour. The comparison of different grafted coupling agents with different maleic acid anhydride levels of 1.0 wt% and 1.4 wt% is shown in Figure 14.16. A coupling agent with 1.4 wt% bounded maleic acid anhydride shows best performance at this dosage level compared to the dosage level of the product with 1.0 wt% bonded maleic acid anhydride. Here you need a dosage of 2.5 wt% for the same performance. A coupling agent with 0.5 wt% bound maleic acid anhydride does not come to this point during this examination. The exception from this rule is a very high grafted PP wax.

The following calculation will give an explanation for the findings: For a wax with a molecular weight of 10 000 Da, the minimum amount of maleic acid anhydride to graft a single monomer on every polypropylene chain is around 1 wt% (M_w of maleic acid anhydride: 98 g mol^{-1}). Therefore, a 7 wt% grafted polypropylene wax would be an effective coupling agent with maximum dosage around 1.4 wt% for comparable performance. Due to its low molecular weight, the entanglement of the chains of wax with the polypropylene matrix would also not be as strong as for a high molecular weight polypropylene coupling agent. Such low

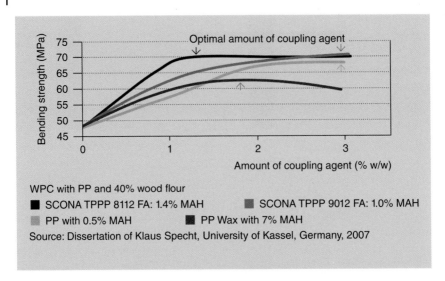

Figure 14.16 Example for search of optimal dosage of a coupling agent at various grafting levels.

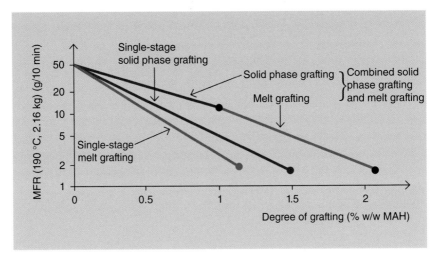

Figure 14.17 Improving the level of grafting by combining solid phase grafting with melt grafting in a two-step synthesis.

M_w coupling agents are recommended if the compound is prepared in a hot cooling mixer system or in a conical twin-screw extruder with very low shearing (Figure 14.17).

To design coupling agents for wood plastic compounds, the target is to get high grafting levels of maleic acid anhydride onto polypropylene with only a minimum of β-scission during the grafting step. This is achievable by using low temperatures during the synthesis step.

In case of polyethylene-based coupling agents, the possible synthesis route is based on a combination of two synthesis steps. Since polyethylene tends to show recombination of radicals before adding a reactive monomer such as maleic acid anhydride, the starting point of the synthesis is a polymer with a melt flow index as high as possible. Therefore, for a very high grafted PE the meltindex (MFR) is low in any case. Single melt grafting or single solid state grafting gives a grafting level of 1.3–1.6 wt% maleic acid anhydride if the MFR should not be less than 1 g cm^{-3} at 190 °C and 2.16 kg. With the same MFR a combination of solid state grafting and melt grafting give a grafting level of 2.1 wt%. In the same way the grafting level of HDPE can be pushed up to higher levels.

For polyethylene-based WPC, coupling agents based on grafted HDPE would be the first choice. But there are also other possibilities for an optimal coupling based on LLDPE-based coupling agents, which are worth mentioning. In general, it is also possible to use a very high grafted LLDPE (2 wt% maleic acid anhydride) if a very good kneading extruder is used for compounding (see Figure 14.18).

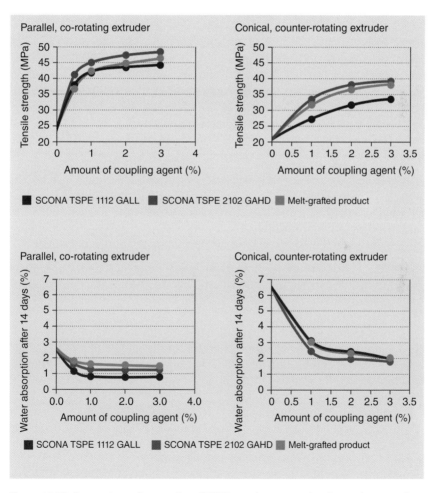

Figure 14.18 Comparison of properties of HDPE wood compounds using various coupling agents based on maleic acid anhydride grafted polyethylene (SCONA TSPE 1112 GALL: LLDPE-*g*-MSA, SCONA TSPE 2102 GAHD: HDPE-*g*-MSA).

In Figure 14.18, the graphs on the left show the results for bending strength and water absorption (tested by 14 days of water storage) for a formulation by a co-rotating twin-screw extruder. The graphs on the right show the results for the same for a formulation compounded with a conical twin-screw extruder. In general, the properties for the co-rotating twin-screw extruder are better for both, mechanical properties and water absorption. The mechanical properties for a conventional melt grafted HDPE-based coupling agent with a grafting level of 1.3 wt% maleic acid anhydride are nearly at the same level; a very high grafted HDPE with 1.7 wt% maleic acid anhydride shows a slightly better performance. The water absorption in this case is lowest with the very high grafted LLDPE due to a more ideal and strong surface coverage of the wood particles. Therefore, the very high grafted HDPE is the best choice in respect to all properties.

14.3.2 Grafted Polyolefins for Polymer Blending

14.3.2.1 Reactive Blending of Polyamides

Most polymers used in a technical scale are immiscible. Blending of such immiscible polymers can be enabled by dispersing one polymer in the main phase. In case of polyamide, the difference in polarity compared to polyolefins is enormous. A fine dispersed mixture of one in the other is only possible by using "detergent molecules." Polyamides are very reactive against acids and acid anhydrides. Therefore, the detergent molecules for reactive blending with a big range of polyolefins can be built *in situ* during the compounding.

The following three groups of reactive components are important for reactive polyamide blending:

1) Soft polyolefins grafted with maleic acid anhydride for getting a high toughness polyamide.
2) PP and PE grafted with maleic acid anhydride to make the polyamide cheaper and to reduce water absorption.
3) Polyolefins grafted with acrylic acid for viscosity regulation of polyamide.

The chemical reaction between an acidic grafted polymer and a polyamide forms a mixed polymer via transamidation. This structure is the true compatibilizer. A schematic reaction equation based on anhydride grafted polymer and PA 6 is shown in Figure 14.19.

The grafted maleic acid anhydride molecules are able to react with end standing NH_2-groups or they react with NH-groups in the middle of the polyamide via transamidation. By using trimellitic acid anhydride as a model substance, Lehmann et al. [25] showed that the splitting of polyamide occurs during reactive compounding.

This splitting reaction is the reason for creating enough detergent molecules if the right reaction conditions are used. Right reaction conditions refer to a strong kneading extrusion process. Figure 14.20 shows an example for a good kneading screw of a twin-screw extruder for polyamide blending.

Important to note here is the use of enough kneading and reverse flow elements in the screw after the melting zone. The load factor of the extruder should be around 80% and a retention time of at least 30 s is needed. A high shear rate is

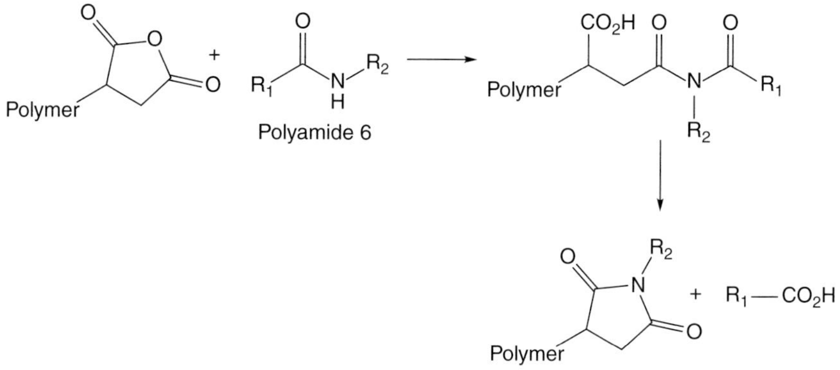

Figure 14.19 Mechanism of *in situ* production of compatibilizer for modified PA during extrusion.

Figure 14.20 Typical screw design for toughness modifying of polyamides.

necessary to have a fine dispersion of the second polymer component in the formulation or of the thermoplastic elastomer (TPE) phase (see Figure 14.20) that needs to be evenly distributed in the polymer blend.

If the compounding is done in a sufficient way, the *in situ* synthesized detergent molecules will decrease the surface energy of the polyolefin phase and stabilize the dispersion. The resulting bubbles of the originally immiscible polymer or the elastomer phase are smaller than 0.5 μm. The small size and the connection via detergent molecules are responsible for the high toughness of the polyamide. This is illustrated in Figure 14.21.

This principle is independent of the kind of polyolefin used. Rapthel *et al.* [27] have demonstrated this effect for the compounding of PA 6.6 with syndiotactic polystyrene (sPS). Figure 14.22 shows the results of compounding PA 6.6 with 20 wt% sPS without modifier (b) and by using additional 10 wt% maleic acid anhydride grafted sPS (grafting level 1.5 wt%) (a).

The compound without any modifier shows phase separation with bigger bubbles and sharp border lines. The compound with 10 wt% modifier shows much smaller bubbles with weak border lines. The authors show that the toughness of the PA 6.6 is stronger if the bubbles are smaller (about 0.5 μm). Some authors describe that the toughness of polyamide will decrease further if particle size is much smaller than 0.5 μm [24]. If properties of reactive polyamide blends are not in the expected range, the dispersion of polyolefin parts is not optimal.

The reason for increasing toughness is not only for the better dispersion of sPS parts. Additional reasons for increasing toughness are small particles connected by "detergent molecules" with the polyamide phase. During stretching, cracks can be built. Figure 14.23 shows building of these cracks in a polyamide 6.6/sPS blend. The increase of toughness in this case is low because the sPS is very rigid. The strength of the toughness depends on the kind of polyolefin used. If an elastomer

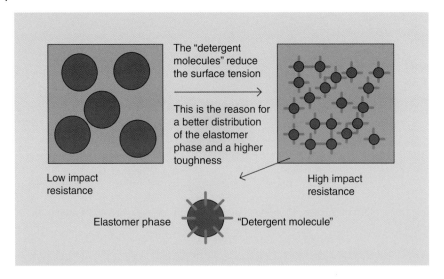

Figure 14.21 Dispersion of polyolefins (immiscible polymers or polymer-elastomer-blend) during reactive compounding.

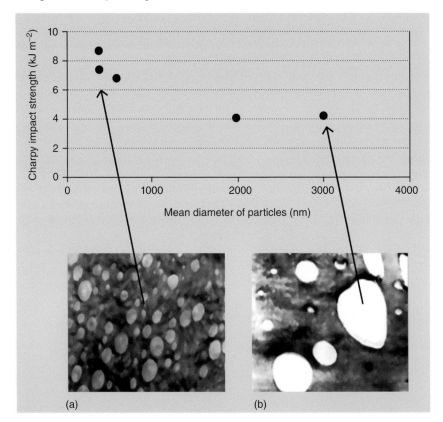

Figure 14.22 Polyamide blended with sPS, TEM-figure. Ultra-thin cut, contrast with RuO_4-formalin solution. The lighter spots show the sPS finer distributed (a) or worse distributed (b). (Seydewitz et al. 2004 [27]. Reproduced with permission of Carl Hanser Verlag GmbH & Co. KG.)

Figure 14.23 Building of cracks during stretching of a polyamide blend with sPS, TEM-figure. Ultra-thin cut, contrast with RuO_4. (Seydewitz et al. 2004 [27]. Reproduced with permission of Carl Hanser Verlag GmbH & Co. KG.)

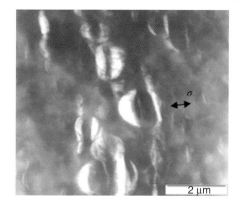

(based on styrene block copolymers (SEBS) or polyolefin elastomers (POE)) is used for blending of polyamide, the result will be an increase in toughness. For high toughness polyamide the use of modified elastomers is necessary. The most used elastomer components are shown in Table 14.3:

The right choice of modifier depends mainly on the goals of the user. Some criteria for choice are:

- Flow ability of grains
- Viscosity of melt, especially for injection molding
- Crystallinity of modifier (shrinking is influenced)
- Toughness at low temperatures (depending mainly from T_g of the backbone of the modifier)
- Price–performance ratio.

The price–performance ratio can be improved by using high grafted products that have to be diluted with neat polymers. To realize such a highly efficient product it is necessary to combine solid state and melt grafting in order to get very high grafting levels. For optimum size of soft particles in the polyamide phase, a grafting level of maleic acid anhydride of around 0.5 wt% is necessary. Common products have a grafting level that is mostly in this range. A higher grafting level causes a further decrease of the melt flow index of the finished compounds and does not increase the toughness of the polyamide.

Table 14.3 Most used backbone elastomers for high toughness modifier.

Raw material	Price	Physical state	Process ability
EPM/EPDM	Medium	Low crystalline/amorphous	Pellets sticky
POE	Medium	Low crystalline	Pellets, free flowing
SEBS/SEEPS	High	Amorphous	Pellets, free flowing

EPM, ethylene/propylene rubber; EPDM, ethylene/propylene/diene monomer rubber.

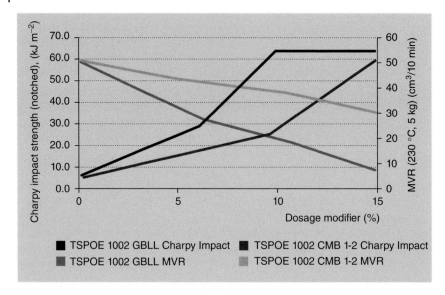

Figure 14.24 Results for testing maleic acid anhydride grafted POE modifier (different SCONA TSPOE 1002-Types) regarding toughness and MVR.

The grafting level of these precursors of compatibilizers is very important for further discussion. The assertion that the high grafting level of 1.5 wt% maleic acid anhydride has a big impact on the properties of the blend was proven by dilution of the material with neat POE. The tests were carried out with polyamide 6 with a relative viscosity of 2.8. The results are summarized in Figure 14.24. The higher grafted SCONA TSPOE 1002 GBLL has a faster increase of toughness and with a maximum level at a dosage of 10 wt%. The MVR of the polyamide blend drops down quickly from 60 to less than 10 cm^3/10 min, which makes this blend unsuitable for injection molding applications.

The diluted version of this modifier SCONA TSPOE 1002 GBLL with two parts of neat POE gives a slight increase in toughness. The value for 15% dosage is already comparable to the first experiment described earlier. However, the MVR for this sample is three times higher and the blend can be used for injection molding. The toughness at low temperatures (down to −40 °C) is in both cases comparable. The reason therefore is the very low glass transition temperature of the POE.

The same behavior shows highly grafted, easy flowing SEBS. The grades available on the market typically have grafting levels of 1.1 up to 1.7 wt% maleic acid anhydride. Dilution of these grades will give a similar effect as described earlier. These SEBS grades are amorphous and could be used to reduce the shrinkage of polyamide.

For reactive PA-PP and PA-PE blends, the same chemistry is used to get the materials compatible. Holsti-Mieittinen *et al.* [29] described how to use various modifiers for this purpose. A high molecular polypropylene grafted with only 0.15 wt% maleic acid anhydride gave good results for compatibilization, but

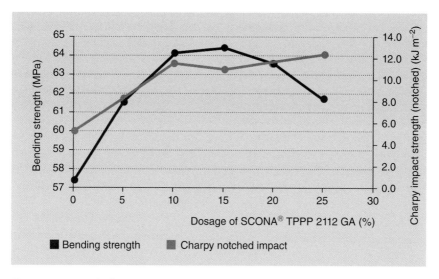

Figure 14.25 Results for testing a maleic acid anhydride grafted polypropylene (SCONA TPPP 2112 GA) (in a mixture 70 % PA, X % SCONA TPP 2112 GA, 30-X % PP) regarding bending strength and Charpy impact strength (notched).

only if this material was used in the polypropylene phase. Best compatibility was received by using a maleic acid anhydride grafted SEBS (Figure 14.25).

For the compatibilizing of polypropylene blends with polyamide the solid state grafting of polypropylene is the method of choice for the synthesis of compatibilizers. The products need a high grafting level and as little degradation of the polypropylene as possible [28].

For high quality blends of polyamide with polyethylene the back bone of the modifier needs to be compatible with the polyethylene phase. In case of blending PA with LDPE or with LLDPE the polymeric backbone of the compatibilizer should be LLDPE as well. A commercially available grade for this application is SCONA TSPE 1112 GALL, an LLDPE with a grafting level of 2.0 wt% maleic acid anhydride. This high grafting level allows a dilution of 1 : 3 or 1 : 4 with a neat polymer. In case of blending PA with HDPE, the modifier should be based on HDPE. Here, SCONA TSPE 2102 GAHD, a HDPE with a grafting level of 1.5 wt% maleic acid anhydride should be the first choice.

Besides maleic acid anhydride, acrylic acid can be used as graftable monomer for reactive compounding components. Unlike maleic acid anhydride, acrylic acid is a self-polymerizing monomer that builds longer side chains within the grafting process. In addition to this difference in grafting behavior, the mode of action is different to the anhydride grafted homologs.

The reaction of acrylic acid grafted polyolefins with polyamides causes also transamidation reactions, comparable to maleic acid anhydride based compatibilizers. Besides this, acrylic acid grafted polyolefins can react with the NH-groups of the amide structures and build out a branched structure as shown in Figure 14.26.

Blending of polyamides with acrylic acid grafted polyolefins can be used for high-toughness polyamides as well as for blends of polyamides with

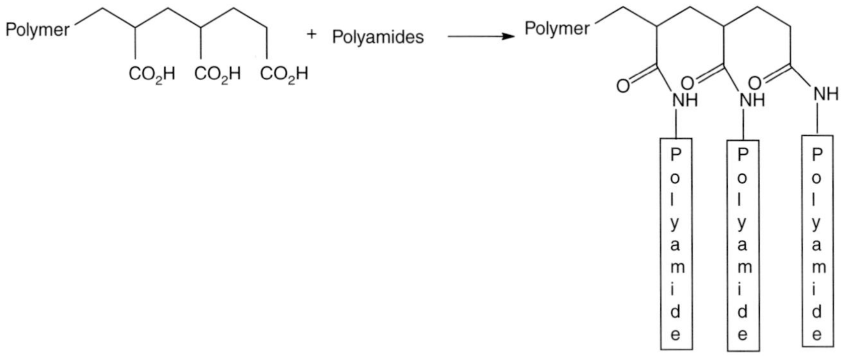

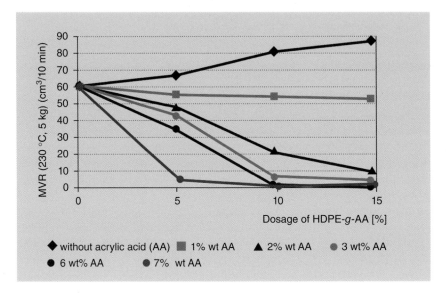

Figure 14.26 Reaction of acrylic acid grafted polyolefins with polyamides.

Figure 14.27 Viscosity regulation of PA 6 by using acrylic acid grafted HDPE (AA: acrylic acid).

polypropylene or polyethylene. The result is in any case a less-colored but heavy-flowing compound compared to compounding of polyamides with maleic acid anhydride grafted polyolefins. This effect is illustrated in Figure 14.27 as an example of use of acrylic acid grafted HDPE [30].

While comparing acrylic acid grafted HDPE with different levels of grafting for blends of polyamide 6 in dosages of 5, 10, and 15 wt%, the MVR (measured at 230 °C and 5 kg) can be taken as an indicator for viscosity of the blend. The control blend with HDPE without grafted acrylic acid results in a slight increase in MVR. The grafted homologs cause a decrease in MVR. The higher the grafting level the greater the drop down of MVR. The strongest effect is obtained with samples with a grafting level of 7 wt% or higher. During preparation of these blends there was an increasing danger that the mixing of the blends was not optimal since the rapid increase of viscosity of the blends causes inhomogeneity. An indicator for

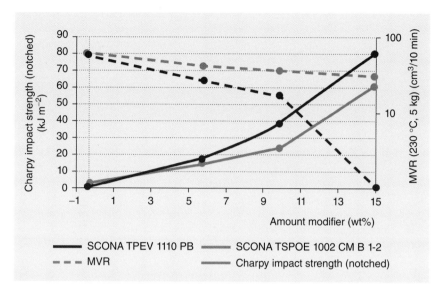

Figure 14.28 Comparison of polyamide 6 blended with maleic acid anhydride grafted POE and acrylic acid grafted EVA regarding toughness and melt flow of the blends.

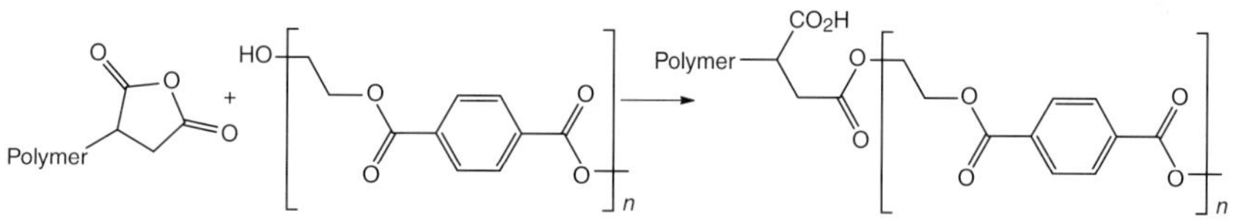

Figure 14.29 Reaction mechanism of PET with maleic acid anhydride grafted polyolefins.

this is the decreasing Charpy impact strength of these compounds. A grafting level of acrylic acid of 5–7 wt% seems to be the optimum for this application.

The increase of toughness is limited when using HDPE as a backbone polymer. Higher toughness can be achieved using grafted EVA.

The differences achievable when changing from HDPE to EVA as a backbone polymer for grafted compatibilizers with reference to the toughness and melt flow of blends are shown in Figure 14.28. Using a higher dosage of acrylic acid grafted onto EVA (with a content of vinyl acetate of 14 wt%) in blends with polyamide 6 gives a comparably high toughness to maleic acid anhydride grafted POE. But the melt flow of the resulting blends is much lower because the viscosity is higher.

Reactive Blending of Polyesters Reactive blending of polyesters follows the same rules indicated for polyamide. For achieving good blends with improved mechanical properties, enough reactions between the polyester and the grafted polyolefin as compatibilizer is needed (Figure 14.29).

In contrast to polyamides the reactivity of polyesters with grafted polyolefins is limited due to limited reaction partners for compatibilizers. Typically, the terminal OH- or COOH-groups of the polyester are available for reactions. In case the polyester contains more OH- chain end groups maleic acid anhydride grafted

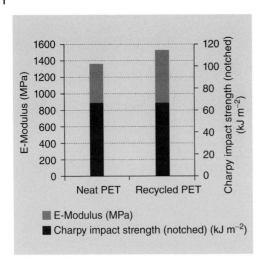

Figure 14.30 Comparison of neat PET and recycling PET compounds with 12% maleic acid anhydride grafted POE.

coupling agents are the best choice for reactive extrusion since more reaction partners for the acidic compatibilizer are present. This fact can be easily shown for PET. Thus, toughness modification of PET can be done by similar products as for toughness modification of polyamides. This topic is of interest for the use of recycled PET in injection molding applications or in strapping tapes. In case of recycled PET the possibility for a reaction with maleic acid anhydride grafted polyolefins is better compared to new PET material because the molecular weight of the recycled PET is lower, and the amount of OH end groups for the reaction with the compatibilizer is higher.

Figure 14.30 shows the results of PET recycling blends with maleic acid anhydride grafted POE comparing new and recycled PET. According to the reaction mechanism showed in Figure 14.29 the recycled PET has more possibilities for reaction with the grafted POE.

In case of PBT the polymer is terminated by acid groups. Therefore, epoxy-functional modifiers are used as toughness modifiers and compatibilizers for polyolefin blends. Until now some epoxy-functional terpolymers are the only commercially available modifiers on the market. For toughness modifying of PBT or PBT-GF compounds this is mostly the right choice.

There is actually an unmet need for products increasing the rigidity of these polyester containing blends and for more unipolar blends with PBT. Some epoxy-functionalized polyolefins made by solid state grafting can fulfill this gap in the near future.

14.3.3 Grafted TPE's for Overmolding Applications

Since the 1960s, TPEs became increasingly more important. They have rubber-like properties and are usable like normal thermoplastic materials. TPEs are styrene blocks containing polyolefins with a structure as shown in Figure 14.31.

A typical application for soft TPEs is overmolding used for many parts in the automotive sector and also in the sports industry and in tool manufacture [31].

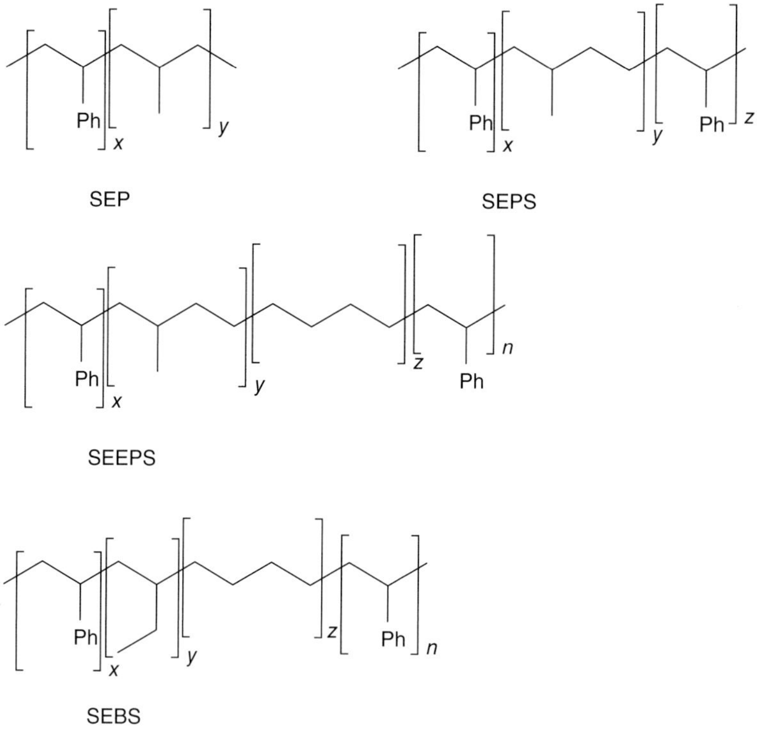

Figure 14.31 Possible structures of TPEs [32].

The hard part of the material combination can typically be PP, PA, ABS, PBT, or other technical polymers or even metal. The adhesion between PP and TPEs is good enough without any modification. The reason is the partial compatibility of both polymers. Most other polymer combinations used need grafted SEBS or SEEPS-types as adhesion promoters.

Grafting of reactive monomers onto the TPE backbone needs a more detailed discussion. Due to its structure, grafting can happen at different blocks of the polymer. The tertiary carbon atoms of the butyl, propyl, and styrene parts are the preferred positions for the first attack of a peroxide-initiated radical. If so, then a partial break of the molecules as described before for polypropylene (β-Scission) is possible. But since there are also longer CH_2 sequences available, radical-induced crosslinking is also possible. Studies on molecular weight before and after grafting have showed that the TPE polymer molecules react in both described ways: the higher the grafting the more a bimodal molecular weight distribution can be seen; molecular weight increase via crosslinking as well as molecular weight reduction via radical-induced degradation.

For low molecular weight SEBS products, this bimodal molecular weight distribution does not occur. Most of these low molecular weight SEBS polymers are melt grafted and have grafting levels from 1.1–1.8 wt% maleic acid anhydride. They are typically used as adhesion promotors.

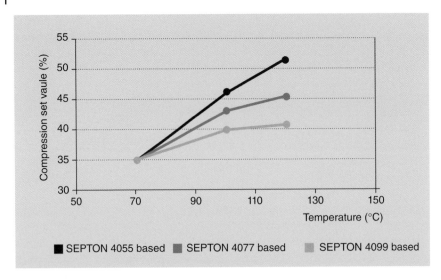

Figure 14.32 Compression set values of commercially available SEBS-based compounds (SEPTON: trade name for SEEPS polymers from Kuraray; with high viscosity oil) [32].

If an improved compression set is necessary, the use of high or ultra-high molecular weight TPE products is necessary. Figure 14.32 shows the improvement of compression set by increasing molecular weight for some SEPTON-type products of Kuraray [32].

If a common low molecular weight maleic acid anhydride grafted TPE is used for adhesion modification the improvement of compression set will disappear. The improvement of compression set will only be possible if a maleic acid anhydride grafted product is used, which has about the same molecular weight as the matrix polymer. High- and ultra-high molecular weight SEBS and SEEPS grades can be grafted in solid state, as well. Even if the products are amorphous the grains are stable enough during the grafting process. Such products can be grafted by using various conditions. It is important to consider the reaction mechanism described earlier. The improvement of compression set will be lost if too much β-scission happens. Otherwise, the processability will be more difficult because of too much crosslinking. Therefore, the grafting level of high molecular weight TPEs is mostly limited (0.9–1.3 wt%).

Some grafted high molecular weight SEBS and SEEPS products are already available on the market (e.g., KRATON FG 1901 G from KRATON Performance Polymers Inc., SCONA TSKD 9103 from BYK-Chemie GmbH, SONA TPKD 8103 PCS from BYK-Chemie GmbH). It seems like there is more potential for these kind of products for many applications in the near future.

A lot of TPEs for over-molding applications contain polypropylene for adjusting the hardness of the product. The polypropylene is partially soluble in the SEBS matrix. This means only a small part of the polypropylene is really molecular solved. The excess of polypropylene is fine dispersed in the compound as shown in Figure 14.33.

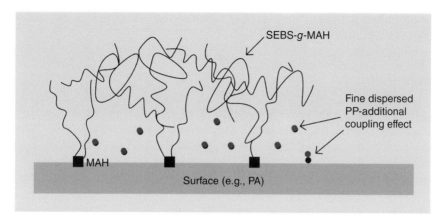

Figure 14.33 Schematic view on surface of SEBS over molded polyamide.

The fine dispersed polypropylene can inhibit the adhesion. To avoid this effect a part of the used polypropylene can be replaced by grafted polypropylene. The minimum content of maleic acid anhydride grafted in the polypropylene phase should be 0.2–0.3 wt%. The possibilities for grafting were discussed earlier. The best choice for this application is a high molecular weight polypropylene. One additional possibility to fulfill the demand for good adhesion could also be to use a polypropylene grade made for multilayer film applications completely as polypropylene phase. Maleic acid anhydride grafted polypropylenes with higher grafting levels can be used as well. But since they are mostly more or less degraded by melt grafting, they have an unintended influence on the MFR of the compound. A better solution is to use a solid state grafted, not degraded grafted polypropylene, such as SCONA TPPP 2112 GA or SCONA TPPP 2003 GB. Both grades have a grafting level higher than 1 wt% maleic acid anhydride and a molecular weight in the range of the neat PP phase.

14.4 Conclusion and Outlook

Grafting of polymers in the industrial scale is mainly done by two different grafting technologies: melt grafting and solid state grafting. Very few grafted products are made by grafting using polymer solutions. Especially, products grafted in the solid phase show outstanding performance for coupling onto glass fibers or wood in polyolefin compounds as adhesion promotors, and also in compatibilizing different immiscible polymers such as polyethylene or polypropylene with polyamide, PET, or TPEs.

Recent upcoming industrial developments for new grafted polymers will broaden the limits in different functional groups that are actually available. First epoxy-functional grafted polymers will be commercially available soon, while multiple different functional groups grafted on polyolefins are under evaluation.

Further, the megatrend of biopolymers in packaging materials and also fiber reinforced polymer compounds makes it necessary to also broaden the polymeric base of grafted polymers. Here, grafted biopolymers will be one of the future pathways for industrial grafted polymers.

References

1 IUPAC (1996), GLOSSARY OF BASIC TERMS IN POLYMER SCIENCE, *Pure Appl. Chem.*, **68**, 2287–2311.
2 Heinen, W. et al. (1998), 13C NMR STUDY OF THE GRAFTING OF 13C LABELED MALEIC ANHDRIDE ONTO PE, PP AND EPM, *Makromol. Symp.*, **129**, 119–125.
3 Gaylord, N.G. (1996), CRC Press Inc. Maleic Anhdride, Grafting, 3988–3991 in *CRC Handbook of Chemistry and Physics*, vol. 9/P (ed. J.C. Salamone).
4 Jia, D., Luo, Y., Li, Y., Fu, W., and Cheung, W.L. (2000), Synthesis and Characterization of Solid-Phase Graft Copolymer of Polypropylene with Styrene and Maleic Anhydride, *J. Appl. Polym. Sci.*, **78**, 2482–2487.
5 Rzayev, Z.M.O. (2011), Grsft Copolymers of Maleic Anhydride and Its Isostructural Analogues: High Performance Engeneering Matrials, *Int. Rev. Chem. Eng.*, **3**, 153–215.
6 Minoura, Y., Ueda, M., Mizunuma, S., and Oba, M. (1969), The reaction of polypropylene with maleic anhydride. *J. Appl. Polym. Sci.*, **13**, 1625–1640.
7 Ide, F., Kamada, K., and Hasegawa, A. (1968) *Kobinshi Kagaku*, **25**, 107–115; *Chem. Abst.* **69** (1968) 11031.
8 Bergström, C. (2003) *A Polyester Based TPV with Excellent Oil Resistance at High Temperatures.* TPE 2003, Brussels, Belgium 16–17. September 2003 87–93.
9 Lazar, M., Hrcková, L., Fiedlerová, A., Borsig, E., Rätzsch, M., and Hesse, A. (1996) Functionalization of isotactic poly(propylene) with maleic anhydride in solid phase. *Die Angew. Makromolekurare Chem.*, **243** (4236), 57–67.
10 Rengarajan, R., Vvicic, M., and Lee, S. (1989) Solid phase graft copolymerization: 2. Effect of toluene. *Polymer*, **30**, 933–935.
11 Picchioni, F., Goossens, J.G.P., and van Duin, M. (2005) Solid-state modification of isotactic polypropylene (iPP) via grafting of styrene. II. Morphology and melt processing. *J. Appl. Polym. Sci.*, **97** (2), 575–583.
12 Gerecke, J. et al., Verfahren zur Herstellung von polarmodifiziertem PP, (1990), DD 275 159 A 3.
13 Asmus, K.D. (1980) *Kunststoffe*, **70** (6), 336–343.
14 Gerecke, J. et al., Verfahren zur Herstellung von carboxylierten PP, (1990) DD 275161 A3.
15 De Roover, B., Sclavons, M., Carlier, V., Devaux, J., Legras, R., and Momtaz, A. (1995) Molecular characterization of maleic anhydride-functionalized polyprpylene. *J. Polym. Sci. A*, **33**, 829–842.
16 Borsig, E., Lazar, M., Fiedlerova, A., Hrckova, L., Rätzsch, M., and Marcincin, A. (2001) Solid-state polypropylene grafting as an effective chemical method of modification. *Macromol. Samp.*, **176**, 289–298.
17 Savignat, B. (2015) Functional polyolefins: maleated polyolefins as versatile HFFR coupling agents; Kabel 2015, Cologne 3–5.

18 Sarami, R., Ebrahimi, N.G., and Kashani, M.R. (2008) Study of polypropylene/polyethylene terephthalate blend fibres compatibilized with glycidyl methacrylate. *Iran. Polym. J.*, **17** (4), 243–250.
19 Walcott, M.P. and Englund, K. (1999) *A Technology Review of Wood-Plastic Composites*. 33rd International Particelboard/Composite Materials Symposium, pp. 103–111.
20 Lu, J.Z., Wu, Q., and McNabb, H.S. Jr., (2000) Chemical coupling in wood fiber and polymer composites: a review of coupling agents and treatments. *Wood Fiber Sci.*, **32** (1), 88–104.
21 Bledzki, A.K. and Gassan, J. (1997) Natural fiber reinforced plastics, in *Handbook of Engineering Polymeric Materials*, Abschn. 52, pp. 787–808 CRC Press, Cheremisiroff, N. P.
22 Kreiter, J. und Golombek, J. (2003) Use of WPC in outdoor-applications. Wood Fibre Polymer Composites Sympo-sium in Bordeau, März 2003.
23 Specht, K. (2007) PhD thesis. Alterungsverhalten von spritzgegossnenen hozfaserverstärkten PP-Verbunden University of Kassel.
24 Sykacek, E., Frech, H., and Mundigler, N. (2007) Eigenschaften hochgefüllter Holz-Polypropylen-Composites mit unterschiedlichen Haftvermittlern. *Österr. Kunststoff-Zeitschrift*, **38**, 12–15.
25 D. Lehmann, K.-J. Eichhorn, D. Voigt und U. Schulze (1996) Cabonsäureanhydridgruppen und Carbonsäregruppen in Polyamidschmelzen; KGK Kautschuk Gummi Kunststoffe; 49.Jahrgang, Nr. 10/96, 658–665.
26 TUFTEC (2010) Broshure April 2010; Asahi Kasei Chemicals Corporation; Synthetic rubber Division.
27 Seydewitz, V., Häußler, L., Rapthel, I., and Michler, G.H. (2004) Morphologie und mechanischen Eigenschaften: Blends aus Polyamid 66 und syndiotaktischem Polystyrol. *Kunststoffe*, **4**, 98ff.
28 Schöne, S. 2014 Graduation work, FH Merseburg, Germany.
29 Holsti-Miettinen, R., Seppälä, J., and Ikkala, O.T. (1992), Effects of Compatibilizers on the Properties of Polyamide/Polypropylene Blends. *Polym. Eng. Sci.*, **3.2** (13), 868–877.
30 Kröll, S. (2014) Technische Universität Dresden. Fakultät Maschinenwesen, Institut für Werkstoffwissenschaften, Lehrstuhl für Polymerwerkstoffe und Elastomertechnik.
31 http://www.kraton.com/about/history.php.
32 Kuraray Europe GmbH (2003) *Rapra TPE*, Kuraray Europe GmbH.
33 CELSTRAN TDS PP-GF30-04CN02/10.
34 A Schulman Inc. (2016) TDS Polyfort® FPP 40 GFC HI.
35 Chandranupap, P., Bhattacharya, S.N., Reactive Processing of Polyolefins with MAH and GMA in the Presence of Various Additives, *Journal of Applied Polymer Sience*, **78**, 2405–2415 (2000).

Index

a
acetylcaprolactam 353
aluminum trihydroxide (ATH) grades 303–306
amide-amide interchange 159
anthracene 76, 77
Arrhenius equations 47
Arrhenius laws 143
attenuated-total-reflection (ATR-IR) technique 111

b
Bagley corrections 119, 121
Banbury mixer 39
Bayer process 301
Beer–Lambert law 111
Bernoullian distribution 153
burning test 323
Buss Kneader technology 331

c
cable trials
    bedding compounds 322
    burning test 323–328
    final cable construction 322
    fire performance, according to EN 50399 322–323
    standard crosshead 322
capillary rheometry 121, 122
caprolactone (CL) 191
ε-caprolactone, polymerization of 55–59
Carreau model 91
Carreau–Yasuda law 50, 57
cationic starches 60
chemometric(s) 112, 115
chemometric model 116
compatibilizer concentration distribution (CCD) 210
conveying elements 16
copolymer 247
cyclopentanone 155

d
degree of exchange (DE) 155
degree of interchange 140
degree of polymerization (DP) 158, 160, 190, 194
degree of randomness (DR) 140, 141, 155
degree of substitution (DS) 59–61
Devaux model 141, 149
devolatilization/degassing 25–26
DGEBA-MDEA 198, 199
diameter of the dispersed phase domain distribution (DDD) 210
di-butyl-tin-laurate (DBTL) 309
diglycidyl ether of bisphenol A (DGEBA) 197, 200
Dirac function 72
distributive mixing
    elements 17
    kinematic modeling of 88–89
D_o/D_i ratio 24
downstream feeding 24
DSC 137, 140, 145, 148, 154, 156
dynamic vulcanization 40

Reactive Extrusion: Principles and Applications, First Edition. Edited by Günter Beyer and Christian Hopmann.
© 2018 Wiley-VCH Verlag GmbH & Co. KGaA. Published 2018 by Wiley-VCH Verlag GmbH & Co. KGaA.

e

epoxy-amine
  phase 198
  stabilization 202
epoxy-comonomer 196
ester-carbonate exchange 144
ester-ester exchange 139
esterification, EVOH 113
ethylene carbonate (ETC) 144
ethylene-propylene-diene
    tripolymers/polypropylene
    (EPDM/PP) 333, 334
ethylene vinyl acetate (EVA) 181
  chemical modification of 113
  copolymer 46
  mechanical properties and fire
    resistance 201
  polymer chains 190
  transesterification 52
ethylene-vinyl-alcohol (EVOH) 113

f

finite element method 42
Fire Performance of Electrical Cables
    (FiPEC) test 317
flame retardant wire and cable
    compound
  aging performance 315–317
  cable trials 322
  compounding line 306–308
  crosslinkable HFFR products
    one-step compounding process
      309–313
    two-step compounding process
      308–309
  crosslinking density 314–315
  heat release rate (HRR) 317
  mechanical properties 315
  non-polar compounds 318–321
  oxygen depletion calorimetry 317
  time to ignition (TTI) 317
flame retardant wire and cable
    compounds
  aluminium trihydroxide (ATH) 331
    crystallisation process 302
    endothermic decomposition
      303
  Martinal® grades 304
  mode of action 302–303
  ethyl vinyl acetate copolymers 300
  jacketing applications 299
  magnesium hydroxide (MDH) 331
  polar formulation 301
  polyethylene blends 300
  silane addition 301
  silane crosslinking
    PEX-b 299
    reactive extrusion 301
  silane grafting principle 302
fluorescence spectroscopy 104
fluorescent detector 74
flush mounted sensors 106
Fourier transform infrared (FTIR)
    spectroscopy 137, 140, 182
  methods 160

g

gel-permeation chromatography (GPC)
    121
glycidyl methacrylate (GMA) 114
graft polymers 271
grafted polyolefins
  coupling agents 375
    polypropylene glass fiber
      compounds 380
    wood plastic composites 385
  grafted TPE's, overmolding
    applications 400–403
  industrial synthesis
    free radical grafting 377
    melt grafting technology 377–378
    solid state grafting 377–380
    solution grafting 377
  maleic acid anhydride 375
  polymer blending
    polyamide reactive blending
      392–400
    polyesters reactive blending 399
grafting ratio 188

h

Haake batch mixer 76
heat distortion temperature (HDT) 28
heat release rate (HRR) 185, 186

HPLC pump 181
hybrid system 385
hydrolytic polymerisation 347–348

i

immiscible polymer blends
    Janus nanoparticles (JNPs)
        in situ mechanism 290
        PLLA/PVDF blends 291
        PMMA/PS and SAN/PPE blends 289
        polymer-polymer interfaces 289–293
    morphology development
        compatibilizer effects 253–254
        melt flow stage 251–253
        solid–liquid transition stage 249–251
    reactive comb compatibilizers 272
in-line RTD
    measuring system 73–74
    of preformance 76–77
in-line slit rheometry 117
inorganic particles, *in situ* sol–gel synthesis of 180
in-process rheometry 116–125
in-situ polymerization and *in-situ*
    compatibilization process 179
    vs. classical polymer blending 255, 257
    nano-structured polymer blends
        PP/PA6 nanoblends 257–264
        PPO/PA6 nanoblends 264
        PA6/core-shell blends 264–267
        polymer nanoblends 255–257
interpenetrating network (IPN) 28
ionic exchange capacity 187
ionomers 152
isophorone diamine (IPD) 197
isothermal thermogravimetric (TGA) 141

j

Janus nanoparticles (JNPs)
    in situ mechanism 290
    PLLA/PVDF blends 290
    PMMA/PS and SAN/PPE blends 289
    polymer-polymer interfaces 289

k

Kenics mixer 95
kinetic equations 44, 45, 56, 65
Kjeldahl method 60
kneading blocks 16–18, 20, 25, 102
    mixing elements 88
    types 82
kneading discs (KD) 47, 61
    description of 90
    distributive mixing performance 93–97
    efficiency 93–97
    flow channel 90
    numerical simulation 89–92
    staggering angle 79
    zone 79–81

l

length-to-diameter ratio 75
light transmission 73
linear polymerization 117
liquid crystalline polymer (LCP) 106
2-lobed kneading discs 20
3-lobed kneading discs 20
low density polyethylene (LDPE) 114
Lubonyl method 299
Ludovic© 44, 45, 57, 61–63, 182

m

machine torque capacity 29
MALDI-TOF MS 151, 163–165, 167
maleic anhydride (MA) 31–32, 113, 194
mass polymerization 3
mass spectrometry (MS) 137, 140
measured off-line (MFR) 110
mechanical energy transfer 25
melt elasticity 117
melt flow index 34
melt flow rate (MFR) 102
    test 120
melt grafting technology 377, 378
melt homogenizing effect 22

mercaptopropyltriethoxysilane (MPTES)   187
methylene diphenyl diisocyanate (MDI)   30
microscale   188, 189, 204
mixing bushings   25
mixing mechanisms   25
molar mass (MM)   135, 137, 139, 143, 150, 157
molar mass distribution (MMD)   138, 160
molar ratio   187
molecular weight (MW)   51, 103, 105, 117, 121
molecular weight distribution (MWD)   103, 117, 121
molten polymers   180, 183
monomer(s)   190
   ε-caprolactam   190
   conversion C(t)   56
   in situ polymerization   188
monomer/polymer ratio   112
Monte Carlo method   160

n
nano-structured polymer blends
   PP/PA6 nanoblends   257
   PPO/PA6 nanoblends   264
nanocomposites   179–188
nanometer scale   104
nanoscale   189, 204
nanostructuration   188–196
near infrared (NIR) spectroscopy   73, 109, 114
NMR   137, 140, 163
normal stress difference   104

o
on-line capillary rheometry   110, 120
optical fibers   109
optical spectroscopy   104
output/screw speed ratio (Q/N)   111
oxygen depletion calorimetry   317

p
PA6/core-shell blends   264
PA6 gas chromatography analysis   355
PA6 oligomers   151
partial Couette flow rheometer   106, 110
partial differential equations   50
partial least squares (PLS)   115
PET/MXD6 blends   152
PET/PEN, interchange reactions of   140
PET/poly(butylene terephthalate) (PBT)   143
piezo axial vibrator   122
plasticizers   196
plastics processing
   processing chain of   3
   supply chain of   4
PMMA/PS and SAN/PPE blends   289
polar monomers   194
polarimetry   105
polarized optical microscopy (POM)   105
poly(ethylene terephthalate) (PET)   32–33, 138, 140
   thermal-degradation mechanism   153
poly(phenylene oxidize) (PPO)   196
poly(vinylidene fluoride)/acrylic rubber (PVDF/ACM) blends   335
polyamide 6 (PA6)   185
   anionic polymerisation   348–352
   hydrolytic polymerisation   347–348
   residual monomers   347
   twin screw extruders
      acetylcaprolactam   353
      cellulose fibers   353
      hexamethylene diisocyanate and CPL   352
      material properties   353
      modified reaction system   353
      reaction system and experimental setup   354–356
      residual monomer content, degassing zones   354–364
      screw configuration and process conditions   353
polyamide 12 (PA12)   203
polyamide copolymer   186

polyamide/polyamide (PA/PA)   138, 159–166
poly-ε-caprolactone (PCL)   190
polycarbonate (PC)
  oligomers   143
  recycling   191
polycarbonate/polyamide (PC/PA)   138, 155–159
polycarbonate/polyester (PC/PEs)   143–148
polydimethoxysiloxane (PDMOS)   186, 187
polyetheramide triblock copolymer   190
polyetherimide (PEI)   196
polyester/polyamide (PE/PA)   138, 148–155
polyester/polycarbonate (PEs/PC)   138
polyester/polyester (PE/PE)   138–143
polyester reactive blending   399
polyethersulfone (PES)   196
polyethoxysilane (PAOS)   181
polyethylene (PE)   188
  glycol   190
polyethylene terephthalate (PET)   8, 113
polyimides (PI)   196
polylactic acid (PLA)   102
poly-l-lactic acid (PLLA)   111
polymer blending processes
  CCD   210
  compatibilizer-tracer, concept of   210
    mixers, types of   221
    screw speed, effects of   220
  DDD   210
  emulsification   210
  reaction kinetics and reactive functional groups   222–223
  reactive compatibilizer-tracers, *see* reactive compatibilizer-tracers
  residence time distributions (RTD)   210
  twin screw extruders   209
polymer chain   136
polymer, low thermal conductivity   107
polymer matrix   101

polyolefins   180, 194, 195
polyoxymethylene (POM), thermal degradation of   112
polypropylene (PP)   33–34
  antibacterial properties of   183–184
  controlled degradation of   41, 44, 50–55
  glass fiber compounds
    applications   380
    exemplary screw design   382
    maleic acid anhydride grafted polyolefins   383
    PET   385
    PP/GF compound   385
    sized glass fiber   383
    stiffness and heat deflection temperature (HDT) values   380
    thermoplastic polymers   383
  peroxide degradation of   52–54
  rheological modifications   40
polystyrene (PS)   76, 197
polystyrene-poly(methylmethacrylate) (PMMA)   203
polytetramethylene ether glycol (PTMEG)   190
polyvinylidene fluoride (PVDF)   203
power law   91
power transmission   14, 15
PS-g-PA6-MAMA-1   219
PS-g-PA6-MAMA-2   219

r

Rabinowitsch corrections   121
Raman band intensity   114
Raman spectroscopy   114
reactive blending   135, 144, 145, 247
reactive comb compatibilizers
  carboxyl-terminated PMMA   273
  compatibilization   272
  molecular structures   272
  PLLA/ABS blends
    carboxyl groups reactions   283
    cocontinuous phase   282–289
    internal pressure and free volume enlargement   288
    morphologies   282
    TEM images   283, 288

reactive comb compatibilizers (*contd.*)
    tensile stress–strain curves, of neat components 284
    thermal properties, of neat components and compatibilized blends 286
  PLLA/PVDF blends 274–282
reactive compatibilization 247
reactive compatibilizer-tracers
  blend composition effects 235–238
  emulsification curve build-up, experimental procedure 216–219
  geometry of screw elements, effects of 238–241
  in-line fluorescent light detection device 214–216
  morphology development 224–229
  non-reactive compatibilizers *vs.* reactive compatibilizers 223–224
  PS-g-PA6-MAMA compatibilizer-tracer 213
  reactive compatibilizer-tracer injection location 233–235
  twin screw extruder 229–233
reactive extrusion
  ε-caprolactone, polymerization of 55–59
  coupling 45
  defined 39–40
  dynamic vulcanization 40
  EVA copolymer, esterification of 46–50
  in-line measurements 106
  in-process rheometry 116–125
  industrial applications 40–41
  kinetic equations 44
  on-line measurements 107
  optimization 61–65
  PA/PA 159–166
  PC/PA 155–159
  PC/PEs 143–148
  PEs/PEs 138–143
  PE/PA 148–155
  PET 8
  polymer matrix 188
  polymerisationin 9
  polypropylene, controlled degradation of 50–55
  principle of 41–46, 189
  process modelling 41
  process parameters 355
  reactive blending 40
  requirements, in-process monitoring of 103–111
  rheokinetic model 44
  scale-up 61–65
  starch cationization 59–61
  twin screw flow module 42–44
  types of 6
reactive linear polymers, compatibilization 272
reagents degradation 46
reflectance mode 107, 112
residence revolution distribution (RRD) 81
  local 86–87
  partial 82–86
residence time distribution (RTD) 71, 81, 104
  deconvolution procedure 79
  distributive mixing, kinematic modeling of 88–89
  experimental validation 92–93
  exutrder configurations 75–76
  feed rate, effect of 77–79
  in-line measuring system 73
  in-line preformance 76
  kneading disc zone 79
  local 86–87
  modeling of 88
  numerical simulation 89–92
  partial 82–86
  pressure measurements 110
  screw configurations 75–76
  screw speed, effect of 77–79
  staggering angle 80
  of theory 72–73
  variance 72
residence volume distribution (RVD) 81
  local 86–87
  partial 82–86

reverse pitch screws 16
rheokinetic model 44, 46
rheometer 104, 107, 120
rotational rheometry 103
Runge–Kutta procedure 50

s

screw elements 16–22
screw mixing element (SME) 22
screw speed 14, 48, 103, 104, 112
  effect of 102
  feed rate 182
SiDOPO 185
signal to noise ratio 112
single-screw extruders 101
size exclusion chromatography (SEC) 113
small amplitude oscillatory shear (SAOS) 110, 117
small angle light scattering (SALS) 105, 109
small-angle X-ray scattering (SAXS) 137, 140
softsensor 110
sol–gel method 66
sol–gel chemistry 180, 181
solid state grafting technology 378, 379
specific mechanical energy (SME) 28
spectroscopy techniques 104
staggering angle 79–81, 86, 97
starch cationization 59–61
starve-fed single screw extruders 102
strain hardening effects 117
strain-gauge 109
stress tensor 91
stretch ratio 88, 89, 94, 95, 97
styrene of, radical polymerization 40
styrene–maleic anhydride 111, 112
surface to volume ratio 24

t

tapered kneading blocks (TKB) 21
Taylor diffusion 107
thermal energy transfer 24–25
thermoplastic
  elastomer 190
  nanostructuration 188–196
  polymer/epoxy-amine miscible blends 197–201
  polymerization and crosslinking 179
thermoplastic elastomer (TPE)
  definition 27, 331
  intrinsic copolymer 331
  reactive blending 331
  thermal reversibility 332
  TPVs, see thermoplastic vulcanizates (TPVs)
thermoplastic phase, in situ polymerization 179
thermoplastic polymers
  flame retardant properties 184–186
  in situ nanocomposites synthesis 181–183
  monomers crosslinking via radical polymerization 202–203
  PP/TiO$_2$, antibacterial properties of 183–184
  stabilization 202
thermoplastic polyurethane (TPU) 9, 27, 29
thermoplastic vulcanizates (TPVs)
  deformation mechanism of 332
  dynamic vulcanization procedure 331
  EVA/EVM blends 338–342
  morphological development of 333–334
  PP/EPDM 334
  PVDF/ACM blends 334–338
thermoset 200, 202
  minor phase 179
  polymerization of 196–203
three-lobed kneading blocks 19
time to ignition (TTI) 185
TiO$_2$, antibacterial properties of 183–184
titanium-alcoxy bond 55
toluene diisocyanate (TDI) 31
transmission electronic microscopy (TEM) 191, 198, 202
trimethyol propane trimetharylat (TRIM) 203
twin screw extruders 353

twin screw extruders (*contd.*)
  acetylcaprolactam   353
  ε-caprolactam, polymerization of   112
  cellulose fibers   353
  chemical reactor   179
  material properties   353
  modified reaction system   353
  residual monomer content, degassing zones
    contents of activator   358
    degassing efficiency   357
    gel permeation chromatography (GPC)   363
    influence of amount and entrainer types   365–367
    mass temperature   359, 360
    polymer influence   367–368
    side reactions   361
    thermal degradation   362
  screw configuration and process conditions   353
  character dimensions of   15
  chemical reactions   26–27
  development for   14
  devolatilization/degassing   25–26
  discharge   26
  downstream feeding   24
  feeding   23
  MAH   31
  mechanical energy transfer   25
  melting mechanisms   24
  mixing mechanisms   25
  octanoic acid   113
  PMMA   42
  PP/TiO$_2$ nanocomposite   181
  problems and challenges   45–46
  styrene/maleimide   113
  thermal energy transfer   24–25
  TPE processing   27–29
  TPU polymerization   29–31
  unit operations   22–26
  upstream feeding   23–24
twin screw flow module   42–44
typical unit operations   22

u

ultraviolet–visible (UV–VIS)   104
upstream feeding   23–24

v

viscometers   122

w

wide-angle X-ray spectroscopy (WAXS)   166
wood plastic composites
  bending strength and Charpy impact strength   387
  coupling agents   388
  coupling mechanism, maleic acid anhydride grafted polypropylene   388
  maleic acid anhydride and silanes   388
  plastic material, processing temperature   385
  polypropylene   385

z

zero length die   119
zero shear viscosity   50, 91, 117
Ziegler–Natta catalysis   50
ZSK barrels   16